# Partially Ordered Systems

W0051418

# Partially Ordered Systems

*Editorial Board*: J. Charvolin · W. Helfrich · L. Lam

---

*Solitons in Liquid Crystals*
Lui Lam and Jacques Prost, Editors

*Bond-Orientational Order in Condensed Matter Systems*
Katherine J. Strandburg, Editor

*Diffraction Optics of Complex-Structured Media*
V.A. Belyakov

Katherine J. Strandburg
Editor

# Bond-Orientational Order in Condensed Matter Systems

Foreword by David R. Nelson

With 163 Illustrations

Springer-Verlag

New York  Berlin  Heidelberg  London  Paris
Tokyo  Hong Kong  Barcelona  Budapest

Katherine J. Strandburg
Argonne National Laboratory
Materials Science Division
Argonne, IL 60439
USA

Library of Congress Cataloging-in-Publication Data
Bond-orientational order in condensed matter systems / Katherine J.
  Strandburg, editor.
    p.  cm. — (Partially ordered systems)
    Includes bibliographical references and index.
    ISBN -13:978-1-4612-7680-7
    1. Condensed matter.  2. Crystals.  3. Glass.  4. Phase
  transformations (Statistical physics)  I. Strandburg, Katherine Jo,
  1957–  .  II. Series.
  QC173.4.C65B66  1992
  530.4′1 — dc20              91-20237

Printed on acid-free paper.

Photocomposed from a LaTex file.

9 8 7 6 5 4 3 2 1

ISBN -13:978-1-4612-7680-7      e-ISBN -13:978-1-4612-2812-7
DOI: 10.1007/978-1-4612-2812-7

# Foreword

Most of us were taught in school that matter comes in three basic forms: solid, liquid, and gas. This classification scheme goes back at least to the Greeks, and has remained essentially unchanged for thousands of years. The important embellishments have come during the last 100 years or so. We now know, for example, that there is no fundamental distinction between a liquid and a gas above the critical point. On the other hand, it is important to distinguish between the repeating arrays of atoms and molecules in a crystalline solid and the nonequilibrium disorderly arrangements in glassy solids. We find additional phases in liquid crystals which, because of the anisotropic shape of their constituent molecules, are in many ways intermediate between liquids and solids. Despite such refinements, a strong basic prejudice remains that simple pointlike atoms or molecules in equilibrium should have just three basic phases.

This prejudice overlooks the fact that two broken symmetries separate low-temperature crystalline solids from high-temperature liquids. The regular rows of atoms in a crystal responsible for Bragg peaks reflect a breakdown of the translational invariance characteristic of a liquid. The crystallographic axes of the solid represent a different, broken orientational symmetry. A broken translational symmetry necessarily implies long-range orientational order. It is possible to imagine phases of matter, however, where the short-range translational order of a liquid coexists with a broken rotational symmetry. The long-range rotational order of this liquid is associated with "bonds" (determined by, say, the Voronoi construction) joining neighboring atoms or molecules. Thirteen years ago, Bertrand Halperin and I discovered that such an intermediate phase exists in a model of two-dimensional melting first proposed by Kosterlitz and Thouless and by Berezinski. The usual latent heat of the crystal-to-liquid melting transition was spread out over an intermediate "hexatic" phase, somewhat similar to a nematic liquid crystal. Unlike a liquid crystal, though, the residual sixfold orientational order of hexatics is not associated with the anisotropy of the constituent particles. This fourth, hexatic phase of matter has now been observed in free standing liquid crystal films, in colloidal crystals, in magnetic bubble arrays, and even in the flux lines of high temperature superconductors.

The idea of bond-orientational order has proven to be more general than the model of two-dimensional dislocations we studied at that time. Bond-orientational order can exist in three dimensions, and in two and three dimensional glasses, and need not be associated with a continuous dislocation unbinding transition. Although a "hexatic phase," with sixfold bond-orientational order, is the most likely possibility in two dimensions, "tetradic," "cubatic," and even icosahedral bond-oriented phases are possible. In this book, Katherine J. Strandburg has brought together an outstanding collection of scientists who have studied bond-orientational order. The focus is on hexatic bond-orientational order in two dimensions, and icosahedral bond order in metallic glasses and three-dimensional quasicrystals.

The subject of bond-orientational order is introduced in the article by Joel D. Brock. Intriguing computer simulations of bond order and quasicrystals are then discussed by Strandburg. Some of the important experiments on liquid crystal phases (where the hexatic was first discovered) are reviewed by C.C. Huang. Cherry A. Murray then presents a very thorough account of experiments on melting and hexatics in colloidal suspensions. Prior to this work, many investigators thought the hexatic phase would not occur in systems of pointlike particles in two dimensions.

The remainder of the book focuses on icosahedral bond order in glasses and quasicrystals. Tin-Lun Ho discusses some fascinating issues which arise in the faceting of quasicrystals. Icosahedral clusters are a natural packing of (slightly distorted) tetrahedra in three dimensions, playing a role similar to hexagonal clusters in two dimensions. Subir Sachdev discusses how local orientational order in these clusters determines the structure of undercooled liquids and metallic glasses. Landau expansions for long-range translational and orientational order in quasicrystals are reviewed by Marko V. Jarić, and the icosahedral glass model for quasicrystals is explained by A.I. Goldman.

Bond-oriented phases seem likely to be with us for quite some time. This book provides an excellent introduction for both theorists and experimenters to the basic concepts in the field.

Cambridge, Massachusets                                    David R. Nelson
                                                              May 1991

# Contents

# Contributors

Brock, Joel D.
  School of Applied and Engineering Physics, Cornell University,
  Ithaca, NY 14853, USA

Goldman, A.I.
  Ames Laboratory—USDOE and Department of Physics,
  Iowa State University, Ames, IA 50011, USA

Ho, Tin-Lun
  Department of Physics, Ohio State University,
  Columbus, OH 43210, USA

Huang, C.C.
  School of Physics and Astronomy, University of Minnesota,
  Minneapolis, MN 55455, USA

Jarić, Marko Vukobrat
  Center for Theoretical Physics, Texas A&M University,
  College Station, TX 77843, USA

Murray, Cherry A.
  Condensed Matter Physics Research Department,
  AT&T Bell Laboratories, 600 Mountain Avenue,
  Murray Hill, NJ 07974, USA

Nelson, David R.
  Department of Physics, Harvard University,
  Cambridge, MA 02138, USA

Sachdev, Subir
  Center for Theoretical Physics, Yale University,
  New Haven, CT 06511, USA

Strandburg, Katherine J.
  Argonne National Laboratory, Materials Science Division,
  Argonne, IL 60439, USA

# 1

# Bond-Orientational Order

*Joel D. Brock*

## 1.1   Introduction

Even though string theorists are currently working on a "theory of every-thing" which they claim will unify all of the basic forces of nature into a single physical theory, some of the most familiar physical processes, physical processes which we experience every single day, remain largely unexplained. The mechanism by which common fluids, for example, water, freeze is still poorly understood. What are the mechanisms by which the constituent atoms organize themselves as the fluid freezes? What causes the randomly positioned atoms to arrange themselves into a perfectly periodic lattice? How did the atoms communicate with each other? At a more fundamental level, is there any reason to believe that a perfectly periodic crystalline lattice should be the ground state of condensed matter?

Although the complete answers to these questions are not yet known, we have begun to develop some insight into these very familiar problems. Advances in the theory of critical phenomena, particularly the concepts of spontaneously broken symmetry and scaling, and the development of renor-malization group theory, provide a framework within which we can at least define the important issues. We now know that symmetry, spatial dimen-sionality, and fluctuations play crucial roles in determining the structure and cooperative behavior of condensed matter. Advances in experimental capabilities have, in many cases, driven these theoretical advances. It is now possible to do detailed experiments studying the statistical physics of a wide variety of two- and three-dimensional systems. Further insight has been gained from computer simulations of model systems.

Probably the most remarkable advance in our conceptual framework for dealing with complex many-body systems is the notion that the ground state of a system does not have to display the full symmetry of the govern-ing Hamiltonian. Consider the basic laws of physics, the laws which govern the behavior of every system: Newton's laws, Maxwell's equations, relativ-ity theory, or Schrodinger's equation. In either the classical or the quantum limit, these laws are translationally and rotationally invariant. At high tem-peratures, matter is isotropic: usually either a gas or a liquid. An isotropic state is clearly compatible with the complete symmetry of the physical laws and needs no explanation. However, as we all know intuitively, cold matter is quite different. All points in space are no longer equivalent. When you hold a rock in your hand, there is no doubt in your mind that the translational invariance of space has been broken by the rock. Not only does the rock occupy a fixed position in space, but the translational symmetry of space is broken everywhere within the rock; the constituent atoms each occupy fixed positions. Furthermore, the macroscopic properties of cold matter may depend on the *orientation* of the sample. That is, the physical properties of matter can depend on which direction you happen to be looking. For example, the elastic modulus, the dielectric function (index of refraction), and the magnetic susceptibility are all frequently anisotropic in cold matter. Clearly, condensed matter frequently breaks both the translational and the rotational symmetry of the physical laws which ultimately control its behavior.

On the experimental front, most likely due to availability of "clean" physical realizations, condensed matter physicsts have tended to study only the two extremes of the generic condensed matter phase diagram (see Figure 1.1). At one limit are the perfectly periodic, crystalline solids. In these materials the atoms or molecules form a perfectly periodic array that extends to infinity in all three spatial directions. The entire field of solid-state physics is ultimately based on the assumption of a perfectly periodic crystal lattice. Solutions to Schrodinger's equation for a single electron in the presence of a periodic potential lead to the Bloch wave functions and electronic band theory. All of this formalism is ultimately based on the perfect periodicity of the underlying lattice. At the other extreme of the phase diagram are fluids and glassses, systems in which the atoms or molecules are randomly positioned and the system is both orientationally and positionally isotropic. That is, the material looks the same when viewed from any position, in any direction. However, states of matter with an intermediate degree of order are possible. In particular, one may have a state of matter in which the constituent particles are distributed at random, as in a fluid or a glass, but the system is orientationally anisotropic on a macroscopic scale, as in a crystalline solid. In other words, the macroscopic properties of the fluid or glass will be different in different directions. As a simple example of a system with long-range orientational order but with short-range positional order, consider a fluid of anisotropic (e.g., cigar or pancake-shaped) molecules. If all of the molecules are somehow made to point in the same direction, the material will still be a fluid but the optical properties of the material will be anisotropic. The nematic liquid crystal phase used in digital watch displays is an example of this type of system. This type of ordering is called *molecular orientational order*. This is *not* the type of orientational ordering which this book is about.

The subject of this book is a second, much subtler, kind of orientational order which does not require anisotropic constituent particles. In a fluid or glass, the constituent atoms or molecules have, on average, a preferred number of nearest neighbors. These nearest neighbors sit, on average, at

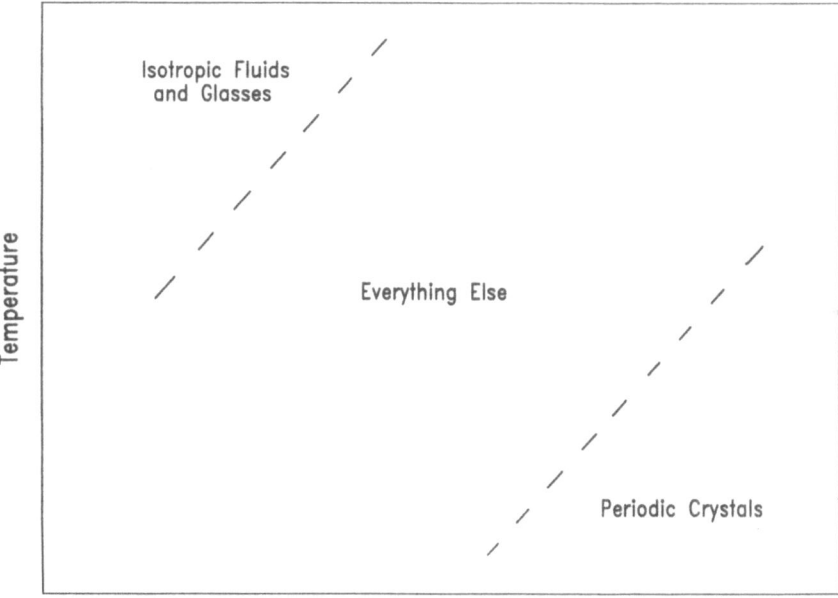

FIGURE 1.1. Generic phase diagram for condensed matter.

some preferred distance from each other. Using these nearest neighbors, one can define a set of local crystallographic axes. In an isotropic fluid, the orientation of these local axes varies randomly with position; however, one can imagine an exotic fluid or glass phase of matter in which *the orientation of the local crystallograhic axes persists over macroscopic distances*. We call this subtle type of orientation order *bond-orientational* (BO) order.

The physics of systems exhibiting bond-orientational order but not periodic positional order is the subject of this book. In the remainder of this chapter, I will present an overview of the subject from an experimentalist's viewpoint. In Section 1.2, I will present all of the essential concepts necessary to discuss bond-orientational order. I will not attempt to review all of the relevant theory. Instead, I will use a couple of very simple models as examples to develop the important physical concepts and define the issues involved. None of the examples are original, I have borrowed them all from others. In Section 1.3, I will discuss a few experimental results which illustrate physical realizations of the concepts developed in Section 1.2 and point out the detailed interpretation of the experimental order parameters. Section 1.4 will cover the extension of the simple concepts to other more complicated systems and coupling between the bond-orientational and positional order.

## 1.2    Elementary Ideas

Initially, as one begins to consider the concept of a system possessing long-range bond-orientational order but not possessing periodic positional order, it is not at all clear that such systems should actually exist. However, at least in simple model systems, the phenomenon is not uncommon. In the mid–1930s, R.E. Peierls [1] and L.D. Landau [2,3] independently developed arguments which concluded that a periodic lattice cannot exist in one- or two-dimensional systems. Peierls's argument is based on the harmonic approximation. Landau, on the other hand, used an order parameter expansion and his theory of phase transitions. The discussion of the two-dimensional harmonic crystal which follows is a variation of Landau's theory [4].

### 1.2.1    Example: Two-Dimensional Harmonic Crystal

Consider an ordered configuration of atoms in a two-dimensional system. The two-dimensional analog of an ordinary three-dimensional crystal is a film in which the constituent atoms occupy the points of a two-dimensional lattice which extends to infinity in both in-plane directions. This atomic configuration can be completely described by a periodic density function, $\rho(x, y) = \sum_{\mathbf{R}} \rho_A(\mathbf{r} - \mathbf{R})$, where $\rho_A(\mathbf{r})$ is the density of a single atom and $\{\mathbf{R}\}$ is the set of lattice sites.

At finite temperatures, we expect each atom to vibrate about its equilibrium lattice position. Let $\mathbf{u}(x, y)$ denote the displacement field vector due to the thermal fluctuations of some small region labeled by the two-dimensional position vector $\mathbf{r} = (x, y)$. The probability $w$ of any given thermal fluctuation $\mathbf{u}(\mathbf{r})$ is given by the formula

$$w \propto \exp(-\delta F / k_B T),$$

where

$$\delta F = \int (F - \overline{F}) \, d^2 r$$

is the deviation of the total free energy of the system from its average value and $F$ is the free energy per unit area. Using standard statistical mechanical techniques [4], we can calculate the mean square fluctuations of the Fourier components of the displacement vector. Specifically, to calculate any thermodynamic average $\langle A \rangle$, we evaluate

$$\langle A \rangle = \frac{\displaystyle\sum_{\{\mathbf{u}(\mathbf{r})\}} A e^{-\delta F[\mathbf{u}]/k_B T}}{\displaystyle\sum_{\{\mathbf{u}(\mathbf{r})\}} e^{-\delta F[\mathbf{u}]/k_B T}}$$

where the sum is over all possible displacement fields $\mathbf{u}(\mathbf{r})$.

In a harmonic crystal, the free energy functional is assumed to have the form

$$\delta F[\mathbf{u}(\mathbf{r})] = \frac{1}{2} \int \lambda_{ijlm} \frac{\partial u_i}{\partial x_j} \frac{\partial u_l}{\partial x_m} \, d^2 r,$$

where $\lambda_{ijlm}$ is the elastic modulus tensor [5]. If we take advantage of the periodicity of the lattice, $\mathbf{u}$ can be represented as a Fourier series,

$$\mathbf{u}(\mathbf{r}) = \sum_{\mathbf{k}} \mathbf{u}_{\mathbf{k}} \, e^{i \mathbf{k} \cdot \mathbf{r}}.$$

The components of the vector $\mathbf{k}$ take both positive and negative values, and the coefficients $\mathbf{u}_{\mathbf{k}}$ are related by $\mathbf{u}_{-\mathbf{k}} = \mathbf{u}_{\mathbf{k}}^*$, since $\mathbf{u}$ is real. The series includes terms with wave numbers, $k \leq 1/d$, where $d$ is the linear dimension of the region previously defined. If we use this Fourier expansion, $\delta F$ takes the form[1]

$$\delta F[\mathbf{u}] = \tfrac{1}{2} \Omega \sum_{\mathbf{k}} \lambda_{ijlm} k_j k_m \, u_{i\mathbf{k}} u_{l\mathbf{k}}^*$$

$$= \tfrac{1}{2} \Omega \sum_{\mathbf{k}} \beta_{il}(\mathbf{k}) u_{i\mathbf{k}} u_{l\mathbf{k}}^*,$$

---

[1] The terms containing products $u_{i\mathbf{k}} u_{i\mathbf{k}'}$ with $\mathbf{k}' \neq \mathbf{k}$ disappear on integration over the volume.

where the elements of the real tensor $\beta_{il}(\mathbf{k})$ are quadratic functions of $\mathbf{k}$. The probability distribution is thus Gaussian and the mean square fluctuations of the Fourier components of the displacement vector are given by

$$\langle u_{ik} u_{lk}^* \rangle = \frac{k_B T}{\Omega} \beta_{il}^{-1}(\mathbf{k}),$$

where $\beta_{il}^{-1}(\mathbf{k})$ are the components of the tensor inverse to $\beta_{il}$. The notation can be made more transparent by rewriting the tensor $\beta_{il}^{-1}$ as $[A_{il}(\hat{n})]k^2$, where the tensor $A_{il}$ depends only on the direction of the vector $\mathbf{k}$ ($\hat{n} = \mathbf{k}/k$).

### Lack of Long-Range Positional Order

The mean square value of the displacement field, $\langle |\mathbf{u}|^2 \rangle$, is now found by summing over $\mathbf{k}$; changing from a discrete sum to an integral over $\mathbf{k}$, we obtain

$$\langle |\mathbf{u}|^2 \rangle = k_B T \int \frac{A_{ii}(\hat{n})}{k^2} \frac{d^2 k}{(2\pi)^2},$$

$$= \frac{k_B T}{(2\pi)^2} \int_0^{2\pi} A_{ii}(\phi) \, d\phi \int_0^{1/d} \frac{dk}{k}.$$

The important point to note after doing all this algebra is that the integral over $k$ diverges logarithmically as $k \to 0$. The divergence of the mean square displacement implies that a given atom [to which a particular value of $\rho(x,y)$ corresponds] is displaced through large distances. As the sample size increases, the size of this displacement increases. Clearly, the density function $\rho(x,y)$ is being "smoothed out" by the thermal fluctuations. When $\sqrt{\langle u^2 \rangle} \approx a$, the ideal lattice spacing, the long-range periodicity is lost. Consequently, no density function $\rho(x,y)$ is allowed except for the trivial case where $\rho = \text{constant}$. Thus, in the thermodynamic limit of infinite samples, a two-dimensional harmonic crystal does not posses long-range positional order. Thermal fluctuations always destroy the long-range periodicity of a two-dimensional lattice at finite temperatures.

It is customary to discuss the nature of the positional ordering in more detail by considering the correlations between thermal fluctuations at two different locations in the system. The first point to note is that, when $T = 0$, a two-dimensional lattice of any size will have long-range order. In this classical harmonic theory, the divergence of the integral is due to thermal fluctuations, which formally vanish at $T = 0$. Taking advantage of this $T = 0$ behavior, let $\rho_0(\mathbf{r})$ be the perfectly periodic atomic density function of the system at $T = 0$. At sufficiently low temperatures, only long-wavelength fluctuations are excited in the lattice. If the displacement due to thermal fluctuations $\mathbf{u}$ varies only slightly over length scales of the order of the lattice constant $a$, the fluctuating density can be written as

$\rho(\mathbf{r}) = \rho_0[\mathbf{r} - \mathbf{u}(\mathbf{r})]$. The correlation between density fluctuations at different positions is then given by the statistical average

$$\langle \rho(\mathbf{r}_1)\rho(\mathbf{r}_2) \rangle = \langle \rho_0[\mathbf{r}_1 - \mathbf{u}(\mathbf{r}_1)]\rho_0[\mathbf{r}_2 - \mathbf{u}(\mathbf{r}_2)] \rangle.$$

As before, the periodic function $\rho_0(\mathbf{r})$ can be expanded in a Fourier series:

$$\rho_0(\mathbf{r}) = \bar{\rho} + \sum_{\mathbf{G} \neq 0} \rho_{\mathbf{G}}\, e^{i\mathbf{G}\cdot\mathbf{r}},$$

where the $\mathbf{G}$ are the reciprocal lattice vectors. Note that the constant term $\bar{\rho}$ representing the mean density of the material has been separated from the sum. Substituting this series expansion into our expression for the correlation function, we obtain an infinite series of terms of the form

$$|\rho_{\mathbf{G}}|^2 \exp[i\mathbf{G}\cdot(\mathbf{r}_1 - \mathbf{r}_2)] \langle \exp[-i\mathbf{G}\cdot(\mathbf{u}_1 - \mathbf{u}_2)] \rangle,$$

where for brevity, I have written $\mathbf{u}(\mathbf{r}_1) = \mathbf{u}_1$.

Again, using the probability distribution derived earlier, we can calculate the statistical average

$$\langle \exp[-i\mathbf{G}\cdot(\mathbf{u}_1 - \mathbf{u}_2)] \rangle = \exp\left(-\tfrac{1}{2}G_i G_l \chi_{il}\right),$$

where

$$\chi_{il}(\mathbf{r}) = \langle (u_{i1} - u_{i2})(u_{l1} - u_{l2}) \rangle$$
$$= 2\langle u_i u_l \rangle - \langle u_{i1} u_{l2} \rangle - \langle u_{i2} u_{l1} \rangle.$$

We already have already found expressions for the mean values $\langle u_{ik} u'_{lk} \rangle$. Using them, we find that

$$\chi_{il}(\mathbf{r}) = k_B T \int \frac{A_{il}(\hat{n})}{k^2} 2(1 - \cos\mathbf{k}\cdot\mathbf{r}) \frac{dk_x\, dk_y}{(2\pi)^2}.$$

Therefore, we find

$$\chi_{il} = \frac{k_B T}{\pi} \overline{A_{il}} \ln k_{\max} r;$$

the bar over $A_{il}$ indicates an averaging over all possible orientations of the vector $\mathbf{k}$ in the plane.

The desired correlation function is now obtained by substituting back into our original expression and summing over $\mathbf{G}$. The sum will be dominated at large distances $\mathbf{r}$ by the least rapidly decreasing term:

$$\langle \rho(\mathbf{r}_1)\rho(\mathbf{r}_2) \rangle - \bar{\rho}^2 \propto \frac{1}{r^{k_B T \alpha_{\mathbf{G}}}} \cos\mathbf{G}\cdot\mathbf{r},$$

$$\alpha_{\mathbf{G}} = \frac{G_i G_l \overline{A_{il}}}{2\pi},$$

where $\mathbf{G}$ is taken to be the reciprocal lattice vector for which $\alpha_\mathbf{G}$ has its least value.

Although the algebraically decaying positional correlations in two dimensions formally exclude true long-range order, our harmonic crystal obviously it still quite well ordered. In the classical limit, the fluctuation dissipation theorem gives

$$\chi(\mathbf{q}) = \beta \int d\mathbf{r} \, e^{-i\mathbf{q}\cdot\mathbf{r}} \langle \rho(\mathbf{O})\rho(\mathbf{r}) \rangle$$

$$= \beta \int d\mathbf{r} \, e^{-i\mathbf{q}\cdot\mathbf{r}} \frac{e^{i\mathbf{G}\cdot\mathbf{r}}}{r^\eta}$$

$$= \beta \int r^{1-\eta} \, dr \int d\theta \, e^{-i|\mathbf{G}-\mathbf{q}|\cos\theta}$$

$$\chi(\mathbf{q}) = \frac{\beta}{|\mathbf{G}-\mathbf{q}|^{2-\eta}} \int_0^\infty x^{1-\eta} \, dx \int_0^{2\pi} d\theta \, e^{-ix\cos\theta}.$$

The remaining integrals are model-independent and equal a constant number. The temperature and wave vector dependence of the susceptibility are all contained in the prefactor. Therefore, in two dimensions, algebraic decay of a correlation function implies a diverging susceptibility to an ordering field. Any *infinitesimal* ordering field will suppress the fluctuations and stabilize the system, producing long-range order. Algebraic decay produces very sharp $1/(q^{2-\eta})$ peaks in the scattered intensity, making it very difficult to distinguish an algebraically decaying correlation function from true long-range order. We will call such systems, *quasi-long-range-ordered*.

Again, let us step back and extract the essential physics of the model from the algebra. The two-dimensional $\mathbf{k}$-space of this model causes correlations between density fluctuations at two different points in space to decay algebraically to zero as the separation $r \to \infty$. In three dimensions, the mean square fluctuation is finite and the correlations between density fluctuations do not decay to zero at infinite separation. The entire effect is completely due to the dimensionality of space.

## Presence of Long-Range BO Order

Even though they demonstrated the absence of long-range periodic positional order in one and two dimensions, Peierls and Landau realized that these systems could still be quite highly ordered. In particular, they understood that a uniform density does not necessarily imply that the system is isotropic. In 1937 Landau wrote [3]

> If the body is isotropic, then $\rho$ = const; however, from $\rho$ = const it does not follow that the body should necessarily be isotropic. If $\rho$ = const, then this means that all positions of an atom, more precisely its centre of mass, in the body are equally probable. Nevertheless in this case different orientations in the

body can be non-equivalent. Namely, when the position of any particular atom No. 1 is given, then the probability of different positions of a neighbouring atom No. 2 is a function of their relative positions (i.e., of the vector $r_{12}$ connecting atom No. 1 and 2). This probability $\rho_{12}$ can depend on the direction of $r_{12}$. Then the body will be anisotropic regardless of the fact that for every atom $\rho = \text{const.}$

In this particular paper, Landau was referring to what I have previously called molecular orientational order. However, Landau and Peierls appear to have been aware of bond orientational order. In a recent visit, Rudolph Peierls commented, "We knew in the 1930's that it was possible theoretically for the positional and orientational order to vanish at different temperatures; we just could not think of a scenario in which it would actually happen." [6]

Almost 40 years later, the question of bond orientational ordering was reopened. In 1968 [7], David Mermin began studying crystalline order in two dimensions when the constituent particles interact only via a pair potential $\Phi(r)$. Assuming only some rather general boundary conditions, he showed that every $\mathbf{k} \neq 0$ Fourier component of the density must vanish in the thermodynamic limit. As did Landau and Peierls, Mermin noted that two-dimensional lattices are still very highly ordered; however, he went on to point out the peculiar nature of the remaining order. The average direction of the vector connecting any two adjacent atoms at finite temperatures does not vary from the direction the vector points when $T = 0$. Therefore, although it does not exhibit long-range periodic translational order, a two-dimensional lattice does exhibit "long-range orientational order."

To better understand Mermin's discovery, let us return to the example of the harmonic crystal. In this simple system, the angle between the local crystallographic axes and the axes of the ideal lattice is given by

$$\theta(x,y) = \tfrac{1}{2}(\partial_x u_y - \partial_y u_x),$$

where, as before, $\mathbf{u}(x,y)$ represents the thermal displacement field vector. If we use the Fourier series representation of $\mathbf{u}$, $\theta$ can be expressed as

$$\theta(x,y) = \tfrac{1}{2} \sum_{\vec{k}} (ik_x u_{y,\mathbf{k}} - ik_y u_{x,\mathbf{k}})e^{i\mathbf{k}\cdot\mathbf{r}}.$$

Consequently, the thermal average $\langle \theta^2 \rangle$ becomes

$$\langle \theta^2 \rangle = \frac{k_B T}{(4\pi)^2} \int_0^{2\pi} f_i(\phi) f_j(\phi) A_{ij}(\hat{n})\, d\phi \int_0^{1/d} k\, dk,$$

where $f_x(\phi) = \cos(\phi)$ and $f_y(\phi) = \sin(\phi)$. The tensor $A_{ij}(\hat{n})$ has the same definition as before.

The derivatives in the definition of $\theta(x, y)$ have the effect of bringing down an extra factor of $k^2$ from the exponential which is more than sufficient to kill the logarithmic divergence in the integral over $k$. Thus, the mean square bond-orientational fluctuation is finite, even in infinite samples. Hence, the *orientation* of the "bond" directions can be transmitted infinitely far.

Again to discuss this in more detail, we want to study the correlation between fluctuations in the bond directions at different positions in the lattice. Using the same arguments I used for the displacement fluctuations, we get

$$\langle \theta(\mathbf{r}_1)\theta(\mathbf{r}_2) \rangle = \frac{k_B T}{(4\pi)^2} \int_0^{2\pi} f_i(\phi) f_j(\phi) A_{ij}(\hat{n})\, d\phi \int_0^{1/d} \cos \mathbf{k} \cdot (\mathbf{r}_1 - \mathbf{r}_2) k\, dk,$$

which is finite. In a two-dimensional harmonic crystal, the bond-orientational correlation function does not decay to zero as $r \to \infty$; the bond-orientational order is long-ranged.

We will find it convenient to define a bond-orientational order parameter by designating $\theta(\mathbf{r})$ to be the angle between the local bond direction at position $\mathbf{r}$ and some arbitrary reference axis. In a two-dimensional crystal with $n$-fold rotation symmetry, the bond-orientational order parameter is defined as

$$\Psi_n(\mathbf{r}) = \langle \exp^{in\theta(\mathbf{r})} \rangle,$$

where the brackets indicate a coarse grain average. $\Psi_n$ has two parts: a magnitude, which is not related to the symmetry, but which is a measure of the degree of order, and a phase. The phase is the symmetry-breaking parameter.

## 1.2.2    Mean Field Phase Diagram and Coupled Order Parameters

In the late 1970s, after the development of both the scaling theory of phase transitions and the renormalization group methods, theorists renewed their attack on the melting/freezing problem. One of the more remarkable theoretical predictions was that, in two dimensions, freezing could occur via two continuous phase transitions. The intermediate phase was predicted to be a bond orientationally ordered phase with triangular symmetry: a *hexatic* phase. Before attacking the full problem, let us begin by considering the simplest type of model describing freezing in two dimensions which includes both bond-orientational and positional order: a mean field model (one which ignores fluctuations).

In our discussion of the harmonic crystal, we found that the energy of the crystal must depend not on absolute displacements of the crystal but only on relative ones, i.e., on the gradient $\partial u_j / \partial x_i$. We would like, however, to describe the free energy in terms of the crystalline order parameter in order to study the phase transition. What is the relationship between the strain field $\mathbf{u}$ and the amount of crystalline order in a sample? On a more fundamental level, what is the order parameter of a crystal?

Imagine a $d$-dimensional crystal just at its inaccessible second-order transition point. Surely the crystalline order would first appear to be a $d$-dimensional density wave,

$$\rho(\mathbf{r}) = \bar{\rho} + \sum_{\mathbf{G}} \rho_{\mathbf{G}} \, e^{i\mathbf{G}\cdot\mathbf{r}},$$

where $\rho_{\mathbf{G}} = \rho^*_{-\mathbf{G}}$ ($\mathbf{G}$ denotes the reciprocal lattice vectors of the structure). The lowest-order term in the Landau free energy must then be of the form

$$F = \tfrac{1}{2} \sum_{\mathbf{G}} \alpha(|\mathbf{G}|)(T - T_c(\mathbf{G}))|\rho_{\mathbf{G}}|^2 + \text{higher-order terms.}$$

The orientation of the actual $\mathbf{G}$'s that appear is, at first, completely arbitrary. The growth of different $\mathbf{G}$'s that are directed and phased appropriately to form a crystal occurs only due to "higher-order terms" in the Landau expansion. A third-order invariant $\rho_{\mathbf{G}_1}\rho_{\mathbf{G}_2}\rho_{\mathbf{G}_3}$, where $\mathbf{G}_1 + \mathbf{G}_2 + \mathbf{G}_3 = 0$, is the next symmetry-allowed term in the expansion.

To illustrate the physical implications of a third-order invariant, consider three density waves whose wave vectors form the sides of a triangle: $\mathbf{G}_1 + \mathbf{G}_2 + \mathbf{G}_3 = 0$. Recall that $\mathbf{G}_i$ specifies both the orientation and the wavelength of the $i$th density wave. We begin by establishing the first density wave. At this point, there is no particular phase which is energetically favored over any other and we are free to place the origin whereever we choose. Next we establish the second density wave. Again, there is no particular phase which is energetically favored. The situation is pictured in Figure 1.2. Finally, we must establish the third density wave. This time our choice of phase does make a difference. By varying the phase of the third wave, we can place its maxima at the joint maxima of the first two density waves or in between these joint maxima. In general, one of these configurations will have a lower energy than the others. Thus, the third-order invariant is the first term in the Landau expansion which picks out a definite phase for $\mathbf{G}_3$ and therefore begins to organize the crystal. Obviously, higher-order invariants will also contribute to the final form.

We can consider the coupled set of $\rho_{\mathbf{G}}$'s which make up our crystal to be the order parameter. Note that this order parameter has both an *amplitude* which is not related to symmetry in any way, but which is a measure of the degree of order, and two symmetry-breaking parameters: the *direction* of $\mathbf{G}$, which determines the orientation of the crystal, and the *phase* of $\rho_{\mathbf{G}}$, which is directly related to the previously defined strain variable $\mathbf{u}$. To see this, set

$$\rho_{\mathbf{G}} = |\rho_{\mathbf{G}}|e^{i\phi_{\mathbf{G}}},$$
$$\phi_{\mathbf{G}} = \mathbf{G} \cdot \mathbf{u},$$

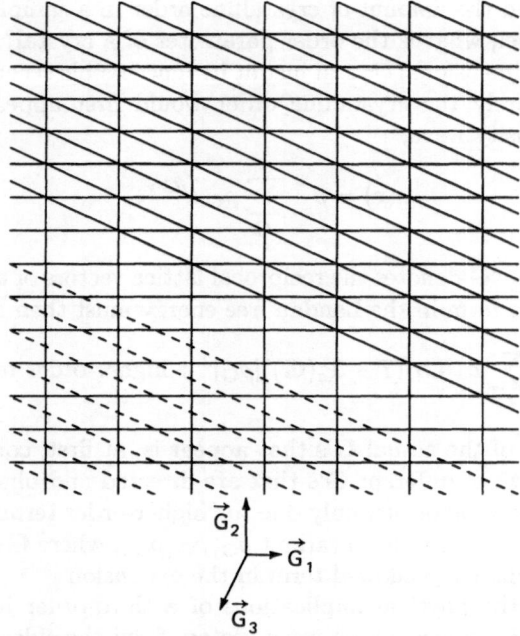

FIGURE 1.2. Origin of Cubic Invariant: A crystal is built by adding up many sinusoidal density waves. The first density wave can be placed in any arbitrary position in space. Similarly, there is no preferred position for the second density wave. The placement of the third density wave, however, is no longer arbitrary. In general, the system may energetically prefer to place the maxima of the third density wave such that they coincide with some definite phase of the first two density waves. For example, the system may energetically prefer to have all the maxima occur at the same position. The solid lines above represent the maximum of the three density waves. On the other hand, the system might prefer to have the minima of the third wave coincide with the maxima of the first two. The dashed line corresponds to this latter case. Whatever the energetic preference, it is clear that the phase of the third density wave is not arbitrary.

which gives us

$$\rho(\mathbf{r}) = 2 \sum_{\mathbf{G}} |\rho_{\mathbf{G}}| \cos\left[\mathbf{G} \cdot (\mathbf{r} + \mathbf{u})\right].$$

Here $\mathbf{u}$ is a vector; thus, we have two acoustic phonon modes in two dimensions. This now connects back to our original expression for the free energy functional of a crystal.

Putting everything together, we can write the excess free energy describing the freezing transition in a system with triangular symmetry in two dimensions as

$$\delta F = \tfrac{1}{2} a_T \sum_{\mathbf{G}} |\rho_{\mathbf{G}}|^2 + b_T \sum_{\mathbf{G}_1 + \mathbf{G}_2 + \mathbf{G}_3 = 0} \rho_{\mathbf{G}_1} \rho_{\mathbf{G}_2} \rho_{\mathbf{G}_3} + \text{higher-order terms}$$
$$+ \tfrac{1}{2} a_6 |\Psi_6|^2 + b_6 |\Psi_6|^4 + \text{higher-order terms}$$
$$+ \gamma \sum_{\mathbf{G}} |\rho_{\mathbf{G}}|^2 \left\{ \Psi_6 e^{-i\theta_{\mathbf{G}}} + \Psi_6^* e^{i\theta_{\mathbf{G}}} \right\}.$$

The phase transition is driven by either $a_T$ or $a_6$ changing sign at $T_c$. If $a_T$ goes negative before $a_6$, the cubic term in $\rho_{\mathbf{G}}$ will drive the phase transition first order. The coupling term will then ensure that the bond-orientational order parameter $\Psi_6$ is phase-locked to $\mathbf{G}$. Positional order thus implies bond-orientational order in this simple model. On the other hand, if $a_6$ becomes negative first, our simple theory predicts a continuous phase transition into a bond orientationally ordered phase. Unlike the solid phase, the bond orientationally ordered phase does not require any of the $\rho_{\mathbf{G}}$ to become finite. The phase diagram associated with this particular free energy is shown in Figure 1.3.

### 1.2.3    Modern Theory of Phase Transitions

As mentioned above, the 1970s witnessed huge advances in our understanding of critical phenomena and phase transitions. The physical concepts of scaling and spontaneous symmetry breaking were united with the computational power of renormalization group theory to create a very powerful theory describing the behavior of systems near a critical point. Thermal fluctuations play a central role in the modern theory of phase transitions. In fact, near ordinary critical points the physics is so dominated by thermal fluctuations that their effects can be ignored only when the dimensionality of space exceeds four. As the spatial dimensionality is reduced, fluctuations become more and more important. If the spatial dimensionality is reduced far enough, fluctuations drive the phase transition temperature to zero, destroying the long-range-ordered phase entirely. This is all in qualitative agreement with the intuition we have developed by studying the two-dimensional harmonic crystal.

The theoretical situation is actually much more rigorously established [8]. In particular, models with *discrete* symmetries [9–11] lose long-range order

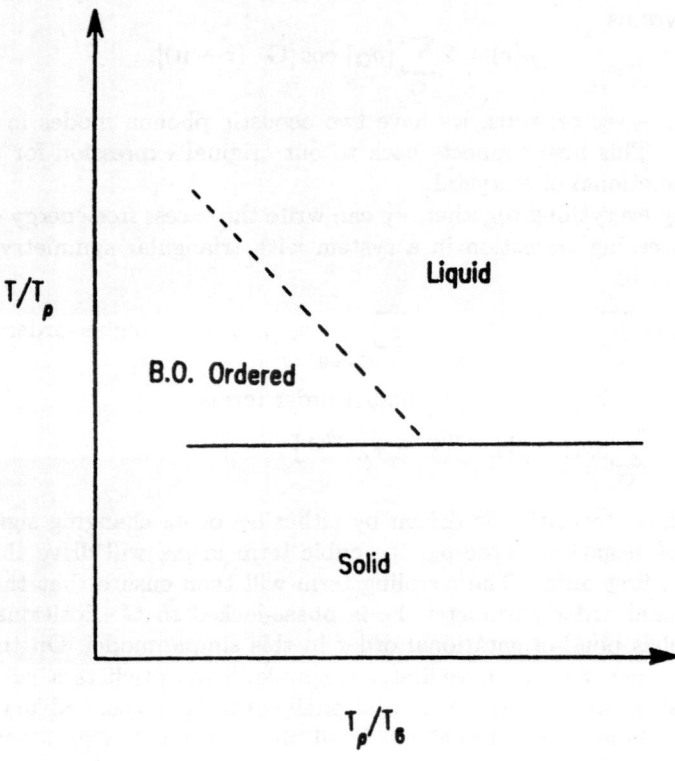

FIGURE 1.3. Mean Field Theory Phase Diagram: $T_\rho$ is the temperature at which $a_T$ changes sign. $T_6$ is the temperature at which $a_6$ changes sign. A solid line indicates a first-order phase transition and a dashed line indicates a second-order phase transition.

in the limit $d \to 1^+$. Models with broken *continuous* symmetries [7,12–14] lose long-range order in two dimensions or less. If the broken continuous symmetry is *Abelian*, a finite-temperature two-dimensional phase transition is still possible. Phase transitions in such systems occur via the unbinding of pairs of point defects.

Kosterlitz, Thouless, and Berezinski proposed a theory of the superfluid transition in two dimensions [15–17] using these ideas. In this theory, below $T_c$, thermally generated vortices in the superfluid wave function exist in bound pairs of opposite vorticity. Above $T_c$, these vortex pairs unbind, destroying the ordered state. Kosterlitz, Thouless, and Berezinski pointed out that similar ideas could be applied to the melting of a two-dimensional crystal, with dislocations taking the place of vortices. Following this initial suggestion, quantitative theories of melting based on these ideas were developed by Halperin and Nelson [18–20] and by Young [21].

The physical model associated with dislocation-mediated melting theory is really rather simple. The elastic strain energy associated with an unbound, isolated dislocation occupying an area $L^2$ is $E = \pi k_B T K \ln(L/a)$, where $K$ is a temperature-dependent elastic constant. Since there are $(L/a)^2$ places to put the dislocation, the entropy gain upon unbinding is $S \approx 2k_B \ln(L/a)$. Therefore, the phase transition occurs when the free energy, $F = E - TS$, changes sign, i.e., when $K(T_c) \approx 2/\pi$.

Naively, one would suspect that introducing an enormous number of *free* dislocations would melt the crystal. However, unbinding the dislocation pairs does not completely melt the crystal; i.e., the system does not become an isotropic fluid when the dislocations unbind. The details of the ordering and the functional form of the correlation function are model-dependent. In the Halperin–Nelson theory, after the dislocations unbind, the positional correlations are short-ranged but the bond-orientational correlations are quasi-long-ranged. The positional correlations decay exponentially while the bond-orientational correlations decay algebraically. A second unbinding transition involving disclinations is required to achieve an isotropic, two-dimensional fluid.

These two types of lattice defects, dislocations and disclinations, are depicted schematically for the case of a two-dimensional hexagonal lattice in Figure 1.4. A dislocation is created by adding an additional half-row of atoms to the previously perfect lattice and is completely described by a vector **b**, called the Burgers vector. Outside the defect core, a dislocation has very little effect on the local orientation of the hexagonal lattice. This is illustrated in Figure 1.4(a), where the relative orientation of the two hexagons highlighted by cross-hatching has not been disturbed by the presence of the dislocation. This is exactly the situation corresponding to a bond orientationally ordered phase. The long-range periodicity of the lattice has been destroyed by the dislocation; however, the long-range coherence of the orientation of the local lattice still persists.

In contrast, a disclination has a dramatic effect on the orientation of the

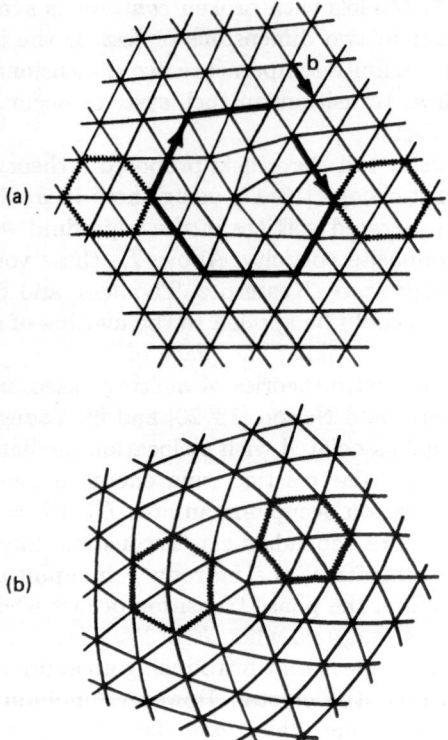

FIGURE 1.4. Lattice defects in a two-dimensional triangular lattice. (a) A dislocation: An additional half-row of atoms has been inserted into the lattice. The Burgers vector **b** is a lattice vector. Notice that the orientation of the local crystallographic axes is not significantly altered by the introduction of the dislocation. The long-range periodicity of the lattice, however, is lost. (b) A disclination: The crystallographic axes rotate by 60° around the defect center, which has fivefold symmetry, destroying the long-range order. Notice that the dislocation is built out of a five- and a sevenfold disclination. [From J.D. Brock, R.J. Birgeneau, J.D. Litster, and A. Aharony, *Contemporary Physics*, **30**, 321 (1989).]

local lattice, rotating the crystallographic axes by 60°. The relative orientation of the cross-hatched hexagons in Figure 1.4(b) illustrates this rotation. Clearly the disclination has destroyed the long-range bond-orientational order of the hexagonal system. In the two-dimensional hexagonal lattice, disclinations may be either fivefold or sevenfold symmetric. Notice that the dislocation in Figure 1.4(a) is comprised of two bound disclinations, one fivefold and one sevenfold symmetric. It is the unbinding of these bound disclination pairs that finally produces the isotropic fluid. The novel intermediate phase with short-range positional but quasi-long-range bond-orientational order was dubbed *hexatic* for two-dimensional systems with hexagonal symmetry.

The important point for our purposes is the discovery of a model which *predicts* a phase with short-range positional order and (quasi-) long-range bond-orientational order. Evidently, a system can fully develop its bond-orientational order without developing any (quasi-) long-range periodicity.

## 1.3   Experimental Results

Although theoretical models and their analytic solutions are very elegant and persuasive, models are not physical theories unless they make a definite statement about some physically measurable quantity. Indeed, there are a large number of models, many of which have analytic solutions, which do not describe any physical system. Appropriately then, after defect-mediated melting theory in two dimensions predicted the existence of a bond orientationally ordered *hexatic* phase, an intense search for physical realizations of such systems began. Several systems exhibiting phases closely related to ideal two-dimensional hexatic phases were found: smectic liquid crystals, rare gases physisorbed on graphite, and colloidal suspensions of polystyrene spheres in water. The experimental work I am interested in for the purposes of this chapter can be grouped into two main categories: studies of the static structure and thermodynamic measurements. This ignores a large body of work on the dynamical and viscous properties of bond orientationally ordered systems, however. I will concentrate on the liquid crystal evidence.

### 1.3.1   Thermodynamic Measurements

The smectic phases of thermotropic liquid crystals are differentiated by varying kinds of ordering. Bond orientational order turns out to be essential ingredients in the smectic classification scheme based on the type of microscopic order present.

Liquid crystals are usually long, rodlike organic molecules. An example is the homologous series of $n$-alkyl=$4'$-$n$-alkoxybiphenyl-4-carboxylate, which is given the more convenient name $nm$OBC. These particular molecules

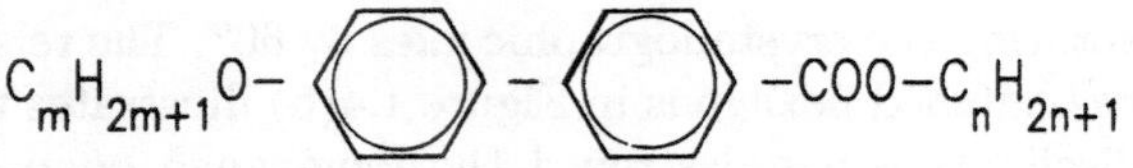

$$C_mH_{2m+1}O-\hexagon-\hexagon-COO-C_nH_{2n+1}$$

FIGURE 1.5. Chemical structure of n-alkyl-4'-n-alkoxybiphenyl-4-carboxylate.

have the structure shown in Figure 1.5. In a smectic phase, the molecules tend to line up parallel and then segregate into layers. The simplest smectic phase is the smectic A phase, denoted $S_A$. This phase has traditionally been described as a system that is a solid in the direction along the director and a fluid normal to the molecular symmetry axis; equivalently, it is a stack of two-dimensional fluid layers. It is more properly described as a one-dimensional density wave, along the molecular symmetry axis, in a three-dimensional fluid. One of two phases originally labeled the smectic B phase is now believed to be an example of a stacked hexatic phase and is denoted $S_{BH}$. In each smectic layer of the $S_{BH}$ phase, the molecular symmetry axis is parallel to the smectic density wave, as in the $S_A$ phase, and there is short-range positional order in the smectic plane; however, there is long-range bond-orientational order in the smectic plane. If the order continues to increase, we leave the true liquid crystal phases. The $S_{BC}$ phase is actually a three-dimensional crystal. The $S_E$ phase is another three-dimensional crystal which has developed a "herring-bone" packing arrangement in the smectic planes.

If the bond-orientational order in the ideal two-dimensional system is quasi-long-range, the susceptibility to an ordering field will be diverging. Thus, we expect the bond orientational order in the $S_{BH}$ phase to be long-ranged, as any infinitesimal coupling between the planes will lock the planes together.

The temperature-concentration plane of the phase diagram of this system is shown in Figure 1.6. One of the intriguing aspects of this system is the similarity of this phase diagram to the mean field model phase diagram shown in Figure 1.3. One sees a two-dimensional fluid phase, a hexatic bond orientationally ordered phase, and a crystal phase in both systems.

By definition, the heat capacity $C_V$ is proportional to the rate of change of entropy with respect to temperature.

$$C_V = T\left(\frac{\partial S}{\partial T}\right)_V = -T\left(\frac{\partial^2 F}{\partial T^2}\right)_V,$$

where $S$ is the entropy and $F$ is the Helmholtz free energy. There is also a clear relationship between the model Hamiltonian used and $C_V$ through the free energy. Clearly, almost all of the entropy "comes out" of the system at the fluid to hexatic phase transition. See Figure 1.7. There is a huge peak in the heat capacity at the $S_A$ to $S_{BH}$ phase transition and only a tiny peak at the $S_{BH}$ to $S_E$ phase transition.

By measuring the details of the divergence of the heat capacity at the

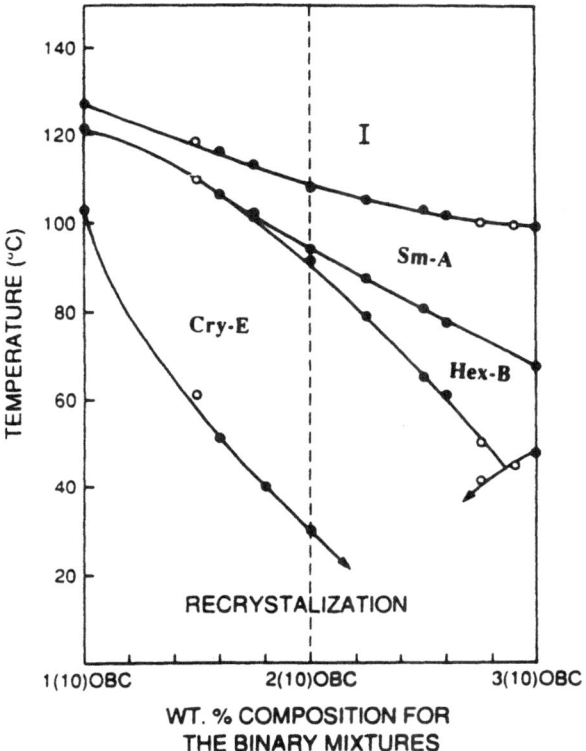

FIGURE 1.6. Temperature-concentration phase diagram displaying the fluid-hexatic-crystal phase sequence in a thermotropic liquid crystal. [From G. Nounesis et al., *Physical Review A* **40**, 5468 (1989).]

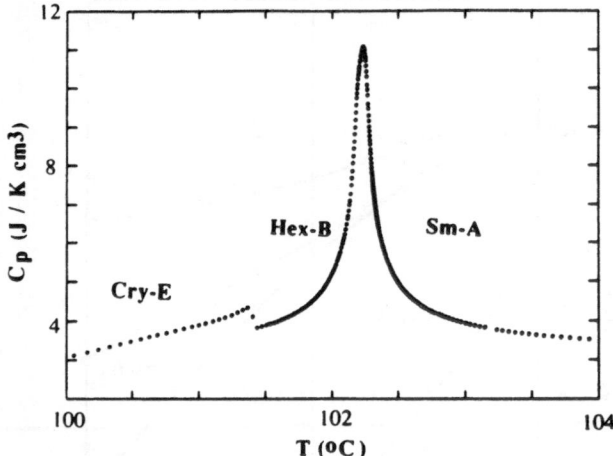

FIGURE 1.7. Heat capacity vs. temperature of the fluid-hexatic-crystal phase sequence in a thermotropic liquid crystal. [From G. Nounesis et al., *Physical Review A* **40**, 5468 (1989).]

fluid to hexatic phase transition, one can measure the heat capacity exponent $\alpha$ and compare it to the predictions of various models.

## 1.3.2   Static Structural Studies

Although the essence of two-dimensional melting theory is the identification of dislocations and disclinations as the elementary excitations out of the ground state, the essential testable prediction of the theory was that a hexatic phase should exist. In particular, the identifying features of a hexatic phase are short-range positional correlations and (quasi-) long-range bond-orientational correlations. Scattering experiments of all types — X-ray, neutron, and electron — are tailor-made to study correlation functions and thus to search for hexatic phases.

In the weak scattering limit (Born approximation), the scattered intensity of all of these probes is given by the expression

$$I(\mathbf{q}, \omega) \propto \frac{1}{2\pi} \int dt \, e^{i\omega t} \int d\mathbf{r}_1 \, d\mathbf{r}_2 \, e^{-i\mathbf{q}(\mathbf{r}_1 - \mathbf{r}_2)} \langle \delta n(\mathbf{r}_1, 0) \delta n(\mathbf{r}_2, t) \rangle,$$

where $\delta n(\mathbf{r}) = n(\mathbf{r}) - \bar{n}$ is the local deviation from the mean electronic number density. The difference between these three scattering probes is all contained in the proportionality constant. In the weak scattering limit, all three probes measure the Fourier transform in both time and space of the density-density correlation function.

In the case of X-ray scattering, the energy resolution of the spectrometer $\Delta\omega$ is typically much larger than $k_B T$. Thus, the spectrometer is integrating over all thermal energy scales. This energy integration has the effect of reducing the X-ray result to the spatial Fourier transform of the equal time density-density correlation function.

In 1978, Birgeneau and Litster [22] suggested that some of the exotic smectic liquid crystal phases might be three-dimensional realizations of a bond orientationally ordered phase. In this view, each smectic layer is an independent, two-dimensional, bond orientationally ordered system. In 1979, Moncton and Pindak [23] recognized that, in the thin film limit, freely suspended films of liquid crystal material should provide ideal substrate-free systems in which to study two-dimensional melting. They then began a series of studies on the melting of very thin films at the $S_A \rightarrow S_B$ phase transition using synchrotron X-ray sources. Subsequently, Pindak and coworkers [24] made the first identification of a hexatic bond orientationally ordered liquid crystal phase. However, these experiments were not able to make quantitative measurements of the bond-orientational order. The variable number and orientation of the hexatic domains present within the area probed by the X-ray beam made quantitative measurements impossible. Single-domain hexatic samples are necessary to obtain quantitative information on bond-orientational order. The first quantitative measurements of bond-orientational order were by Brock and co-workers [25]. These measurements were also synchrotron X-ray scattering studies of freely suspended films; however, the films were very thick and were magnetically aligned tilted hexatic phase smectic liquid crystal.

This experiment yielded a surprise. Figure 1.8 illustrates the expected in-plane diffraction patterns for two-dimensional isotropic and hexatic fluid structures. Consider first the normal-fluid isotropic scattering function. The scattering peaks at a momentum transfer $Q$ of roughly $4\pi(\sqrt{3}\,a)$, where $a$ is the average interparticle separation. Since the fluid is isotropic, the intensity and shape of this peak must be independent of direction. Therefore, if one scans the angular variable $\chi$, defined in the figure, one should not see any variation in the intensity of the scattering. The width $\Delta Q$ of the ring of scattering should be roughly $2/\xi$, where $\xi$ is the length scale on which positional correlations between the constituent particles decay.

For a fluid (i.e., short-range positional order) hexatic phase, the peak position and width $\Delta Q$ of the ring of scattering are the same as for the isotropic fluid phase. However, the fluid develops a sixfold modulation in the angular variable $\chi$; i.e., the angular isotropy is broken. For a quasi-long-range ordered (i.e., algebraically decaying positional order) hexatic phase, the scattered intensity is difficult to distinguish from that of a perfect crystal. The delta function peaks associated with a crystal are replaced by algebraic singularities. In the dynamical diffraction theory appropriate near a Bragg peak, the scattered intensity falls off roughly as $1/|\mathbf{G} - \mathbf{q}|^2$. This is to be compared to the expected $1/|\mathbf{G} - \mathbf{q}|^{2-\eta}$ for the quasi-long-range system.

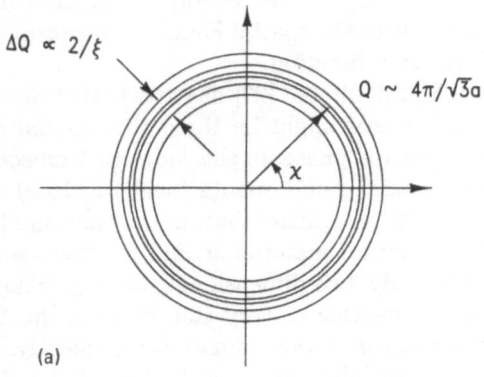

(a)

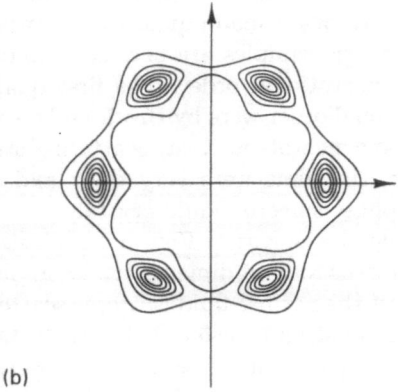

(b)

FIGURE 1.8. Scattering functions for isotropic and bond orientationally ordered systems. (a) In-plane scattering function for an isotropic system. The six circles are lines of constant scattered intensity. The inverse of the half-width at half maximum $\Delta Q$ measures the distance $\xi$ over which the molecules are positionally ordered. The momentum transfer $Q$ is related to the average molecular separation $a$ as indicated. (b) The same scattering function for a bond orientationally ordered system. The hexatic scattering spots are more intense because the total scattered intensity is conserved.

## Experimental Definition of the BO Order Parameter

The amount of $6n$-fold order present in the system can be measured quantitatively by fitting the $\chi$ scan data to the form

$$S(\chi) = I_0 \left[ \tfrac{1}{2} + \sum_{n=1}^{\infty} C_{6n} \cos 6n(\chi) \right].$$

Each of the $C_{6n}$ is a measure of the $6n$-fold order in the sample: i.e., $C_6$ is the hexatic order parameter. $C_{6n}$ is completely analogous to the magnetization $M$ of a ferromagnet.

Note that as defined in a scattering experiment, the bond-orientational order parameter $C_6$ is slightly differently than the definition used in the Halperin–Nelson model. This definition is originally due to Aeppli and Bruinsma [26]. The fluid is divided into microscopic cells of characteristic length $\Lambda_0^{-1}$, where $\Lambda_0^{-1}$ is large compared to the length scale on which positional correlations decay, $\xi$. In each of these cells $V_{\mathbf{r}}$, where $\mathbf{r}$ is the position vector, we calculate the Fourier transformed density $\rho_{\mathbf{q}}(\mathbf{r})$ for $|\mathbf{q}| > \Lambda_0$. Using $\rho_{\mathbf{q}}(\mathbf{r})$, which has an infinite number of components indexed by $\mathbf{q}$, one can define a two-component vector field representing the hexatic order parameter:

$$\Psi_6(\mathbf{r}) = \int d\chi \, e^{i6\chi} |\rho_{q_0}(\mathbf{r})|^2,$$

$$= \psi_6(\mathbf{r}) e^{i6\theta(\mathbf{r})},$$

where $|\mathbf{q}_0|$ is the peak of the fluid structure factor. In a scattering experiment with $q = |\mathbf{q}| > \Lambda_0$, each cell scatters independently. The corresponding structure factor is

$$S(\mathbf{q}) = \int d^d r \, \langle |\rho_{\mathbf{q}}(\mathbf{r})|^2 \rangle.$$

Therefore,

$$\int d\chi \, e^{i6\chi} S(\mathbf{q}_0) = \int d^r r \int d\chi \, e^{i6\chi} \langle |\rho_{\mathbf{q}}(\mathbf{r})|^2 \rangle$$

$$= \int d^d r \, \langle \Psi_6(\mathbf{r}) \rangle.$$

The $C_{6n}$ defined above are simply $C_{6n} = \mathrm{Re} \int d^d r \, \langle \Psi_{6n}(\mathbf{r}) \rangle$, the real component of the spatial average over the illuminated spot size of $\mathrm{Re}\langle \Psi_{6n}(\mathbf{r}) \rangle$ [25].

In a scattering experiment, $\Psi_6(\mathbf{r})$ is defined in reciprocal space. In a real space imaging experiment or in the Halperin–Nelson theory, $\Psi_6(\mathbf{r})$ is defined in real space.

## Bond-Orientational Order Parameter Scaling

After agreeing on a definition for the complex-order parameter $\psi(\mathbf{x})$, it is relatively straightforward to develop a phenomenological model describing the development of bond-orientational order [27]. We begin with the Ginzburg–Landau Hamiltonian [28],

$$\overline{H} = \int d^d x \left\{ \tfrac{1}{2}|\nabla\psi|^2 + \tfrac{1}{2}r|\psi|^2 + u_4|\psi|^4 + u_6|\psi|^6 + h\,\mathrm{Re}\,\psi \right\}.$$

This Hamiltonian is the same as in our original mean field problem, with the addition of a gradient term to take into account to lowest order the effects of fluctuations. For $h = 0$, this model exhibits XY-model critical behavior, provided that $u_4$ is greater than the tricritical value, $u_{4t}$. At $u_r = u_{4t}$, this model has a tricritical point.

Our goal is to study the temperature and $\mathbf{q}$ dependence of the $C_{6n}$. To do this, we must have a systematic way of treating thermodynamic functions near a critical point. As in the mean field problem, we begin by writing the singular part of the free energy in terms of the "distance" from the critical point. The scaling hypothesis then claims that, as the distance from the critical point is varied, the free energy changes its scale but not its functional form. In order to apply these ideas to the problem at hand, we need to add field terms $\overline{H}_n = g_n \int d^d x\,\mathrm{Re}(\psi^n)$ to the free energy functional. The essential point to note here is that each of these terms has a different symmetry. If one writes $\psi = x + iy$, one sees that successive terms $n = 2, 3, 4$, etc., have uniaxial $(x^2 - y^2)$, threefold $(x^3 - 3xy^2)$, cubic $4(x^4 + y^4) - 3|\psi|^4$, etc., symmetry. Near the critical point, in the scaling region, the free energy should scale as $F(t, g_n) = |t|^{2-\alpha} f(g_n/|t|^{\phi_n})$, where $t = (T - T_c)/T_c$, $\alpha$ is the XY-model specific-heat exponent, and $\phi_n$ is the appropriate crossover exponent. Thus,

$$C_{6n} = (\partial F/\partial g_n)_{g_n=0} \sim |t|^{2-\alpha-\phi_n} \sim C_6^{\sigma_n},$$

where

$$\sigma_n = \frac{2 - \alpha - \phi_n}{2 - \alpha - \phi_1} = \frac{2(d - \lambda_n)}{d - 2 + \eta},$$

and $\lambda_n = \phi_n/\nu$ ($\nu$ is the correlation-length exponent of the XY-model $\xi \sim |t|^{-\nu}$). Obviously, this scaling argument is valid only near the XY transition. However, detailed microscopic calculations reveal that, even far from the critical region, the correction to the mean field result ($\sigma_n = n$) scales like a temperature-dependent constant raised to the power $n(n-1)$. Thus, one may write

$$\sigma_n = n + \lambda(T)n(n - 1).$$

The single experimental measurement of $S(\mathbf{q})$ thus yields a simultaneous measurement of many exponents $\sigma_n$, or alternatively many crossover exponents $\phi_n$. The power of the notion of universality is reflected again in the agreement of these results with previous measurements of crossover exponents.

# 1.4  More Complicated Systems

## 1.4.1  Crystals

Until very recently, we believed that all materials exhibiting long-range positional order were perfectly periodic crystals. In the idealized version of these systems, the positional ordering is completely described by a periodic arrangement of unit cells. Furthermore, the orientation of these unit cells is restricted to a discrete set of directions. Due to the periodicity, the allowed orientations of the unit cell are restricted to the set of "crystallographic" subgroups of the rotation group: the 5 two-dimensional and 14 three-dimensional Bravais lattices [29]. The essential point here is that a periodic lattice greatly reduces the number of allowed symmetry operations which take the crystal into itself.

For example, consider a two-dimensional lattice. If we plot the lattice translations which take the crystal into itself and then select the "shortest" of these vectors, they will form a "star." Now assume that a rotation by an angle $\theta$ takes the crystal into itself: that is, **a** goes into **b** (which must also be a lattice translation vector since the lattice and thus the lattice translation vectors, are assumed to be invariant under such rotations). The difference vector **b** − **a** must also be a lattice translation vector. Furthermore, since **b(a)** is assumed to be the shortest of the lattice translation vectors, $|\mathbf{b-a}| \geq |\mathbf{b}|$. Therefore, $\theta \geq 60°$. Successive rotations of the lattice by the angle $\theta \geq 60°$ should continue to take the crystal into itself. Thus, the star can have 1, 2, 3, 4, 5, or 6 arms.

For each vector in the star, there must be a vector in the opposite direction, since for every lattice translation there exists an inverse lattice translation. 1, 3, 5 are eliminated by the lack of an inverse lattice translation.

The most general two-dimensional lattice is the *oblique lattice*. The general oblique lattice is invariant only under rotations of $\pi$ and $2\pi$. However, special lattices of the oblique type can be invariant under rotation of $2\pi/4$, or $2\pi/6$, or under mirror reflections. The $2\pi/4$ rotations lead to a square lattice, the $2\pi/6$ rotations lead to a hexagonal lattice, and the mirror reflections lead to either a rectangular lattice or to a centered rectangular lattice [30].

## 1.4.2  Glasses

A glass has neither long-range positional nor long-range bond-orientational correlations. Glasses are differentiated from fluids by their properties under shear. Glasses posses a finite elastic response to shear, whereas fluids only have a viscous response. Our current theoretical picture of a glass is that it is not an equilibrium but a metastable state. One can think of the free energy surface of a glass in configuration space as being resplendent with sharp, deep valleys separated by towering peaks and ridges. As the system cools, it gets "trapped" in one of these deep valleys. The ridges are so high that the system never has adequate time to "tunnel" into the true ground state of the system.

## 1.4.3  Incommensurate Crystals

If we use these definitions, incommensurate crystals are neither crystals nor glasses. Like ordinary crystals, incommensurate crystals possess long-range positional order and long-range bond-orientational order belonging to one of the allowed crystallographic groups. The distinction is that the long-range positional order is *quasiperiodic* rather than periodic. Such structures are characterized by two or more incommensurate periodicities. In many cases the structure may be considered to consist of two interpenetrating lattices with incommensurate lattice constants. In other cases, the structure may be thought of as two interpenetrating lattices which have been rotated with respect to each other by an angle incommensurate with the rotational symmetry of the lattice. A third scenario involves an incommensurate modulation of the atomic positions superimposed on a perfect crystal.

Incommensurate phases are routinely found at the interfaces and surfaces of materials.

## 1.4.4  Quasicrystals

Quasicrystals are a fundamentally different class of structures from all those we have considered so far. They exhibit

- long-range quasiperiodic positional order, and

- long-range bond-orientational order with a *disallowed* crystallographic rotation symmetry.

The term *quasicrystal* has been used to refer both to the physical systems exhibiting these properties and to the ideal mathematical representations of these properties. The term is an abbreviated form of *quasiperiodic crystal* [31].

Both of these properties are required to distinguish quasicrystals from all other known phases of condensed matter. However, it is the disallowed

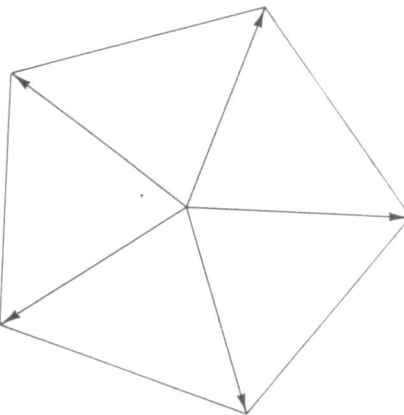

FIGURE 1.9. Fivefold symmetry of the pentagon. The lengths of vectors formed by creating various colinear vectors out of these five basis vectors produce the following incommensurate ratios: $\tau$:1; $\tau^2$:1; $\tau^3$:1 and so on.

crystallographic orientational symmetry which is their essential feature. As mentioned above, two-dimensional quasiperiodic structures are actually quite common at interfaces and surfaces of materials. The quasiperiodicity in these systems is driven by the incommensurate length scales of two different regions. The quasiperiodicity is not required by any symmetry however and hence the ratio of the incommensurate length scales may vary continuously.

Noncrystallographic rotational symmetry changes this. It forces the system to be quasiperiodic: The incommensurate length scales are fixed by geometrical constraints. For example, consider the five vectors shown in Figure 1.9 which point from the center to the vertices of a pentagon. Colinear vectors can be constructed by forming various sums and differences of these five vectors. The lengths of these colinear vectors have the following incommensurate ratios: $\tau$:1, $\tau^2$:1; $\tau^3$:1, etc., where $\tau$ = golden mean = $(1 + \sqrt{5})/2$. Hence, pentagonal orientational symmetry requires quasiperiodic order with incommensurate length scale ratios equal to powers of $\tau$.

## 1.4.5    Quasilattice

Constructing a geometric pattern of lines with a noncrystallographic rotational symmetry and quasiperiodic translational order is quite simple. I begin by drawing a *periodically* spaced set of lines. I then rotate my paper by $2\pi/5$ and repeat the same procedure. Repeating this last step three more times, I complete the geometric pattern. Directing my attention to one of these five periodically spaced set of lines, I find that the lines in two of the remaining four sets intersect this primary set at an angle $2\pi/5$. The remaining two sets of lines intersect the primary set at an angle $\pi/5$. As I go down any given line in the primary set, I find that the intersections from lines at $2\pi/5$ and $\pi/5$ produce two different periodicities. The ratio of these periods is the golden mean.

This simple geometric pattern satisfies all the criteria necessary to be classified as a quasicrystal. It has long-range translational order and it has a noncrystallographic rotation symmetry. However, this pattern is not good as an atomic model of a quasicrystal. The problem is that if I place an atom at every intersection of two or more lines, I eventually have to place two atoms infinitesimally close together. The finite size of atoms makes this procedure unphysical.

A particularly elegant solution to this difficulty is provided by the concept of a *quasilattice*. As in my geometrical example, the lattice point of a quasilattice are quasi-periodically spaced and have a noncrystallographic rotation symmetry. However, the lattice points do not get arbitrarily close together. Rather, the quasilattice can be thought of as a "tiling" of the plane by a finite number of *different* polygons. The atoms are then imagined to sit at the vertices of the polygons. The most famous example of such a quasicrystalline tiling is "Penrose tiling" which inspired the theoretical notion of quasicrystals. The vertices of the tiles have the coordinates

$$\mathbf{r} = n_1\mathbf{e}_1 + n_2\mathbf{e}_2 + n_3\mathbf{e}_3 + n_4\mathbf{e}_4 + n_5\mathbf{e}_5\,,$$

where the $n_j$ are integers and the $\mathbf{e}_j$ are five basis vectors oriented every $2\pi/5$ radians.

## 1.4.6    Quasicrystalline Glass

A competing model for the atomic structure of quasicrystals is the *quasicrystalline glass* model. In this model, icosahedral units are packed together in such a fashion that the icosahedral bond-orientational order is enforced throughout the sample. Such a packing will produce sharp diffraction maxima; however, there are no periodic planes to produce coherent Bragg diffraction. The sharp diffraction peaks can, however, be understood in an analogy to a model developed by Hendricks and Teller [32] to describe the interference patterns produced by partially ordered layered structures found in clays.

The physical distinction between a quasicrystalline glass and a quasicrystal is the lack of long-range positional order in the glass. At sufficient resolution, one ought to be able to measure an intrinsic width to the diffraction maxima from the glass but not the quasicrystal.

## 1.5    Extensions to Three Dimensions

Up to this point, all of my model systems have been two-dimensional. This is not a key ingredient to many of the discussions, however. One can define a three-dimensional bond-orientational order parameter by letting $\mathbf{Q}(\hat{n})$ be the density of bonds in the $\hat{n}$ direction. Analogous to our Fourier expansion of the two-dimensional bond density, we can instead expand the three-dimensional bond density in spherical harmonics, $Y_{lm}(\hat{n})$. The expansion coefficients $\mathbf{Q}_{lm}$ are then the three-dimensional order parameters.

Again, we can construct a mean field model of a system having coupled crystalline and bond-orientational order parameters. We have already seen that in mean field theory the excess free energy describing the phase transition from a disordered fluid to an ordered phase is given by

$$\delta F = \tfrac{1}{2} a_T \sum_{\mathbf{G}} |\rho_{\mathbf{G}}|^2 + b_T \sum_{\mathbf{G_1}+\mathbf{G_2}+\mathbf{G_3}=0} \rho_{\mathbf{G_1}} \rho_{\mathbf{G_2}} \rho_{\mathbf{G_3}}$$
$$+ \sum_{\mathbf{G_1}+\mathbf{G_2}+\mathbf{G_3}+\mathbf{G_4}=0} C_T(\mathbf{G_1} \cdot \mathbf{G_2}, \mathbf{G_3} \cdot \mathbf{G_4}) \rho_{\mathbf{G_1}} \rho_{\mathbf{G_2}} \rho_{\mathbf{G_3}} \rho_{\mathbf{G_4}} + \cdots$$
$$+ A_{bo} \mathbf{Q}^2 + B_{bo} \mathbf{Q}^4 + \cdots$$
$$+ \text{coupling terms},$$

where $\mathbf{Q}$ is the lowest-order orientational order parameter present.

## References

[1] R.E. Peierls, *Helv. Phys. Acta* **7**, Supplement II, 81 (1934); *Ann. Inst. Henri Poincare* **5**, 177 (1935).

[2] L.D. Landau, *Phys. Z. Sowjet.* **11**, 26 (1937); *JETP* **7**, 19 (1937).

[3] L.D. Landau, *Phys. Z. Sowjet.* **11**, 545 (1937); *JETP* **7**, 627 (1937).

[4] E.M. Lifshitz and L.P. Pitaevskii, *Statistical Physics*, 3rd Ed., Part 1 (Vol. 5, *Landau and Lifschitz Course of Theoretical Physics*), translated by J.B. Sykes and M.J. Kearsley, Pergamon Press, New York, 1980, pp. 432–438.

[5] L.D. Landau and E.M. Lifshitz, *Theory of Elasticity* (Vol. 7, *Landau and Lifshitz Course of Theoretical Physics*), translated by J.B. Sykes and W.H. Reid, Pergamon Press, New York, 1970.

[6] R.E. Peierls, private communication.

[7] N.D. Mermin, *Phys. Rev.* **176**, 250 (1968).

[8] An excellent review of the theoretical situation is D.R. Nelson, "Defect-Mediated Phase Transitions," in *Phase Transitions*, Vol. 7, Academic Press, London, 1983, p. 1.

[9] A.A. Migdal, *Zh. Eksp. Teor. Fiz.* **69**, 1457 (1975); *Sov. Phys. JETP* **42**, 743 (1976).

[10] L.P. Kadanoff, *Ann. Phys.* (New York) **100**, 359 (1976).

[11] D.J. Wallace and R.K.P. Zia, *Phys. Rev. Lett.* **43**, 808 (1979).

[12] N.D. Mermin and H. Wagner, *Phys. Rev. Lett.* **17**, 1133 (1966).

[13] P.C. Hohenberg, *Phys. Rev.* **158**, 383 (1967).

[14] N.D. Mermin, *J. Math. Phys.* (New York) **8**, 1061 (1967).

[15] V.L. Berenzinskii, *Zh. Eksp. Teor. Fiz.* **61**, 1144 (1972); *Sov. Phys. JETP* **34**, 610 (1972).

[16] J.M. Kosterlitz and D.J. Thouless, *J. Phys. C* **5**, L124 (1972).

[17] J.M. Kosterlitz and D.J. Thouless, *J. Phys. C* **6**, 1181 (1973).

[18] B.I. Halperin and D.R. Nelson, *Phys. Rev. Lett.* **41**, 121 (1978).

[19] B.I. Halperin and D.R. Nelson, *Phys. Rev. Lett.* **41**, 519 (1978).

[20] D.R. Nelson and B.I. Halperin, *Phys. Rev. B* **19**, 2457 (1979).

[21] A.P. Young, *Phys. Rev. B* **19**, 1855 (1979).

[22] R.J. Birgeneau and J.D. Lister, *J. Phys. Lett.* (Paris) **38**, 1399 (1978).

[23] D.E. Moncton and R. Pindak, *Phys. Rev. Lett.* **43**, 701 (1979).

[24] R. Pindak, D.E. Moncton, S.C. Davey, and J.W. Goodby, *Phys. Rev. Lett.* **46**, 1135 (1981).

[25] J.D. Brock, A. Aharony, R.J. Birgeneau, K.W. Evans-Lutterodt, J.D. Litster, P.M. Horn, G.B. Stephenson, and A.R. Tajbakhsh, *Phys. Rev. Lett.* **57**, 98 (1986).

[26] G. Aeppli and R. Bruinsma, *Phys. Rev. Lett.* **53**, 2133 (1984).

[27] A. Aharony, R.J. Birgeneau, J.D. Brock, and J.D. Litster, *Phys. Rev. Lett.* **57**, 1012 (1986).

[28] R. Bruinsma and G. Aeppli, *Phys. Rev. Lett.* **48**, 1625 (1982).

[29] N.W. Ashcroft and N.D. Mermin, *Solid State Physics*, Holt, Rinehart and Winston, New York, 1976.

[30] C. Kittel, *Introduction to Solid State Physics* (5th edition), Wiley, New York, 1976.

[31] D. Levine and P.J. Steinhardt, *Phys. Rev. Lett.* **53**, 2477 (1984).

[32] S. Hendricks and E. Teller, *J. Chem. Phys.* **10**, 147 (1942).

# 2

# Computer Simulation Studies of Bond-Orientational Order

*Katherine J. Strandburg*

Computer simulation techniques have been used to study the behavior of the bond-orientational order in many systems and to investigate the effects of imposing bond-orientational order on the properties of physical systems. In this chapter we review the techniques used in computer simulation for the study of bond-orientational order and describe the advantages offered by these techniques. Several illustrative examples of the application of simulation techniques to the study of bond-orientational order are described and some inherent difficulties in the simulation approach to this study are also explored.

The chapter by Joel Brock in this book gives an excellent overview of the concept of bond-orientational order and the issues it raises for understanding the properties of condensed matter systems. Here we will confine ourselves to a very brief introduction to bond-orientational order and the systems in which its importance has so far been investigated before proceeding to the discussion of computer simulation probes of this type of ordering.

## 2.1 Introduction

The canonical descriptions of crystal and glass are familiar to all of us from introductory solid-state physics. There we learned that crystals are stable, periodically ordered arrangements of atoms possessing long-range translational order (by which we mean, loosely speaking, that the relative positions of atoms are determined out to large distances) and that glasses are metastable, disordered arrangements of atoms in which, while some short-range correlations in atomic positions remain, the positions of widely separated atoms are uncorrelated. The recognition that ordering in the relative orientations of nearest-neighbor bonds between atoms might be of importance in determining the structure of condensed matter has led to new ideas about the reasons for glass formation and to the recognition of the possibility of novel condensed matter phases. Figure 2.1 lists some of the ways in which translational and bond-orientational order may be combined in a condensed matter phase and gives experimental realizations where

| PHASE | TRANSLATIONAL ORDER | BOND-ORIENTATIONAL ORDER | EXPERIMENTAL REALIZATION |
|---|---|---|---|
| 2D Solid | Algebraically decaying | Long-range sixfold | Adsorbed gases, Thin liquid crystal films Electrons on helium, etc. |
| 2D Hexatic | Short-range | Algebraically decaying sixfold | Colloidal suspensions Thin liquid crystal films |
| 2D Long-range Hexatic | Short-range | Long-range sixfold | Adsorbed gases? Layered superconductors? |
| 3D glass | Short-range | Short-range icosahedral | metallic glasses |
| 3D Icosahedral glass | Short-range | Long-range icosahedral or stacked decagonal dodecagonal, octagonal, etc. | Al-Mn quasicrystal? |
| 3D Random Tiling or Quasiperiodic Crystal | Long-range | Long-range icosahedral or stacked decagonal, dodecagonal, octagonal, etc. | Al-Cu-Fe quasicrystal? Al-Cu-Co quasicrystal? |
| 3D hexatic | Short-range | Long-range sixfold | Thick liquid crystal films |
| 2D Random Tiling | Algebraically decaying | Long-range decagonal, dodecagonal, octagonal, etc. | |
| 2D Quasiperiodic crystal | Long-range | Long-range decagonal, dodecagonal, octagonal, etc. | |
| Liquid | Short-range | Short-range | normal liquids |

FIGURE 2.1. The various possibilities for bond-orientational order in condensed matter phases.

appropriate. As described elsewhere in this book, the concept of bond-orientational order has been of great importance in the interpretation of experiments on liquid crystals [1], colloidal suspensions [2], metallic glasses [3], and quasicrystals [4–6], and it has also been suggested as an important factor in the behavior of flux lines in high-temperature superconductors [7–9].

In two dimensions the hexagonal preferred nearest-neighbor bond orientation (in the absence of angle-dependent interactions or size mismatches) is consistent with the periodicity of the triangular lattice crystal structure. In this context, bond-orientational ordering may serve as an intermediate type of ordering between solid and liquid, allowing the system to increase its entropy over the crystal at a moderate energy cost. In three dimensions (for simple central potentials) the situation is more complicated. Here the preferred icosahedral nearest-neighbor bond-orientational order is inconsistent with periodicity and the local energetic preference for icosahedral order competes with the standard global preference for periodically ordered structures. This "frustration" leads to a variety of possibilities for stable and metastable phases (see Figure 2.1).

Computer simulation techniques for the investigation of these issues (described in more detail below) include molecular dynamics studies in which the simulated atoms move according to classical Newtonian mechanics, Monte Carlo studies in which the equilibrium statistical mechanics of model systems are probed, and nonequilibrium growth simulations in which the consequences of ad hoc rules for particle aggregation are explored. The basic idea in all of these studies is to choose a simple "toy" model designed so as to embody those attributes hypothesized to be fundamental in eliciting a particular behavior and then to use the computer to run a "gedanken experiment" in which the effects of those particular choices can be explored. This type of numerical investigation is to be distinguished from attempts to model or calculate the specific properties of real materials which, while equally interesting, are not described here. The distinction is crucial both to understanding and appreciating the contributions made by the numerical studies described here and to the design of appropriate future simulations of this type. The standard by which to judge these simulations is not "How realistic is it?", but "How much has it illuminated our understanding of the basic physics behind what we observe?"

In the particular arena of the study of bond-orientational order, simulation techniques allow us the following "handles" on the system which are unavailable to experiment and whose effects are generally too difficult to calculate by analytical techniques:

- choice of the atomic potential, including the ability to turn on and off angle-dependent forces

- direct measurement of bond-orientational correlation functions which, since they are not two-particle correlations, are not accessible by standard diffraction techniques

- complete control over the dimensionality of the system, especially the ability to perform substrate-free two-dimensional explorations

- variation of the dynamics, allowing the exploration of the consequences of a variety of growth rules and the possibility of the more rapid equilibration of slow motions

- possibility of imposition of a (nonphysical) field coupling directly to the bond-orientational order parameter

- direct observation and/or simulation of the topological defects which determine the long-range bond-orientational order

The limitations on the usefulness of these simulations are due, to some extent, to the limited system sizes and time scales which can be investigated within our computer resources. Some ways to cope with these difficulties are discussed below. Continued improvements in these aspects due to the design of faster computers with more clever architectures are almost certainly to be expected. The importance of these restrictions on size and speed, can be overestimated, however. The fastest computers will not tell us how to design our simulations so as to extract the important physics in a given phenomenon. In this arena our continued progress is decidedly in our own hands.

# 2.2 Numerical Simulation Techniques

A number of excellent books and reviews exist for those who would like a serious and practical introduction to the techniques used in the simulations described in this chapter [10–16]. In this section we attempt simply a brief description of these techniques intended for the reader who would like to understand what has been done and to feel some confidence in evaluating the validity and significance of simulation results. The reader who would like to become a practitioner is referred to the resources mentioned above.

## 2.2.1 Atomistic Simulations

In these simulations the basic objects are particles, generally taken to represent atoms or molecules. The molecular dynamics or Monte Carlo method is used to generate successive configurations (or computer-stored lists of positions and, for molecular dynamics, momenta) of these particles under the influence of particular choices of external parameters and interaction potential. Energies, correlation functions, and other quantities are defined for these systems as appropriate averages over these configurations. Important choices to be made for these simulations are the list of measured quantities, the choice of potential, and the choice of external parameters and boundary conditions. The Monte Carlo method allows further choices in the allowed rearrangements in moving from one configuration to the next.

## Molecular Dynamics

The molecular dynamics method is basically a numerical integration of the classical Newtonian equations of motion. Positions and velocities for each particle are stored in the computer. At each update step the net force on each particle due to its interaction with the other particles (and with any external fields) is calculated. Then each particle is moved an amount appropriate for a small "time step" $\Delta t$ chosen to be an appropriate microscopic time scale for the system. Various numerical algorithms may be employed in the integration. One common method is the Verlet algorithm [17,18] in which the positions and velocities at time step $n$ are iterated according to:

$$r_i^{n+1} = 2r_i^n - r_i^{n-1} + F_i^n \frac{(\Delta t)^2}{m}$$

$$v_i^n = (r_i^{n+1} - r_i^{n-1})/2(\Delta t),$$

(2.1)

where $r_i^n$ is the position of the $i$th particle at step $n$, $v_i^n$ is the velocity of the $i$th particle at time step $n$, and $F_i^n$ is the net force on the $i$th particle at step $n$ due to its interactions with the other particles.

Equilibrium quantities are calculated as time averages (usually after an initial equilibration time) using the fact that for an ergodic system in equilibrium, long time averages are equivalent to ensemble averages. Time-dependent quantities are also obtainable as averages over a number of different "runs" from different initial conditions, or, once the system is in equilibrium, from averages over pairs of times separated by a given time interval in the same run.

A primary limitation of the molecular dynamics method is the necessity to keep the time-integration interval $\Delta t$ small. Since the computer time required is proportional to the number of time steps explored, the total length of time corresponding to a given run is severely limited (generally corresponding to about 100 psec). However, because the motion of all particles in a given step depends only on the previous positions of the others, the molecular dynamics algorithm is highly parallelizable and powerful new parallel computers should significantly enhance the effectiveness of molecular dynamics by extending the physical times which can be simulated.

In its simplest form molecular dynamics is a constant energy technique. In order to perform averages appropriate to other ensembles (such as constant temperature or pressure), modifications of the technique are required.

Constant temperature, for example, may be provided by periodic rescaling of the velocities of the particles in order to maintain the appropriate magnitude of the kinetic energy. Other methods are also used [14,16]. A number of methods for incorporating constant pressure into molecular dynamics simulations have been used, including hybrid molecular dynamics–Monte Carlo simulations where volume changes are performed according to a Monte Carlo technique, coupling of the system to a hydrostatic "piston" of arbitrary mass, and length rescaling so as to match the virial pressure to the desired external pressure [14,16]. Controversy has ensued over the correctness of these various techniques, which we will not go into here [19–23]. However, it is wise, particularly in the case of the constant pressure ensemble, to exercise caution in the use of these methods and in interpreting results based on them when questions depending strongly on the details of the fluctuations are under consideration.

### Monte Carlo

The Monte Carlo simulation method, as described here, is a method of numerical evaluation of the configuration sums of equilibrium statistical mechanics. (Monte Carlo methods are also applied in other contexts, e.g., for dealing with quantum mechanical systems [24]. Here we describe only their application in classical statistical mechanics.) The objective of the method is to generate configurations (in this context, lists of particle positions) with a probability $P_s$ (where $s$ labels a particular configuration) given by the appropriate statistical ensemble (constant temperature, pressure, etc.) Note that here, in contrast to the case for molecular dynamics, the implementation of different ensembles arises in a natural way.

The algorithm generally used to generate configurations according to the appropriate probability distribution is the Metropolis algorithm [25]. Transition rates $T_{ss'}$ from configuration $s$ to configuration $s'$ are chosen so as to lead to this probability distribution. The time rate of change of the probability for a given configuration is given by

$$dP_s/dt = \sum_{s'} (T_{s's} P_{s'} - T_{ss'} P_s).\qquad(2.2)$$

If $dP_s/dt$ is set to zero for the desired probability distribution for all $s$ and if all configurations are accessible by a sequence of transitions from any given configuration (ergodicity), then the process will eventually converge to the desired probability distribution and statistical mechanical averages may be calculated. One way to ensure that the sum in Eq. (2.2) vanishes is to set it to zero on a term-by-term basis, i.e., $T_{s's} P_{s'} = T_{ss'} P_s$ for all $s$ and $s'$. This so-called "detailed balance" condition is imposed in all of the Monte Carlo simulations described here.

Consider the case of a canonical ensemble. The appropriate probability is given by $P_s = \exp(-\beta E_s)$, where $\beta$ is the inverse temperature. Detailed balance requires that

$$T_{ss'}/T_{s's} = P_{s'}/P_s = \exp[-\beta(E' - E)]. \tag{2.3}$$

One common way to fulfill these conditions is to generate the configuration $s'$ by moving one particle to a position chosen randomly within a sphere of radius $d$ (where $d$ is an empirically chosen parameter) around its initial position. The acceptance probability is then

$$\begin{aligned} T_{ss'} &= 1, & E' < E \\ T_{ss'} &= \exp[-\beta(E' - E)], & E' > E \end{aligned} \tag{2.4}$$

for configurations generated as described and $T_{ss'} = 0$ for all others. The transition probabilities are implemented numerically by comparing the quantity $T_{ss'}$ to a pseudo-random number $r$ generated uniformly in the range 0 to 1 and "accepting" the configuration $s'$ if $T_{ss'} \geq r$.

Note that thus far we have said little about the choice of "moves" which take configuration $s$ into configuration $s'$. In principle, only the ergodicity condition constrains these choices. However, in the real world of limited computer budgets a balance must be struck between choosing moves which are reasonably likely to be accepted (and therefore will probably not represent radical changes) and moves which carry the system further in sweeping out configuration space (and are therefore less likely to be accepted). In the procedure described above, for example, the value of $d$ is arbitrary, but if it is chosen too large, few moves will be accepted. If, on the other hand, it is chosen too small, almost all moves will be accepted but no significantly different configurations will be sampled in the simulations. If we swing too far in either direction, we will not make efficient use of our computational resources and will not explore phase space effectively.

Within the constraints of practicality, though, there still remains a great deal of freedom in the choice of Monte Carlo moves. A clever choice of a Monte Carlo move can render the Monte Carlo procedure significantly more effective than molecular dynamics for calculating equilibrium properties. The exploitation of this freedom is a topic of current interest in a number of contexts (see, e.g., [10,26,27]). One simple example of such a choice of move is described in Section 2.3.2.

In summary, the admittedly biased view of this author is that, unless dynamical information is required (a nontrivial restriction, of course), the Monte Carlo method is more powerful than the molecular dynamics method and is, in general, to be preferred.

## Measurable Quantities

Given a set of equilibrium configurations generated by either the Monte Carlo or molecular dynamics method, one can then proceed to calculate virtually any quantity which can be written as a statistical average in the

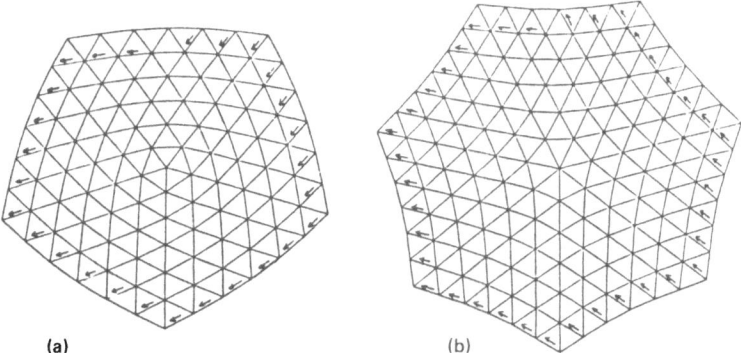

FIGURE 2.2. Positive and negative disclinations in a triangular lattice. The small arrows indicate the rotation of the triangular cells by 60°: (a) clockwise and (b) counterclockwise, as a clockwise path around the disclination is traveled. Note that these disclinations may also be described as particles having (a) five and (b) seven neighbors, respectively, rather than six.

given ensemble, simply by computing that quantity for each configuration generated and then averaging over a large number of configurations. Quantities of particular interest in the context of bond-orientational order include:

- the bond-orientational order parameters for $n$-fold order

$$\Psi_n = \frac{1}{N} \left\langle \left| \sum_j e^{in\theta_j} \right|^2 \right\rangle, \tag{2.5}$$

where $\theta_j$ is defined as the angle the $j$th nearest-neighbor bond makes with an arbitrary axis, the sum is over all nearest-neighbor bonds, and $N$ is the number of such bonds.

- the bond-orientational correlation functions

$$g_n(r) = \langle \psi_n(r)\psi_n^*(0)\rangle, \tag{2.6}$$

where $\psi_n(r)$ is the local bond-orientational order parameter for $n$-fold order.

- the Voronoi polyhedron decomposition of the configuration into a cell for each particle which contains the volume which is closer to that particle than to any other. Analysis of these polyhedra allows the identification of defects in the bond-orientational order [28]. In the two-dimensional case, for example, particles with five or seven neighbors are associated with disclinations (see Figure 2.2).

We describe results for some of these quantities in the examples in Section 2.3.

## 2.2.2    Simulations of Abstract Objects

The simulations described in the previous section might be described, in some loose sense, as "realistic." That is, while they may make no serious attempt to model specific materials, they begin at a reasonably microscopic level with particles that resemble real atoms and interactions qualitatively like those which real atoms experience. Another category of simulation which has been of importance in the study of bond-orientational order begins by abstracting the "fundamental" or "important" structures believed to control the behavior of interest. The behavior of these objects themselves is then explored in simulation in order to see whether the initial assumptions about what is fundamental or important are correct. The familiar Ising model for ferromagnetism is such a model, as are the numerous spin models for the structures of adsorbed systems, the solid-on-solid model for interface roughening, and the many, many others which are the bread and butter of statistical mechanical modeling.

In the context of bond-orientational order, examples of such objects (some of which are described in detail in the examples which follow) are the icosahedral building blocks used in the icosahedral glass studies of quasicrystals [5], the tiles used in the Penrose-tiling-based studies [6], and the dislocations used in some simulation studies of two-dimensional melting. In many of these models the bond-orientational order is "built-in" and the objective of the simulation is to determine the consequences of building in bond-orientational order in a particular way. In that sense, then, these studies are less "fundamental" than the atomistic studies. They may not provide any insight into the origin of bond-orientational order in condensed matter, but they can be extremely important in guiding our physical insight and in teaching us how to interpret the behavior of more complex systems. In addition, because these systems are simpler, they can often be studied in more detail using available computer resources than the more complicated atomistic models.

In studies such as these, the choice and definition of the fundamental objects to be simulated are the primary issues. Questions of quantities to measure, simulation technique, and boundary conditions are still present as well.

The numerical studies of abstract objects described in this chapter fall into two categories. In some of these studies, the abstract objects themselves (e.g., dislocations) may be treated as interacting particles by the methods described in the previous section, or at least (in studies of interacting Penrose tiles, for example) the model includes a Hamiltonian and, while molecular dynamics is no longer appropriate, standard Monte Carlo methods may be used to study the phase diagram of the abstract model. We should mention at this point that while in this chapter we describe numerical *simulation* methods, for many of these abstract models other numerical approaches (such as transfer matrix calculations) are useful and

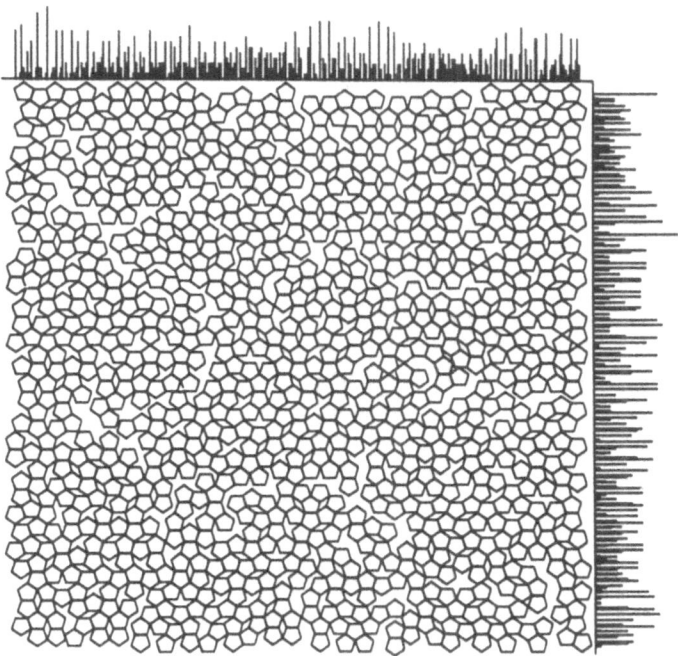

FIGURE 2.3. A cluster aggregation simulation for a two-dimensional decagonal quasicrystal. (This figure is taken from the chapter by A. Goldman in this volume.)

such calculations have a similar "flavor" to the simulations described here.

The second category of computations are growth models in which the basic statistical mechanical concepts such as temperature and energy are, if not exactly abandoned, themselves abstracted to their postulated essential features so that they are hidden in physically motivated but rather arbitrary "growth rules" describing the competition between geometry and chance in determining the growth of condensed matter structures. Well-known examples of such models include the various versions of diffusion-limited aggregation [29].

Growth studies of both perfect quasicrystals [30–33] and icosahedral glass [5] are members of this category. In these studies bond-orientational order is assumed to be preserved through the aggregation of clusters, modeled for these purposes as geometrical objects. The simulation procedure consists of adding these objects randomly to a growing cluster according to some postulated probabilities for "sticking" at particular sites (see Figure 2.3). In spite of their apparent arbitrariness, these models have provided considerable insight into the possible origins of quasicrystals.

An example of the interplay between atomistic and "abstract object" simulations is provided by the search for the two-dimensional bond orientationally ordered hexatic phase. The theoretical idea that the melting behavior of two-dimensional systems might be controlled by the behavior of certain topological defects (dislocations and disclinations) was proposed by Kosterlitz, Thouless, and Bereszinskii [34–37] and more detailed calculations were performed by Halperin, Nelson, and Young [38–40]. These calculations predicted a novel bond orientationally ordered phase (the "hexatic") The outcome of experiments and atomistic simulations was (and remains) far from clear.

An important question then arose: Was the *assumption* of the important role played by topological defects incorrect, or were the *approximations* necessary for the theoretical calculations inappropriate for the experimental and atomistic systems studied? At that point, the usefulness of the abstract object simulations is clear—simulations of the defects themselves are able to provide a more complete idea of the consequences of the postulate that the defects "drive" the behavior of the system, thus providing the potential for distinguishing between an inadequate calculation and an incorrect physical idea.

### 2.2.3    Basic Problems with Numerical Simulation Studies: Short Times, Small Sizes, and the Importance of Boundary Conditions

As mentioned earlier, serious limitations on the usefulness of simulation as a tool arise from the heavy computational requirements of these studies. While experiments and theories are often "in the thermodynamic limit" (or at least very large), the sizes of our simulated systems are often pitifully small (a simulation involving as many as $10^4$ degrees of freedom is still rare). Because of the small system sizes available, the question of choice of boundary conditions becomes relevant. In addition, the simulation times involved are typically very short compared to real laboratory time scales.

Recent advances, both in computer technology and in methods to cope with limited system sizes and simulation times, however, leave room for considerable optimism that simulation methods will continue to provide more and more useful input in the deciphering of nature's puzzles and, in particular, of the bond-orientational order puzzle considered in this book. (However, experience leads one to expect that our ambitions for these methods will also grow!) The power of parallel computation for large-scale scientific number-crunching is only beginning to be realized. While molecular dynamics algorithms adapt rather naturally to parallel implementation, the efficient implementation of Monte Carlo algorithms on parallel computers is a subject of current research among computer scientists and the scientific user community.

Of perhaps more interest to readers of this book, however, is the issue of how most effectively to use the information available from our limited system sizes and simulation times. It is increasingly clear that a brute force approach (run the largest system you can until your computer budget is gone) is not the most effective use of limited computer resources and that far more clever methods exist for the efficient generation and effective use of simulation data.

In this subsection, I will describe a few of these techniques, some of which have been applied in the study of bond-orientational order and some of which have not. Here, as elsewhere in this chapter, the list is intended to be illustrative and certainly not exhaustive. The choice of topics is somewhat arbitrary and is intended to inspire rather than to recommend the "best approaches"—indeed often the best approach is specific to the particular system studied.

## Finite Size Scaling

The basic insight of finite size scaling is that, in trying to determine the behavior of a system in the limit of infinite size, more may be learned by studying the way in which the system approaches larger sizes than by simply looking at the properties of the largest system accessible. This basic procedure allows an extrapolation of the behavior observed for small systems to a "best guess" for the behavior in the thermodynamic limit. There can be no guarantee, of course, that important changes do not occur in the range of inaccessible system sizes. Like any extrapolation, the study of system size dependence relies on assumptions of smooth behavior. Often, however, something is known about the expected behavior in the thermodynamic limit so that one can have a high degree of confidence in extrapolating from available system sizes. For a recent review of finite-size-scaling approaches see [41].

The term finite size scaling refers specifically to the study of the dependence of physical properties on system size at a phase transition. At temperatures away from the transition temperature, quantities such as the specific heat or magnetic susceptibility are size-independent for moderate system sizes. At a second-order phase transition, however, there are a number of properties which diverge in the thermodynamic limit. These divergences are controlled by the existence of a diverging length scale in the system—the correlation length $\xi$. The correlation length measures the size of ordered patches in the disordered phase or of disordered patches in the ordered phase.

In the finite systems accessible by computer simulation true divergences cannot occur, since the system dimension is the largest available length scale. The behavior of these diverging quantities in a finite system is determined by the ratio of the correlation length to the system size, and at the transition temperature quantities like the susceptibility increase with system size in a way which depends on the particular critical exponents associated with that transition.

A new method for distinguishing first- and second-order transitions (an issue which has plagued the study of two-dimensional melting) has recently been advanced by Lee and Kosterlitz [42]. This method uses the behavior of histograms of, e.g., the occurrence of particular values of the order parameter in the generated configurations, to define the order of the transition. At a first-order transition, two-phase coexistence dictates the appearance of a double-peaked structure in such histograms. At a first-order transition the two peaks become increasingly distinct as the system size is increased even for relatively small system sizes (less than the correlation length $\xi$). Thus, the size dependence of the histogram structure is a sensitive probe of the nature of the transition. This technique is currently being applied to the problem of two-dimensional melting [43]. Its application in distinguishing a hexatic phase from a region of two-phase coexistence would seem to be natural.

Even away from a phase transition the study of system size effects is important and can be quite illuminating. In the simplest case, one simply tracks the dependence of thermodynamic quantities or correlation functions on system size to ascertain whether the system is large enough that size and boundary effects are negligible. In other cases, the presence of long-range translational order is determined by the divergence with system size of the calculated Bragg intensity at the appropriate scattering wave vectors. In some two-dimensional systems, critical behavior is observed throughout an entire phase rather than at an isolated point and finite size scaling may be used to identify the phase. The two-dimensional hexatic and random-tiling quasicrystal (see Section 2.3.2) phases are both in this category and, as will be described below, the study of system size dependence is a primary tool for understanding these phases. In complicated systems, especially when different types of ordering are in competition, important effects may occur at large length scales even away from the vicinity of a phase transition. In Penrose-tiling-type quasicrystals the effective "unit cell" is infinite and important contributions from long-length scales are unavoidable. The study of size dependence is crucial for these systems as well.

In spite of the well-known importance of finite size effects, the study of size dependence is neglected in a surprising number of simulation studies (probably because of limited computer resources). In some simple systems the systematic study of these effects is not crucial because, if phase transitions are not under investigation, the size effects become negligible at accessible system sizes. However, in the complicated studies of bond-orientational order described here these effects cannot be neglected and must be addressed in any serious simulation.

## Histogram Method

A beautifully simple method for dealing with the necessity of very long Monte Carlo runs in systems where equilibration is slow has been invented by Ferrenberg and Swendsen [44] and elaborated by others [45–48]. While this method has yet to be applied in any of the studies described here, it is certain to be of importance in such studies in the future. The method allows the use of Monte Carlo data generated at one value of external parameters (such as temperature) to be used in computing averages at another nearby value and can thus significantly reduce the total required computation time.

Given a probability distribution $P_T(E)$ obtained at temperature $T$, the distribution $P_{T'}(E)$ may be obtained in the following way:

$$P_T(E) = Z^{-1}W(E)\exp(-\beta E), \tag{2.7}$$

where $Z$ is the partition function, $W(E)$ is the temperature-independent density of states of energy $E$, and $\beta$ is the inverse temperature. Then

$$P_{T'}(E) = P_T(E)\exp[-\beta(E' - E)], \tag{2.8}$$

may be obtained by a simple transformation of $P_T(E)$. Numerically this technique is limited to a range of temperatures in which the important energies are represented with good statistics in $P_T(E)$. However, combining information from several histograms can significantly extend the temperature range of applicability of the method [45,46].

### Special Monte Carlo Moves

As we have said already, the practical efficiency of the Monte Carlo method depends strongly on the choices of allowed transitions between configurations. Thus, while straightforward single-degree-of-freedom moves are still the standard and most common approach, there is an increasing trend toward the design of special Monte Carlo moves to improve the simulation efficiency for particular systems. These moves range from simple particle rearrangements such as the one used in the quasicrystals simulation described in Section 2.3.2 [49,50], to elaborate schemes for transforming the system from one representation to another in which an extended change in configuration is made before transforming back again (as in the Swendsen–Wang Potts-percolation algorithm [27]). All of these moves deal with the same basic difficulty: the simulated system often gets "stuck" in configurations which are separated from other configurations which are important in the statistical mechanical average by high energy barriers to single-particle changes. The special moves are used to allow the system to hop directly from one energy basin to the next.

An example is shown in Figure 2.4. Here we see an Ising spin configuration below the critical temperature. Most of the system has oriented in the "up" direction. Near the center, however, is a large cluster of "down" spins. At this temperature this configuration is not desirable—the cluster should turn over and point "up." However, single-spin flips will be slow to accomplish this since they are only favorable near the edge of the cluster.

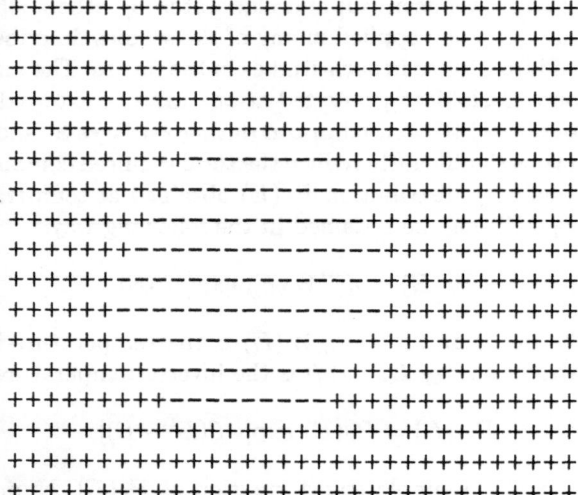

FIGURE 2.4. An Ising model with a large cluster of incorrectly oriented spins.

A many-particle move which flipped the whole cluster would be advantageous. A word of caution is in order here, however. A procedure which involved *searching* for clusters of this type and then flipping them would not be allowable since it would violate the detailed balance condition [Eq. (2.3)]. There must be a nonzero probability for a move which returns the system to the initial state (Figure 2.4). A procedure in which large blocks of spins were chosen *at random* before attempting to flip them according to the Metropolis method would, however, be allowed.

If appropriate care is taken in satisfying the necessary conditions (ergodicity and detailed balance), cleverly designed Monte Carlo moves can greatly enhance the power of the method.

## Boundary Conditions

For the most part simulations are performed using periodic boundary conditions. The basic reason for this choice is that a system with periodic boundary conditions has no edges—all degrees of freedom are in the "bulk." Thus, while "size effects" remain in a simulation with periodic boundary conditions, the boundary conditions have the effect of eliminating "edge effects." Edge effects can be substantial in small systems and are often not well understood. The general procedure in simulations with free or pinned boundary conditions is to throw out the information from some edge layer (usually defined empirically by seeing where the measured properties begin to deviate significantly from those observed in the central "bulk" region). This necessity significantly reduces the effective size of the simulated system and requires the simulation of significantly larger systems.

It would seem, therefore, that periodic boundary conditions are the more attractive choice. However, one cannot be completely sanguine about the imposition of these boundary conditions for systems with important bond-orientational degrees of freedom. The periodic boundary conditions themselves pick out directions in space and thus do not allow the system complete freedom in choosing its rotational symmetry. Periodic boundary conditions are particularly problematical for systems whose preferred orientational symmetry (e.g., fivefold) is inconsistent with periodicity.

Thus, in simulations of quasicrystals care needs to be taken in the choice of boundary conditions. One approach is to use free boundary conditions and put up with the limitations on system size. Another approach is to use the "best periodic approximants" to quasiperiodic structures. These periodic approximants are the finite structures which are closest to quasiperiodic for their system size. They may be defined using the "projection method" for finding quasiperiodic structures. A one-dimensional example is shown in Figure 2.5. A one-dimensional sequence can be obtained by projection from a two-dimensional lattice in the following way: A projection line is chosen and so is a projection strip of width encompassing a unit square. The sequence is obtained by projecting all lattice points within the projection strip onto the projection line. If the slope of the projection strip is irrational, then the projected sequence will never repeat. The Fibonacci sequence is obtained by choosing the slope equal to the golden mean. A close approximation to the quasiperiodic sequence may be obtained by modifying the slope slightly away from its irrational value to a nearby rational value. In that case, a periodic sequence will be obtained but it will be "close" to the quasiperiodic sequence. If the slope is chosen to be one of the "best rational approximants" to the golden mean, then the projected structure will be a best periodic approximant to the Fibonacci sequence. Periodic approximants to higher-dimensional quasiperiodic structures are obtained in an analogous way. (More detailed discussion of the projection method and periodic approximants may be found in some of the articles reprinted in [6].) One study of the effects of boundary conditions in a quasiperiodic system is reported in [51].

Periodic boundary conditions can also cause difficulties in simulations of phase transformations of more standard periodic systems in which the density and shape of the unit cell are kept fixed. In those cases, the behavior of the system may depend in unphysical ways on the extent to which the thermodynamically stable structure "fits" nicely into the unit cell, artificially stabilizing structures which fit and making it hard to cool into structures which do not fit. These problems can be addressed by using constant pressure boundary conditions or boundary conditions which allow for shape changes or by testing the effects of using boundary conditions with varying degrees of fit. While one study did show a significant effect of the dimensions of the unit cell on the melting behavior of a two-dimensional

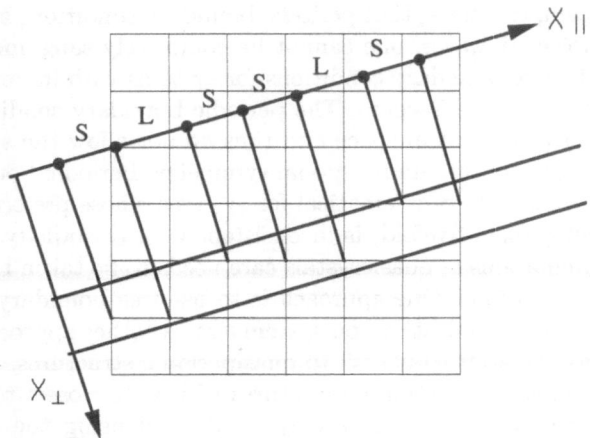

FIGURE 2.5. The projection method of obtaining a one-dimensional quasicrystal by projection from a two-dimensional square lattice.

system [52], the effects of periodic boundary conditions on the questions of bond-orientational ordering remain largely unexplored.

# 2.3    Examples of Computer Simulation Studies of Bond-Orientational Order

In the remainder of this chapter we present several examples from our own work and that of others of computer exploration of bond-orientational order. Many of these topics have been extensively reviewed in other chapters of this book and in other arenas. The examples are simply intended to illustrate some of the approaches which have been taken. In this book the reader is directed to the chapters written by Sachdev and by Goldman for other examples.

## 2.3.1    Measurement of Bond-Orientational Order in a Two-Dimensional Atomistic System

The publication of the defect-mediated melting theory with its prediction of a novel, bond orientationally ordered "hexatic" phase in two-dimensional systems prompted a large number of computer simulations [53]. The theory suggests that two-dimensional melting proceeds in two stages. In the first stage dislocation pairs unbind to form a bond orientationally ordered hexatic liquid. In the second the disclination pairs which make up the dislocations unbind to form an isotropic liquid. Studies of bond-orientational order in two-dimensional atomistic systems are available for a variety of interaction potentials. The results have been quite controversial. Here we describe one of these studies in some detail. The reader is referred to [53] for a reasonably up-to-date discussion of this controversy.

In 1987 Udink and van der Elsken [54] and Udink and Frenkel [55] published studies of two-dimensional melting in a large system of 12,480 particles interacting with a Lennard–Jones potential at a density $\rho\sigma^2 = .873$. This work included a detailed study of the size dependence of the bond-orientational order in this system. The hexatic phase is predicted to have algebraically decaying bond-orientational order. Such power law correlation function decay is analogous to that observed at a critical point and specific finite-size-scaling predictions can be made. The detailed size-dependence study allows the strong conclusion that a bond orientationally ordered phase is observed in this system.

While the authors hesitate to call this region a hexatic phase because of the high density of defects in this phase (the theory was derived in the low defect density limit), the agreement of the exponents describing the finite size scaling of the bond-orientational order with the predictions of the defect-mediated melting theory suggests that hexatic is an appropriate description of this region. Also, although the melting transition is predicted to occur at the point when the first dislocation pair unbinds, local defects which renormalize the interaction between a widely separated pair are expected even below melting and will certainly be seen in the hexatic phase itself. Thus, the presence of even a high density of local defects (such as grain boundary loops) near melting does not preclude a dislocation-unbinding transition.

The observed bond-orientational order is perhaps surprising in light of the fact that the appearance of the configurations is consistent with their description as crystallite patches separated by grain boundaries. Such visual inspection of the configurations has led a number of workers to conclude that they exhibit two-phase coexistence [53]. Interestingly, the melting line above which the orientationally ordered phase is observed is consistent with the melting line predicted by a free energy calculation which assumes a first-order transition. However, bulk phase separation is not observed. One might comment in addition here that (as has been pointed out by Fisher, Halperin, and Morf [56]) a tendency of dislocations to cluster in grain boundary loops is consistent with predicted dislocation correlations in the hexatic phase.

From the point of view of this book, of course, the label which is attached to this region of the phase diagram is less important than the very interesting observation of a bond orientationally ordered phase in this simple two-dimensional system. Here we give a brief description of the simulation evidence for this bond-orientational order.

The bond-orientational order parameter is defined in the simulations as

$$\psi_6(r_i) = \frac{1}{N_i} \sum_{j=1}^{N_i} \exp(6i\theta_{ij}), \tag{2.9}$$

where the sum is over all neighbors $j$ of the particle $i$ as determined in a Voronoi polygon construction and $\theta_{ij}$ is the angle between the bond linking particles $i$ and $j$ and a fixed reference axis. The bond-orientational correlation function

$$g_6(r) = \langle \psi_6(r)\psi_6^*(0) \rangle \tag{2.10}$$

is also obtained. A translational order parameter was also calculated. In the solid phase the translational order parameter should go as $N^{-\eta}$, where $\eta$ can reach no value larger than $1/3$. In the hexatic phase the bond-orientational order parameter should have a similar dependence $N^{-\eta_6}$ (where $\eta_6$ should go to zero in the solid). The defect-mediated melting theory would predict that $\eta_6 = 1/4$ at the hexatic-isotropic transition. Power-law-size dependence in either the translational or bond-orientational order parameter indicates quasi-long-range order of the corresponding type. The bond-orientational order correlation length was also computed using a fit to an exponential form for the bond-orientational correlation function.

A divergence of this correlation length indicates the onset of orientational ordering. The results of these fits are shown in Figure 2.6. The translational exponent $\eta$ goes through $1/3$ at a temperature $T_m$ well separated from the temperature $T_i$ at which the exponent $\eta_6$ goes through $1/4$ and the bond-orientational correlation length diverges. The melting temperature $T_m$ is consistent with the transition temperature obtained by other means. The solid curve through the orientational correlation lengths is obtained by a fit to the prediction of the defect-mediated melting theory

$$\xi_6 = \exp\left[\frac{b}{\sqrt{|T - T_i|}}\right]. \tag{2.11}$$

Thus, the region between $T_m$ and $T_i$ is clearly a region where translational order is lost but bond-orientational order is retained. In addition, the results of the finite-size-scaling analysis are quantitatively in accord with the predictions of the defect-mediated melting theory.

This study provides an excellent illustration of the power of the study of finite size effects in approaching these issues.

## 2.3.2    Behavior of a Two-Dimensional Atomistic System in the Presence of a Hexatic Field

A rather different approach to the study of the two-dimensional melting transition has been taken by Morf [57] in a study of a two-dimensional

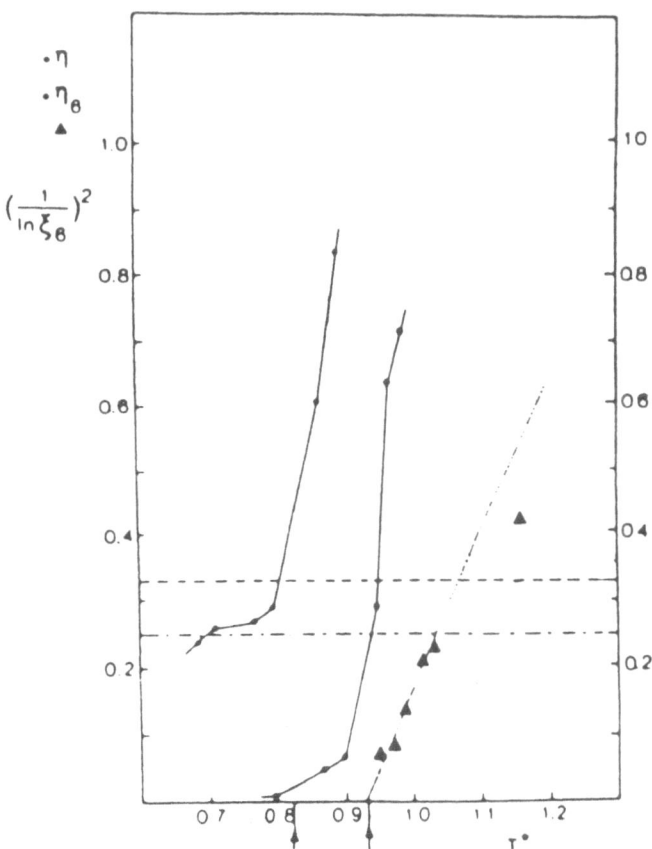

FIGURE 2.6. Scaling results for the algebraic exponent for translational order $\eta$ (closed circles) and for the algebraic exponent for orientational order $\eta_6$ (open circles). The dotted horizontal line is the critical value $\eta = 1/3$ and the dash-dotted horizontal is the critical value $\eta_6 = 1/4$. The triangles are the values of $[1/\ln(\xi_6)]^2$, with $\xi_6$ the orientational correlation length. (From Ref. 55.)

atomistic system interacting with a $1/r$ potential. Because of long-time-scale fluctuations, Morf was never convinced that equilibrium was obtained in his simulations of melting in these systems. There was the additional difficulty in recognizing a hexatic phase in a simulation at finite system size and distinguishing it from a two-phase coexistence region (as discussed above).

Morf got around both of these difficulties by the introduction of a hexatic field coupling directly to the hexatic order parameter. Although the graphite substrate in experiments on physisorbed systems may be thought of as introducing such a field, experimental control over the hexatic field is not available. By looking at the behavior of the phase transition at finite hexatic field, Morf hoped to be able to extrapolate his observations to provide insight into the zero-field behavior. Thus, these computer simulation results were able to provide an important contribution to the debate about the melting transition in this system by tuning a parameter unavailable to experiment.

In these simulations the bond-orientational order parameter was defined as

$$\Psi_6 = \frac{1}{6N} \sum_{i,j} f(r_{ij}) \cos 6\theta_{ij}, \qquad (2.12)$$

where $f(r_{ij})$ is a smooth cutoff function which cuts off between nearest and next-nearest neighbors and the sum is over all pairs, and the interaction term

$$V_6 = -\frac{1}{6} h_6 \sum_{i,j} f(r_{ij}) \cos 6\theta_{ij} \qquad (2.13)$$

was added to the Hamiltonian. This interaction term provides long-range bond-orientational order at all temperatures and thus eliminates the possibility of a hexatic-istropic fluid transition. The dependence of the melting temperature on hexatic field was then explored.

For a first-order transition, the Clausius–Clapeyron equation relates the slope of the melting curve $dT_m/dh_6$ to the ratio of the discontinuity in order parameter $\Delta\Psi_6$ and the melting entropy $\Delta S$. $\Delta\Psi_6$ and $\Delta S$ were available from previous simulations. These calculations predict a slope of $2.2 k_B^{-1}$. A slope can also be predicted from the dislocation-mediated melting theory using values for elastic constants and defect core energies measured in the solid. These calculations yielded $dT_m/dh_6 = .9 k_B^{-1}$.

Simulations for systems of size 780 and 1560 were carried out for fields in the range $1/24 T_m < h_6 < 1/3 T_m$. In this range of hexatic field the melting was reversible and no hysteresis was observed. The slope of the melting curve observed was in good agreement with the prediction of the dislocation-mediated melting theory. Thus, these simulations show that, at least in the presence of small hexatic fields, the melting behavior of this system with a $1/r$ potential is consistent with the dislocation-mediated melting theory rather than the first-order melting theory. Thus, unless the

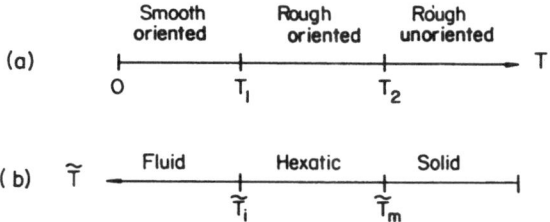

FIGURE 2.7. The duality relation between the interface phases of the Laplacian roughening model and the phases predicted by the defect-unbinding theory of two-dimensional melting.

transition behavior changes at some lower critical value of $h_6$, the melting is likely to be described by the defect-mediated theory at $h_6 = 0$ as well.

### 2.3.3    Study of the Hexatic Phase in a Defect-Based Simulation

In order to explore more fully the consequences of the concept of defect-mediated melting and to clarify the range of parameters in which the approximate theoretical calculations of Halperin and Nelson are applicable, direct simulations of the defect system were performed [58–63]. The defect-unbinding theory predicts two phase transitions—a solid-hexatic transition driven by the unbinding of dislocation pairs followed by a hexatic-isotropic phase transition driven by the unbinding of disclination pairs. In these defect simulations the bond-orientational order is not directly measurable, but is assumed to be implicit in the status of the defects. Unbound dislocations should indicate algebraically decaying bond-orientational order and unbound disclinations should indicate short-range bond-orientational order.

Simulations of dislocations alone were performed by Saito [60,61]. In these simulations the hexatic-to-isotropic transition was inaccessible. Here we describe our simulations of the Laplacian roughening model—a model related by a mathematical duality transformation to the defect model and one in which both transitions may be observed [62,63].

A duality transformation is an operation performed on the partition function which relates the high-temperature phase of one model to the low-temperature phase of another. Nelson [64] showed that the Laplacian roughening model described here is related by duality to the interacting defect system relevant for two-dimensional melting. A major computational advantage of working with the roughening model is that the logarithmic defect interactions are transformed to short-range interactions by this procedure.

The Laplacian roughening model describes a solid-vapor interface in

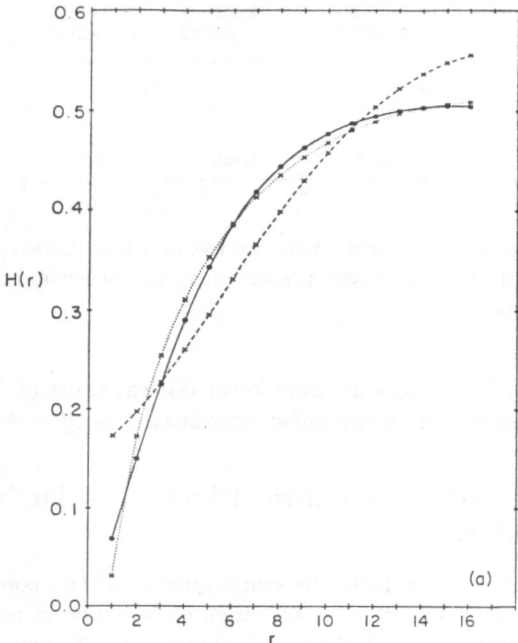

FIGURE 2.8. Height-height correlation functions for the Laplacian roughening model in the (a) intermediate and (b) high-temperature phases. Shown in the figure are the Monte Carlo data (·——·——·) and fits to the predicted forms for the intermediate (x· · ·x· · ·x) and high-temperature (x—x—x) phases.

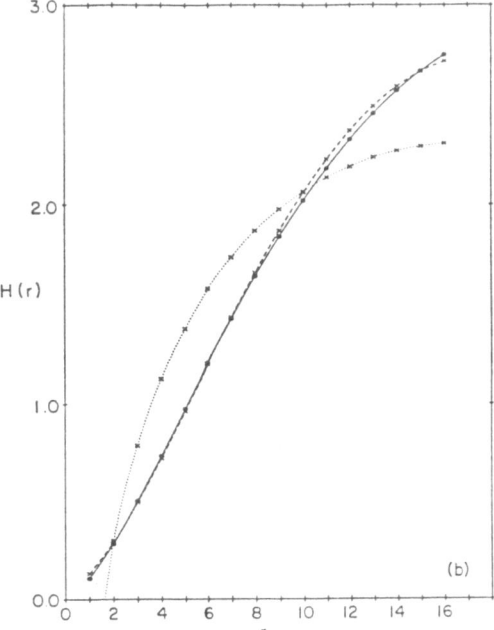

FIGURE 2.8. (b)

terms of the well-known solid-on-solid representation in which the interface is represented by columns of height $h(\mathbf{r})$ arranged here in a triangular lattice. The interaction

$$H = -J \sum_{\mathbf{r}} \left[ 6h(\mathbf{r}) - \sum_{j} h(\mathbf{r} + \hat{\delta}_j) \right]^2 \qquad (2.14)$$

(where $\delta_j$ are the nearest-neighbor vectors) is invariant under uniform translations of the interface by an integer and appropriate uniform tilts of the interface. The system thus has the possibility of three phases (see Figure 2.7) including an intermediate phase dual to the hexatic phase. Simulations of this system revealed the existence of the predicted three phases with correlation function behavior as expected from the defect-unbinding theory. Figure 2.8 shows the height-height correlation function

$$H(r_{ij}) = \langle [h(\mathbf{r}_i) - h(\mathbf{r}_j)]^2 \rangle \qquad (2.15)$$

in the intermediate- and high-temperature phases along with fits to the forms predicted by the defect-unbinding theory. These predicted forms include the finite size dependence appropriate to the periodic boundary conditions.

The Halperin–Nelson calculations were performed in the limit of large chemical potential or "core energy" for defect creation. In order to investigate the effects of relaxing this restriction, a modified version of the Laplacian roughening model was also simulated. It was found using a finite-size-scaling analysis that the two continuous transitions combined at lower core energy into a direct first-order transition between solid and liquid phases (see Figure 2.9). Thus, these simulations showed that, even if the defects drive the melting behavior, the possibilities of both first-order and continuous transitions exist and the existence of the hexatic phase depends on the value of the defect core energy. The simulations also provided some information about the critical exponents characterizing the crossover point. A more extensive simulation study of the critical properties of this system might provide a point of comparison for experimental studies of the phase transitions in liquid crystal hexatic phases.

## 2.3.4    Effects of Frustration on Glass Formation in a Two-Dimensional Lennard–Jones System

The suggestion that glass formation in simple metallic glasses may be due to the competition between the local preference for icosahedral ordering [65] and the global cubic crystalline ordering [66] leads to the hypothesis that two-dimensional systems, in which the preferred local atomic arrangement is compatible with periodicity, will not be glass-formers. A Monte Carlo study by Wong and Chester [67] of a two-dimensional system of Lennard–Jones atoms confirmed this speculation. In addition, they proceeded to

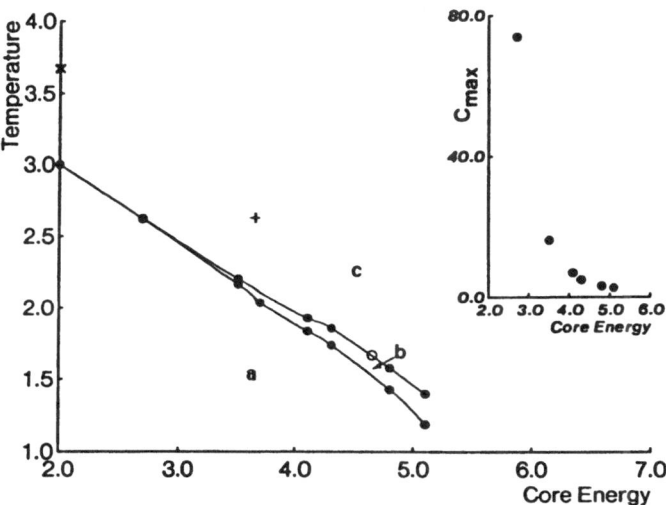

FIGURE 2.9. Phase diagram of the modified Laplacian roughening model as a function of temperature and of the core energy of the corresponding disclination system. Also shown in the inset is the maximum specific heat for a 1024-site system as a function of disclination core energy.

investigate the effect of introducing a frustration similar to that which occurs in three-dimensional monoatomic systems into their two-dimensional systems through the introduction of two different particle sizes. Their results support the idea of frustration as a mechanism for glass formation. In systems with size ratio above about 1.3 and with fraction of large atoms above about .2, glassy states were achieved (see Figure 2.10).

In addition, they investigated the hexatic bond-orientational order of their systems. Earlier work by Nelson, Rubinstein, and Spaepen [68,69] suggested that defects with relatively small size ratios (about 1.1 − 1.2) might destroy the long-range solidlike bond-orientational order, but leave bond-orientational correlation lengths which are significantly longer than the translational correlation length. The Wong and Chester studies of the bond-orientational order correlation functions found slightly longer orientational correlations but no interesting hexatic order was observed. The question of whether the size disparity in these systems might induce some other (non-sixfold) bond-orientational order was not pursued.

## 2.3.5  Square-Triangle Analysis

Glaser et al. [70–73] have used an analysis of local bond-orientational order and modeling based on that analysis to suggest that two-dimensional melting is driven primarily by a condensation of local, geometrical defects. In their picture (related to earlier modelling by Collins, Kawamura, and others

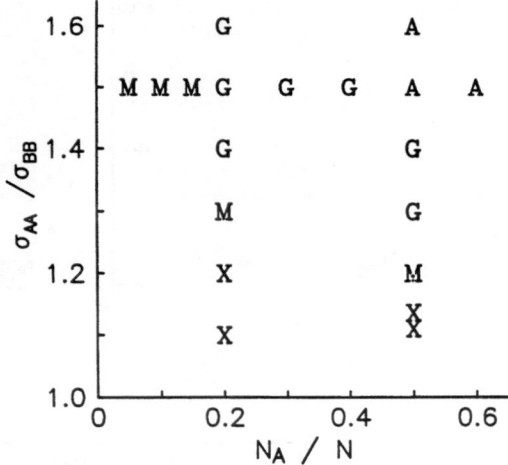

FIGURE 2.10. Summary of the results of quenches to a final reduced temperature of 1.0 of the two-component Lennard–Jones systems at the average reduced density 1.143 as a function of the concentration of $A$ atoms and the ratio of the $A$ and $B$ atomic sizes. The different final states obtained are glasses (G), amorphous but nonglasslike solid (A), microcrystalline state (M), and fairly well-ordered solid (X). (From Ref. 67.)

[74–76]) the local bond-orientational order is used to divide the structure into triangles and squares (with occasional larger polygons). The procedure consists roughly of performing a Voronoi polygon analysis to define near-neighbor bonds and then deleting those bonds which are geometrically closer to being the diagonal of a square than the side of an equilateral triangle. An example of the tiling analysis of a configuration generated by molecular dynamics simulation of a simple pair potential is shown in Figure 2.11.

In this tiling analysis the perfect solid is a tiling composed of triangles alone and the squares represent defects in the structure. Unlike the dislocations and disclinations discussed so far, which are long-range and topological in character, these defects are local and geometrical. Such defects cost energy because they require bonds of the wrong length, but provide a gain in entropy due to the possible tilings available. In order to qualitatively describe the configurations found in the molecular dynamics simulations "tiling faults" (untileable regions) are also required. Based on their analysis of many solid and liquid configurations generated by molecular dynamics, Glaser et al. have proposed a class of statistical mechanical models including a creation energy for nontriangular tiles, interactions between these "defect" tiles, and an energy for the creation of tiling faults. Monte Carlo simulations of these models show that a first order "condensation" phase transition is observed for many values of the input parameters.

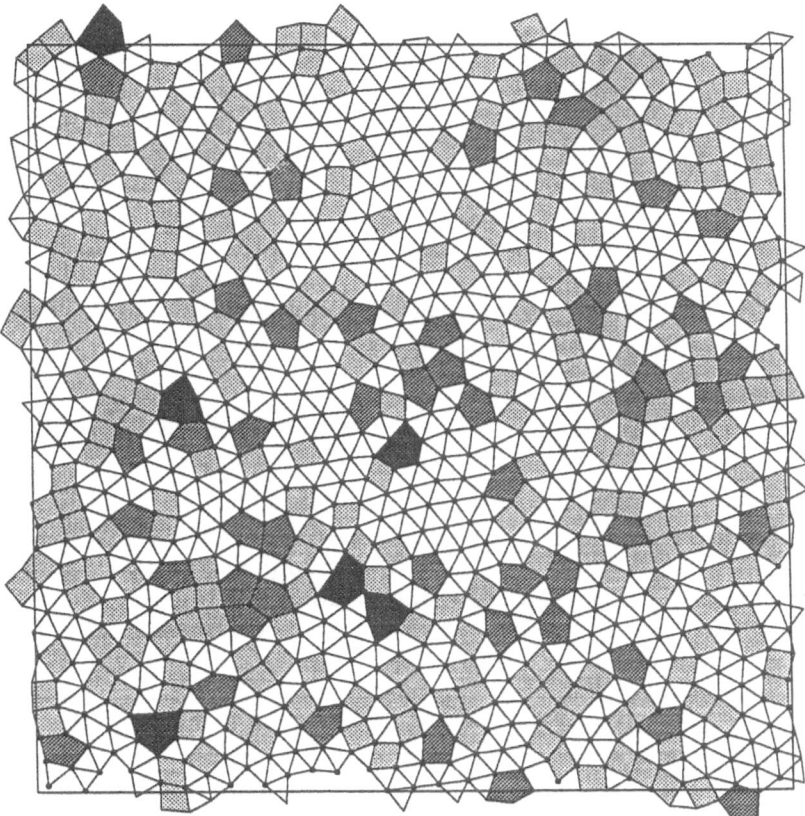

FIGURE 2.11. Polygon construction for a representative configuration of the liquid obtained by molecular dynamics simulation. Polygons with four or more sides are shaded according to the number of sides. Intact nearest-neighbor bonds are indicated by solid lines, and the particle positions are marked by solid dots. (From Ref. 73.)

Specific comparisons of the statistics of the model configurations with the molecular-dynamics-generated particle configurations show a general qualitative similarity in the local structure. However, it is clear that topological defects are not adequately accounted for in the models so far. Even when the topology of the underlying lattice is allowed to vary, the long-range defect interactions are not adequately represented. As a result, unphysical configurations were observed (see Figure 2.12) containing isolated disclinations. Quite apart from the success or failure of these specific attempts to incorporate the observation of a square-triangle structure into a statistical mechanical model, however, the extent to which these local geometrical defects, which are clearly present near the melting transition, participate in or precipitate melting is an intriguing, open question.

### 2.3.6     A Simple Atomistic Model Possessing an Equilibrium Quasicrystal Phase

In order to explore the possibilities for bond-orientational order and translational ordering in a system with a local preference for a nontraditional local bond-orientational order, Widom, Strandburg, and Swendsen [50,77] designed a two-dimensional atomistic model interacting with simple central forces but with a local preference for tenfold bond-orientational order. Two different types of atoms were required for this study since two-dimensional monatomic systems interacting with central forces always form a triangular lattice. The parameters of the interaction were determined so that, at least as far as nearest-neighbor interactions, the atoms would experience a local minimum of the potential at the vertices of the Penrose tiles (see Figure 2.13). The positions of the atoms during the simulations were not constrained to the tile vertices.

Specifically, the two types of atoms [large ($L$) and small ($S$)] interacted with the potential

$$V_{\alpha\beta}(\mathbf{r}_{ij}) = \varepsilon_{\alpha\beta} \left[ \left( \frac{\sigma_{\alpha\beta}}{r_{ij}} \right)^{12} - 2 \left( \frac{\sigma_{\alpha\beta}}{r_{ij}} \right)^{6} \right], \tag{2.16}$$

where $\alpha$ and $\beta$ are $L$ and $S$ for large and small atoms, respectively, $r_{ij}$ is the distance between particles $i$ and $j$, and the model parameters are given by

$$\varepsilon_{LL} = \varepsilon_{SS} = \tfrac{1}{2}, \qquad \sigma_{LL} = 2\sin\tfrac{\pi}{5} = 1.17557\ldots, \qquad \sigma_{LS} = 1$$

$$\varepsilon_{LS} = 1, \qquad\qquad \sigma_{SS} = 2\sin\tfrac{\pi}{10} = 0.61803\ldots. \tag{2.17}$$

In these simulations free boundary conditions were utilized to avoid any interference of the boundary conditions with the bond-orientational ordering. The system was equilibrated at a high initial temperature and then gradually cooled until a freezing transition occurred. An important part of

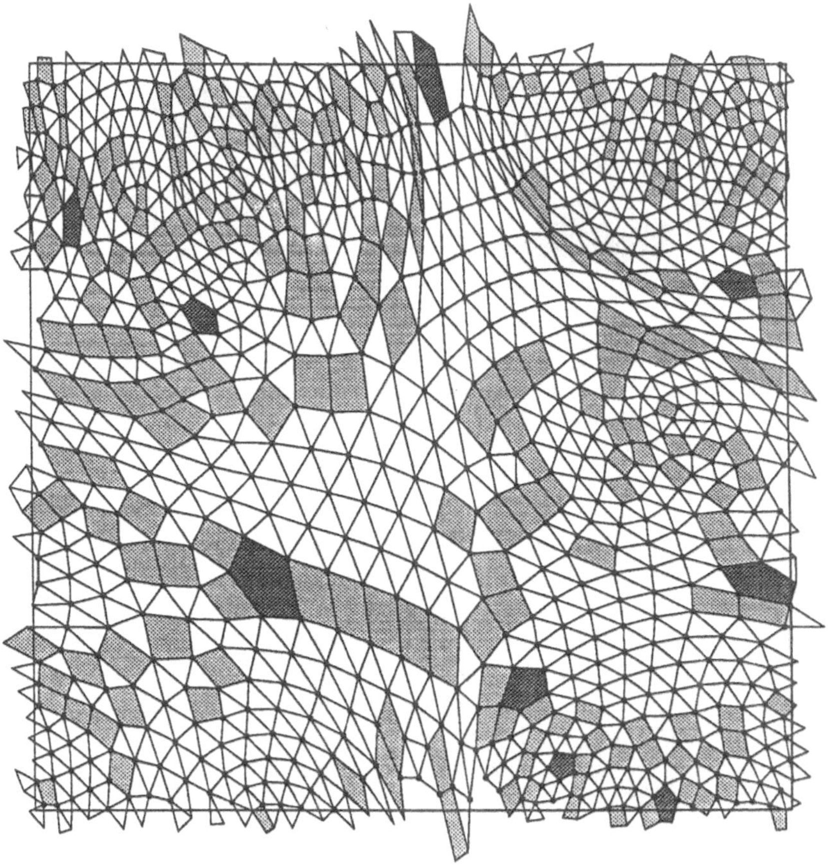

FIGURE 2.12. Liquid-phase configuration resulting from Monte Carlo simulation of a specific version of the square-triangle-tiling-based model. Note the similarity of the local structure to that of Figure 2.11. Note also the unphysical structure on longer distance scales. (From Ref. 73.)

FIGURE 2.13. The preferred separations of large and small atoms in the two-dimensional atomic quasicrystal model.

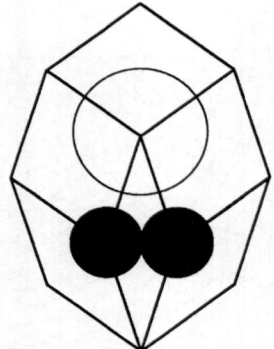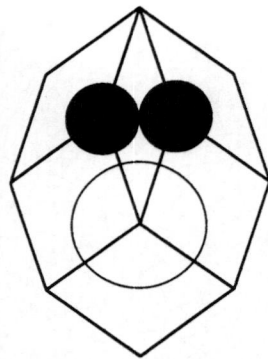

FIGURE 2.14. The particle rearrangement move employed in the simulation. The corresponding tiling rearrangement (when the system is in the quasicrystal phase) is also shown, but the simulation does not require that the atoms sit on tiles in order to make the rearrangement move.

this simulation was the use of a specially designed Monte Carlo particle rearrangement move (see Figure 2.14) which had the effect of allowing the system to move out of locally metastable, but globally unfavorable, triangular arrangements of three atoms of the same type. Without the inclusion of this move equilibrium could not be achieved in feasible computer runs. (The same model was investigated independently by molecular dynamics simulations [78,79] but equilibrium was not achieved.)

The nature of the solid phase was then explored. We measured the bond-orientational order parameter

$$\theta_n^{\alpha\beta} = \frac{1}{N^{\alpha\beta}} \left\langle \left| \sum_j e^{in\theta_j^{\alpha\beta}} \right|^2 \right\rangle, \tag{2.18}$$

where $\theta_j^{\alpha\beta}$ is defined as the angle the $j$th nearest-neighbor bond makes with an arbitrary axis, the sum is over all nearest-neighbor bonds of type $\alpha\beta$ (i.e., $LL$, $SS$, $LS$), and $N_{\alpha\beta}$ is the number of such bonds. We define nearest-neighbor bonds to be those within a cutoff distance $r_c = 1.1\sigma_{LS}$. Clearly the precise choice of $r_c$ is not crucial in a well-ordered configuration.

By computing this order parameter (see Figure 2.15) and by observing that the configurations were, except for surface defects, tileable with Penrose-type tiles (see Figure 2.16) the tenfold bond-orientational order of the solid phase was confirmed. In addition, this phase was shown, by a size-dependence study of "phason fluctuations" (a measure of the deviation of the tile arrangement from perfect quasiperiodicity) (see Figure 2.17), to have algebraically diverging Bragg peaks indicating not only bond-orientational order, but quasi-long-range translational order as well. Here logarithmic divergence of the phason fluctuations indicated quasi-long-range translational order, saturation to a constant would have indicated

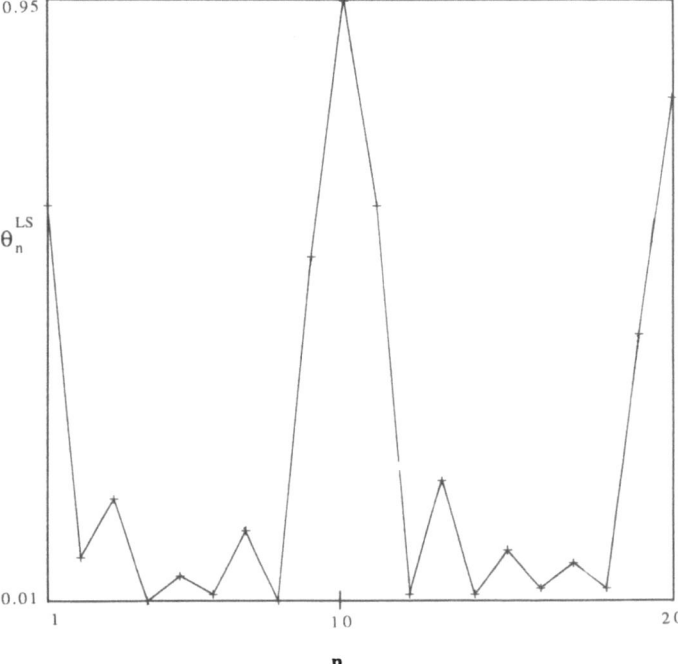

FIGURE 2.15. The bond-orientational order parameter in the quasicrystal phase of the simple atomic model as a function of $n$. Note the strong peak at $n = 10$, indicating decagonal bond-orientational order.

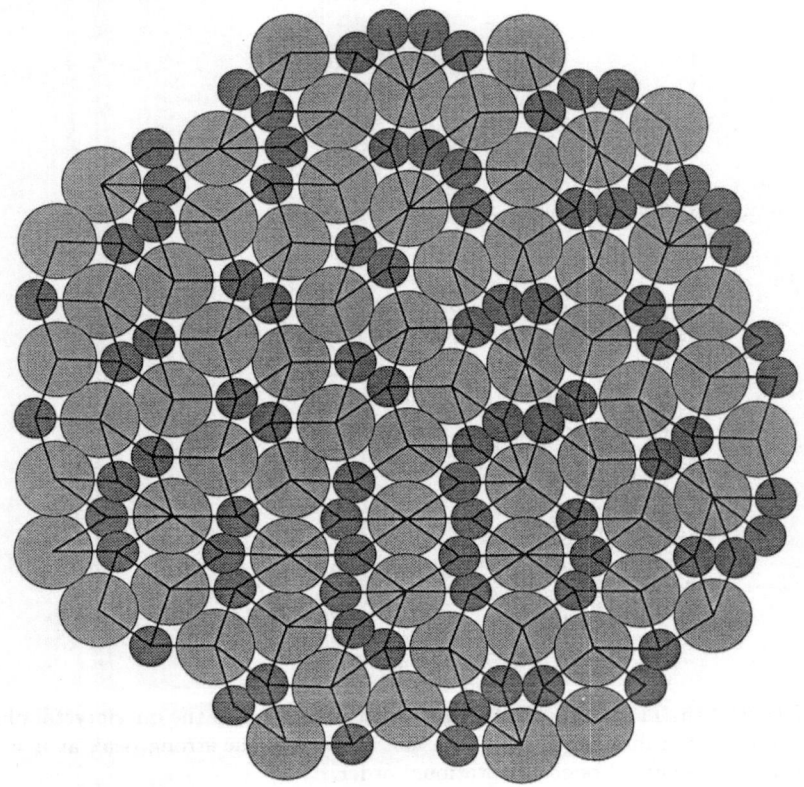

FIGURE 2.16. A sample atomic configuration obtained from the simulation with tiles superimposed.

long-range translational order, and faster divergence would have indicated a loss of translational order.

Thus, this simple atomistic simulation demonstrated the possibility of an equilibrium quasicrystal phase in a system interacting with very simple potentials (and, in particular, without angle-dependent interactions).

Observation of the low-temperature behavior of the particle configurations highlighted an interesting property of this quasicrystal phase. While the system showed a distinct preference for tileable configurations, there

FIGURE 2.17. Size dependence of the phason fluctuations for two temperatures in the quasicrystal phase.

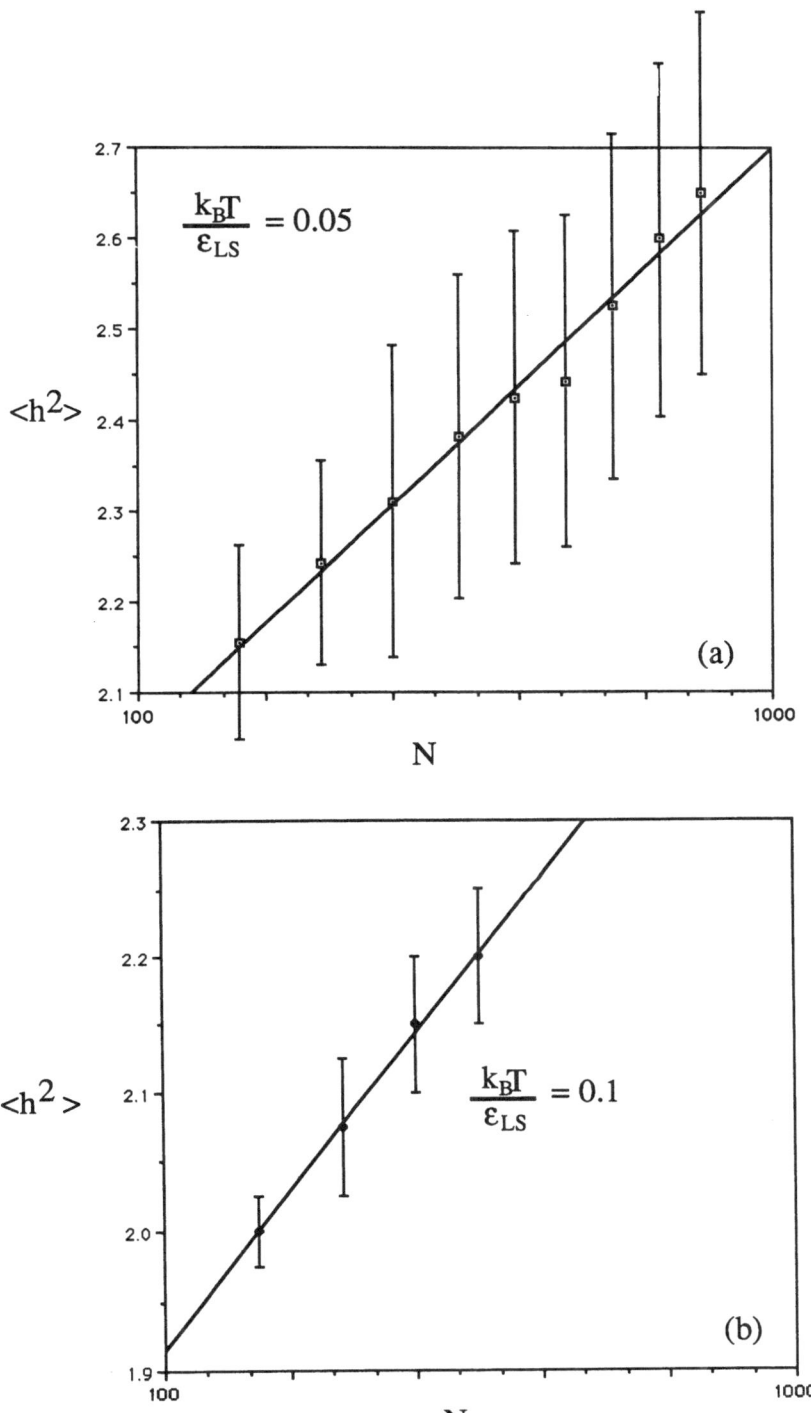

was no noticeable preference for a particular tiling arrangement. This observation gave support to the idea that the entropy provided by the numerous tile arrangements might be an important factor in stabilizing the quasicrystal phase. This "random tiling model" was further explored in the simulations described in the next subsection.

## 2.3.7  Consequences of Nontraditional Bond-Orientational Order in a Random Tiling Model

In order to investigate more fully the consequences of the "random tiling model" suggested by the atomistic simulations above, simulation studies of abstract tiling models in both two [49,50,80] and three dimensions [81,82] have been performed. In these studies we forget about the atoms, and the Penrose tiles (or, in three dimensions, rhombohedra) themselves are selected as the "abstract objects" to be studied. In order to simplify the model as much as possible, all configurations of tiles are taken to have equal energy and thus the acceptance probability for any Monte Carlo move is unity. The Monte Carlo moves which are used in the simulations are illustrated in Figure 2.18. Because we want to explore the ensemble of space-filling tilings, the moves are tile rearrangements which take one space-filling tiling into another. These moves satisfy both the detailed balance (trivially) and ergodicity [83] conditions. Periodic boundary conditions using the periodic approximants described above were used in these simulations.

In spite of the large amount of disorder allowed in these tilings when compared to a perfect quasiperiodic tiling, they were shown numerically to maintain the Bragg scattering associated with quasiperiodicity. More specifically, in two dimensions the delta-function scattering peaks associated with the perfect tiling are exchanged for algebraically decaying peaks, and in three dimensions the Bragg peaks are maintained with diminished intensity and with the addition of some diffuse scattering. Intuitively, one can understand this result to imply that, among space-filling tilings, the quasicrystal is entropically stabilized. In other words, most space-filling tilings are "near" the quasiperiodic tiling.

A more quantitative measurement of distance from the quasiperiodic tiling is provided by the concept of "phason fluctuations" mentioned briefly above. As is illustrated in Figure 2.19, small deviations of the projection strip lead to tile rearrangements. A measure of the fluctuations in the projection strip is obtained by projecting each point in the tiling back up to the higher dimensional space and then measuring the average deviation of the projection strip from the strip corresponding to the perfect quasiperiodic structure. A conjecture by Elser and Henley that a purely entropic free energy proportional to the square gradient of these phason fluctuations would be associated with the space-filling tilings was confirmed by these

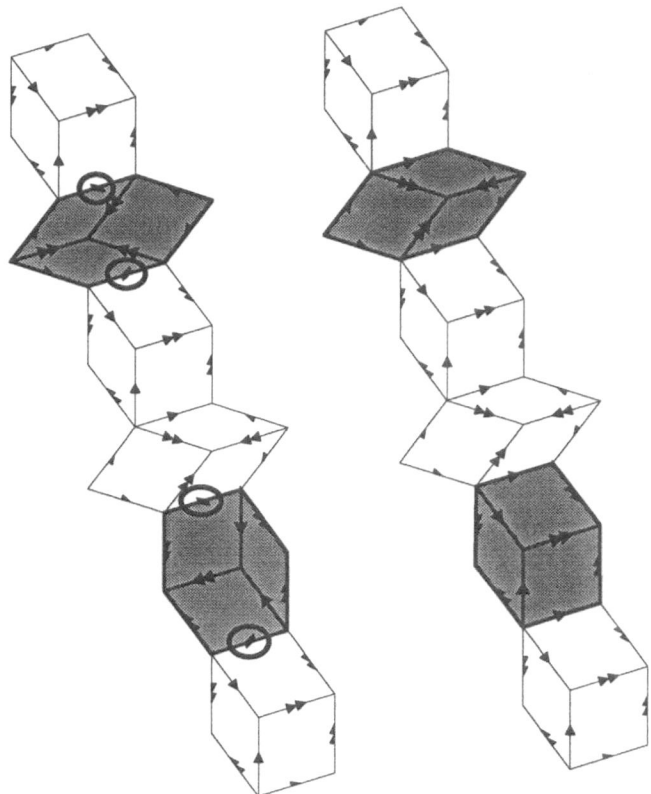

FIGURE 2.18. The flip move used in tiling simulations in (a) two dimensions and (b) three dimensions.

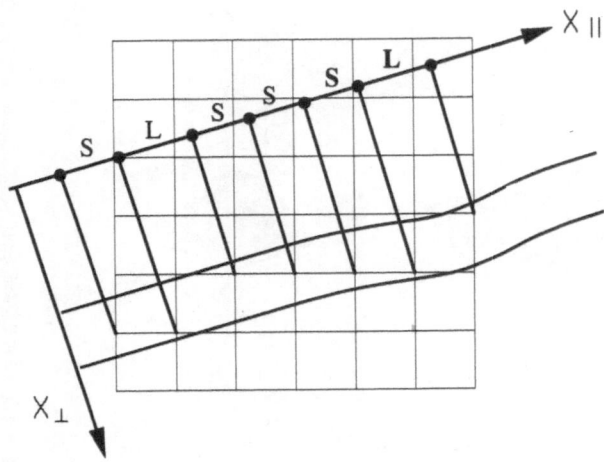

FIGURE 2.19. How strip fluctuations correspond to tile rearrangements.

numerical simulations. The effective elastic constant associated with phason fluctuations was obtained for the random tiling and was shown to agree quantitatively with that obtained in the atomistic simulations described in the previous subsection (see Figure 2.17).

Simulations of three-dimensional random tilings have been used to calculate the entropy associated with the phason fluctuations, to explore the nature of the diffuse scattering associated with the random tiling model, and to predict the dependence of high-resolution electron microscopy observations on the sample thickness. These latter calculations are particularly important since the thermodynamic behavior of the best real quasicrystals seems to indicate that the quasicrystal is a high-temperature (and, therefore, entropically stabilized) phase, but the electron microscopy results do not show evidence of the disorder that might naively be expected in a random tiling quasicrystal. The simulations have shown that, although the random tilings contain large fluctuations around the average quasiperiodic lattice, these fluctuations tend to be averaged out except in very thin samples.

Simulations of this very simple abstract model have thus proved very useful in exploring the consequences of the idea that the bond-orientational order observed in quasicrystals might be due to the large entropy associated with bond orientationally ordered tilings. Even these highly idealized simulations are able to provide qualitative experimental predictions.

## 2.3.8    Growth of a Perfect Quasicrystal

Like the icosahedral glass modeling described in Goldman's chapter in this book, the work of Onoda et al. [30] is an attempt to model quasicrystal growth using cluster aggregation of simple geometrical objects. Here again a particular implementation of the bond-orientational order is assumed and the implications explored. In the Onoda et al. work it is shown that large perfect quasiperiodic crystals (with translational as well as bond-orientational order) can be grown using "almost local" rules. While the results described here have been established analytically, computer implementation of various versions of the rules played a crucial role in the discovery of this growth procedure and can be expected to continue to play an important role in this area of research.

As anyone who has ever played with Penrose tiles can attest, attempts to create large patches of Penrose tiling by aggregating tiles according to the arrow matching rules (shown in Figure 2.20) are continually frustrated by mistakes which cannot be fixed up by local rearrangements of the tiles. This difficulty is often cited as a problem with the quasiperiodic crystal model for quasicrystal structure. If, however, more complicated rules requiring that each vertex created by addition of a tile be one of the eight vertices allowed in a Penrose tiling are used (see Figure 2.21), an almost deterministic procedure results. Specifically, one scans the boundary of the cluster to locate places where the next legal tile is uniquely determined and fills those first. This procedure is continued until a "dead surface" is reached at which no forced tiles exist. At this point, a specific rule is invoked for adding a tile at a corner of the cluster. The procedure then continues.

As pointed out by Jarić and Ronchetti [31,32], these rules are not strictly local since inspection of the entire surface is required to look for forced tiles. However, the rules may be enforced locally by the assignment of local sticking probabilities. The actual structures obtained for a model assigning sticking probabilities $p_f$, $p_0$, and $p_{108}$ to the forced, unforced, and corner sites depend on the ratios of these probabilities. Jarić and Ronchetti point out that $p_0 << 1$, and $p_f/p_{108} >> 1$ (indeed Onoda et al. estimate that $p_f/p_{108} > N$ is needed to grow a perfect quasicrystal of size $N$) are necessary to get reasonably defect-free structures. However, Onoda et al. have also shown that certain defects ensure perfect growth with no dead surfaces when they are used as growth seeds (if $p_0 = p_{108} = 0$). Further work by Ingersent and Steinhardt [33] explores the possible quasiperiodic tilings which can have such local growth.

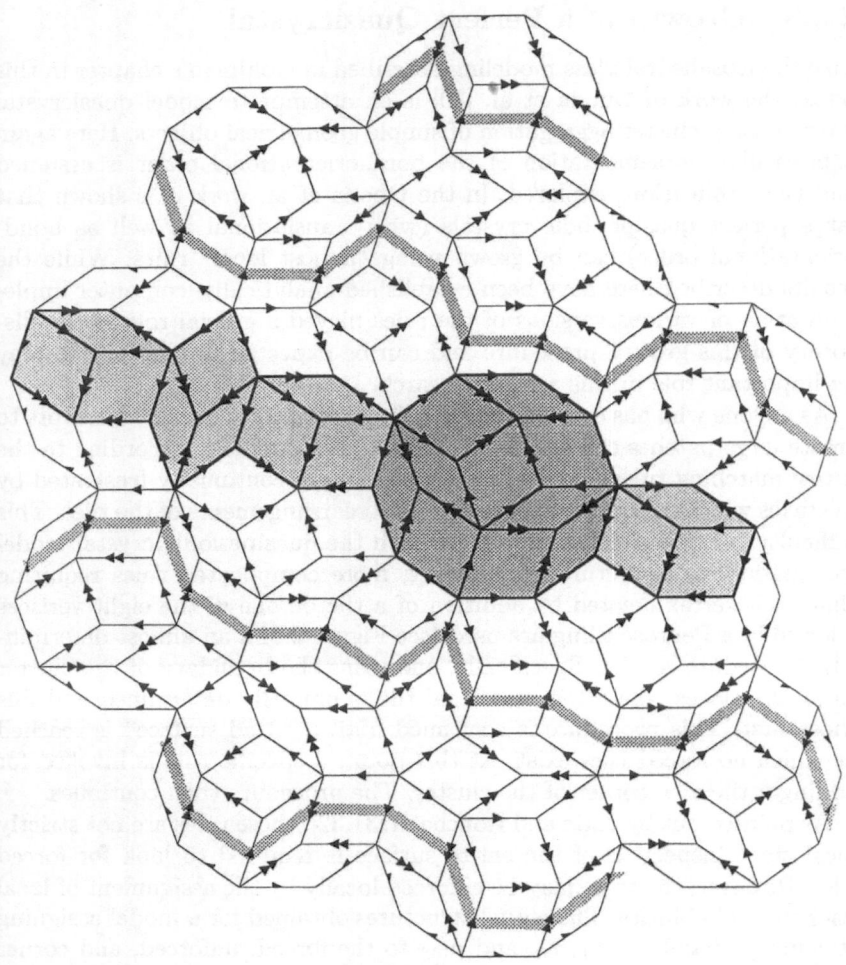

FIGURE 2.20. A portion of a Penrose tiling illustrating the arrow matching rules and highlighting the presence of strings of hexagons within the tiling.

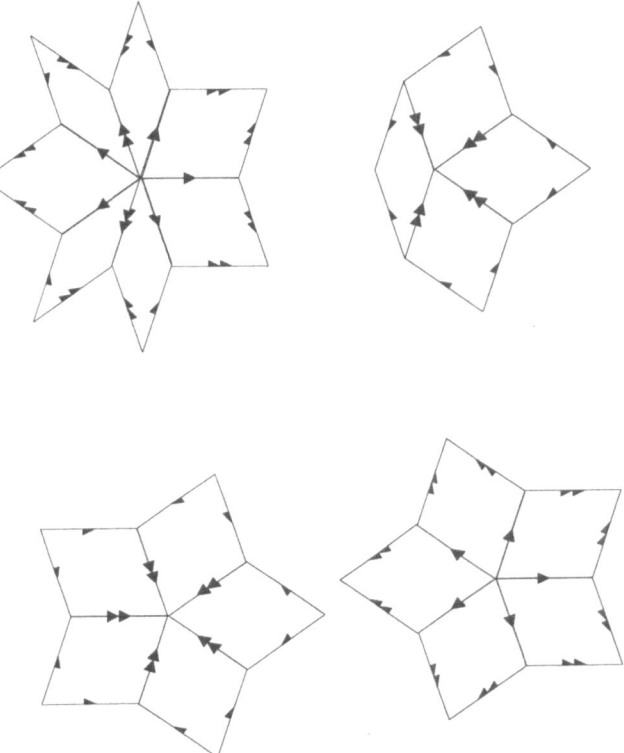

FIGURE 2.21. The eight vertices allowed in a Penrose tiling (figure continued on the next page).

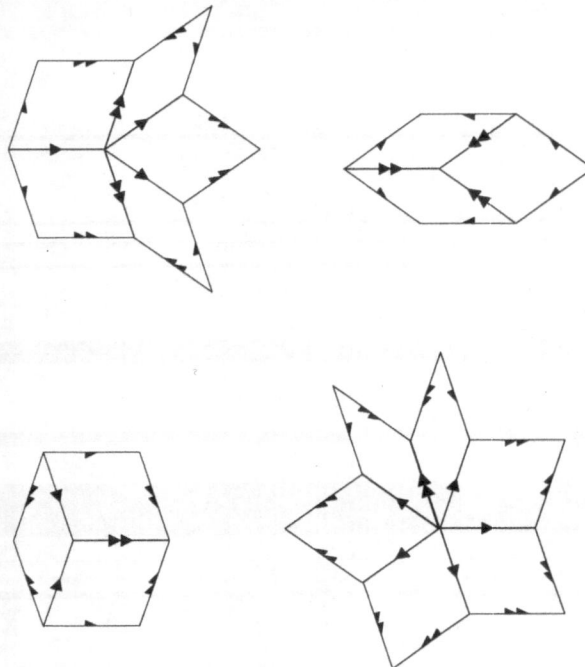

FIGURE 2.21. (*continued*)

## 2.4   Conclusions

The above examples serve to illustrate a variety of ways in which the origins and consequences of bond-orientational order have been explored using numerical simulation. These methods will continue to be of use in such studies, particularly with the implementation of recent advances in dealing with the finite size and time limitations and with continued creativity in simulation design. Possible areas for future work include atomistic simulations of three-dimensional quasicrystals, further tiling simulations of quasicrystal growth and interface roughening, simulations exploring the possibility of bond-orientational order in vortex phases in layered superconductors, and application of the Lee–Kosterlitz finite-size-scaling technique to the problem of two-dimensional melting.

## References

[1] C.C. Huang, in *Bond-Orientational Order in Condensed Matter Systems*, edited by K.J. Strandburg, Springer-Verlag, New York, 1992.

[2] C.A. Murray, in *Bond-Orientational Order in Condensed Matter Systems*, edited by K.J. Strandburg, Springer-Verlag, New York, 1992.

[3] S. Sachdev, in *Bond-Orientational Order in Condensed Matter Systems*, edited by K.J. Strandburg, Springer-Verlag, New York, 1992.

[4] M.V. Jarić, in *Bond-Orientational Order in Condensed Matter Systems*, edited by K.J. Strandburg, Springer-Verlag, New York, 1992.

[5] A. Goldman, in *Bond-Orientational Order in Condensed Matter Systems*, edited by K.J. Strandburg, Springer-Verlag, New York, 1992.

[6] P.J. Steinhardt and S. Ostlund, *The Physics of Quasicrystals*, World Scientific, Singapore, 1987.

[7] M.C. Marchetti and D.R. Nelson, *Phys. Rev. B* **41**, 1910 (1990).

[8] M.C. Marchetti, *Phys. Rev. B* **42**, 9938 (1990).

[9] E.M. Chudnovsky, *Phys. Rev. B* **40**, 11355 (1989).

[10] D.P. Landau, K.K. Mon, and H.-B. Schuttler, *Computer Simulation Studies in Condensed Matter Physics*, Springer-Verlag, Berlin, 1988.

[11] K. Binder, *Monte Carlo Methods in Statistical Physics*, Springer-Verlag, Berlin, 1979.

[12] K. Binder *Applications of the Monte Carlo Method in Statistical Physics*, Springer-Verlag, Berlin, 1984.

74     Katherine J. Strandburg

[13] K. Binder and D.W. Heermann, *Monte Carlo Simulation in Statistical Physics*, Springer-Verlag, Berlin, 1988.

[14] D.W. Heermann, *Computer Simulation Methods in Theoretical Physics*, Springer-Verlag, Berlin, 1986.

[15] M.H. Kalos and P.A. Whitlock, *Monte Carlo Methods: Basics*, Wiley-Interscience, New York, 1986.

[16] M.P. Allen and D.J. Tildesley, *Computer Simulation of Liquids*, Oxford Science Publications, Clarendon Press, Oxford, 1987.

[17] A. Rahman, *Phys. Rev.* **136**, A405 (1964).

[18] L. Verlet, *Phys. Rev.* **159**, 98 (1967).

[19] L.F. Rull, J.J. Morales, and F. Cuadros, *Phys. Rev. B* **32**, 6050 (1985).

[20] F.F. Abraham and S.W. Koch, *Phys. Rev. B* **29**, 2824 (1984).

[21] J.Q. Broughton, G.H. Gilmer, and J.D. Weeks, *J. Chem. Phys.* **75**, 5128 (1981).

[22] S. Tøxvaerd, *Phys. Rev. Lett.* **53**, 2352 (1984).

[23] S. Tøxvaerd, *Phys. Rev. B* **29**, 2821 (1984).

[24] M. Suzuki, *Quantum Monte Carlo Methods in Equilibrium and Nonequilibrium Systems*, Springer-Verlag, Berlin, 1987.

[25] N. Metropolis, A.W. Rosenbluth, M.N. Rosenbluth, A.H. Teller, and E. Teller, *J. Chem. Phys.* **21**, 1087 (1953).

[26] D. Kandel, E. Domany, D. Ron, A. Brandt, and E. Loh, *Phys. Rev. Lett.* **60**, 1591 (1988).

[27] R.H. Swendsen and J.-S. Wang, *Phys. Rev. Lett.* **58**, 86 (1987).

[28] J.P. McTague, D. Frenkel, and M.P. Allen, in *Ordering in Two Dimensions*, edited by S.K. Sinha, North-Holland, Amsterdam, 1980.

[29] T.A. Witten and L.M. Sander, *Phys. Rev. Lett.* **47**, 1400 (1981).

[30] G.Y. Onoda, P.J. Steinhardt, D.P. DiVincenzo, and J.E.S. Socolar, *Phys. Rev. Lett.* **60**, 2653 (1988).

[31] G.Y. Onoda, P.J. Steinhardt, D.P. DiVincenzo, and J.E.S. Socolar, *Phys. Rev. Lett.* **62**, 1210 (1989).

[32] M.V. Jarić and M. Ronchetti, *Phys. Rev. Lett.* **62**, 1209 (1989).

[33] K. Ingersent and P. Steinhardt, *Phys. Rev. Lett.* **64**, 2034 (1990).

[34] V.L. Berenzinskii, *Sov. Phys. JETP* **34**, 610 (1972).

[35] V.L. Berenzinskii, *Zh. Eksp. Teor. Fiz.* **61**, 1144 (1972).

[36] J.M. Kosterlitz and D.J. Thouless, *J. Phys. C* **5**, L124 (1972).

[37] J.M. Kosterlitz and D.J. Thouless, *J. Phys. C* **6**, 1181 (1973).

[38] B.I. Halperin and D.R. Nelson, *Phys. Rev. Lett.* **41**, 121 (1978).

[39] D.R. Nelson and B.I. Halperin, *Phys. Rev. B* **19**, 2457 (1979).

[40] A.P. Young, *Phys. Rev. B* **19**, 1855 (1979).

[41] V. Privman, *Finite Size Scaling and Numerical Simulations of Statistical Systems*, World Scientific, Singapore, 1990.

[42] J. Lee and J.M. Kosterlitz, *Phys. Rev. Lett.* **65**, 137 (1990).

[43] J. Lee and K.J. Strandburg, private communication.

[44] A.M. Ferrenberg and R.H. Swendsen, *Phys. Rev. Lett.* **61**, 2635 (1988).

[45] P.B. Bowen, J.L. Burke, P.G. Corsten, K.J. Crowell, K.L. Farrell, J.C. MacDonald, R.P. MacDonald, A.B. MacIsaac, S.C. MacIsaac, P.H. Poole, and N. Jan, *Phys. Rev. B* **40**, 7439 (1989).

[46] A.M. Ferrenberg and R.H. Swendsen, *Phys. Rev. Lett.* **63**, 1195 (1989).

[47] J.M. Rickman and S.R. Philpot, *Phys. Rev. Lett.* **66**, 349 (1991).

[48] S.R. Philpot and J.M. Rickman, *J. Chem. Phys.* **94**, 1454 (1991).

[49] M. Widom, in *Proceedings of the Anniversary Adriatico Conference — Quasicrystals*, Trieste, Italy, edited by M.V. Jarić and M. Ronchetti, World Scientific, Singapore, 1989.

[50] K.J. Strandburg, *Phys. Rev. B* **40**, 6071 (1989).

[51] E. Sørensen, M. Jarić, and M. Ronchetti, to be published.

[52] J.Q. Broughton, G.H. Gilmer, and J.D. Weeks, *Phys. Rev. B* **25**, 4651 (1982).

[53] K.J. Strandburg, *Rev. Mod. Phys.* **60**, 161 (1988); **61**, 747 (1989).

[54] C. Udink and J. van der Elsken, *Phys. Rev. B* **35**, 279 (1987).

[55] C. Udink and D. Frenkel, *Phys. Rev. B* **35**, 6933 (1987).

[56] D.S. Fisher, B.I. Halperin, and R. Morf, *Phys. Rev. B* **20**, 4692 (1979).

[57] R.H. Morf, *Helv. Phys. Acta* **56**, 743 (1983).

[58] W. Janke and D. Toussaint, *Phys. Lett. A* **116**, 387 (1986).

[59] W. Janke and H. Kleinert, *Phys. Lett. A* **105**, 134 (1984).

[60] Y. Saito, *Phys. Rev. B* **26**, 6239 (1982).

[61] Y. Saito, *Phys. Rev. Lett.* **48**, 1114 (1982).

[62] K.J. Strandburg, S.A. Solla, and G.V. Chester, *Phys. Rev. B* **28**, 2717 (1983).

[63] K.J. Strandburg, *Phys. Rev. B* **35**, 7161 (1986).

[64] D.R. Nelson, *Phys. Rev. B* **26**, 269 (1982).

[65] F.C. Frank, *Proc. R. Soc. London*, Ser. A **215**, 43 (1952).

[66] D.R. Nelson, in *Topological Disorder in Condensed Matter*, edited by L. Garrido, Springer-Verlag, Berlin, 1985.

[67] Y.J. Wong and G.V. Chester, *Phys. Rev. B* **35**, 3506 (1987).

[68] D.R. Nelson, M. Rubinstein, and F. Spaepen, *Phil. Mag. A* **46**, 105 (1982).

[69] M. Rubinstein and D.R. Nelson, *Phys. Rev. B* **26**, 6254 (1982).

[70] M.A. Glaser, N.A. Clark, A.J. Armstrong, and P.D. Beale, in *Dynamics and Patterns in Complex Fluids*, edited by A. Onuki and K. Kawasaki, Springer-Verlag, Berlin, 1990, p. 141.

[71] M.A. Glaser and N.A. Clark, in *Geometry and Thermodynamics*, edited by J.-C. Toledano, Plenum Press, New York, 1990, p. 193.

[72] M.A. Glaser and N.A. Clark, *Phys. Rev. A* **41**, 4585 (1990).

[73] M.A. Glaser, PhD thesis, University of Colorado, 1991.

[74] R. Collins, *Proc. Phys. Soc.* **83**, 553 (1964).

[75] H. Kawamura, *Prog. Theor. Phys.* **70**, 352 (1983).

[76] Y.M. Yi and Z.C. Guo, *J. Phys: Condens. Matter* **1**, 1731 (1989).

[77] M. Widom, K.J. Strandburg, and R.H. Swendsen, *Phys. Rev. Lett.* **58**, 706 (1987).

[78] F. Lancon, L. Billard, and P. Chardhari, *Europhys. Lett.* **2**, 625 (1986).

[79] F. Lancon and L. Billard, *J. Physique* **49**, 249 (1988).

[80] K.J. Strandburg, L.-H. Tang, and M.V. Jarić, *Phys. Rev. Lett.* **63**, 314 (1989).

[81] L.J. Shaw, V. Elser, and C.L. Henley, *Phys. Rev. B* **43**, 3423 (1991).

[82] L.-H. Tang, *Phys. Rev. Lett.* **64**, 2390 (1990).

[83] L.-H. Tang and M.V. Jarić, *Phys. Rev. B* **41**, 4524 (1990).

# 3

# Nature of Phase Transitions Related to Stacked Hexatic Phases in Liquid Crystals

*C.C. Huang*

## 3.1 Introduction

Between the crystalline and the isotropic liquid many liquid crystal compounds exhibit two or more distinct mesophases with various degrees of translational and orientational order. These materials provide new thermodynamically stable phases and are excellent systems for studying melting processes, phase transitions, and critical phenomena. So far, more than 25 different mesophases have been identified among the liquid crystal compounds. Considerable experimental and theoretical effort has been aimed toward understanding the variety of physical phenomena related to various mesophases. Many monographs (e.g., deGennes, 1974; Chandrasekhar, 1977; Liebert, 1978; Luckhurst and Gray, 1979; Gray and Goodby, 1984; Prost, 1984; Pershan, 1988) have been written to present various aspects of liquid crystal phases and/or their phase transitions. In this chapter, we will concentrate on the thermal properties of stacked hexatic phases and phase transitions related to these phases which have been the center of numerous research activities in the past ten years.

To facilitate our discussion of the various kinds of ordering in the relevant liquid crystal mesophases, let us consider the ordering in general first. One of the best ways to characterize the degree of ordering is the correlation function which can be written as

$$P(\mathbf{r}) = \langle \theta(\mathbf{r})\theta(0) \rangle.$$ $(3.1)$

Here the angular brackets denote an average weighted by the Debye–Waller factor and $\theta(\mathbf{r})$ is the order parameter at position $\mathbf{r}$ associated with a given phase transition. For three-dimensional systems in the limit of large $|\mathbf{r}|$, there exist three general cases for the spatial dependence of $P(\mathbf{r})$. First, in the disordered phase the correlation function $P(\mathbf{r})$ decays exponentially in space, so that

$$P(\mathbf{r}) \sim e^{-|\mathbf{r}|/\xi(T)}.$$ $(3.2)$

This is defined as short-order order (SRO). Here $\xi(T)$ is a temperature-

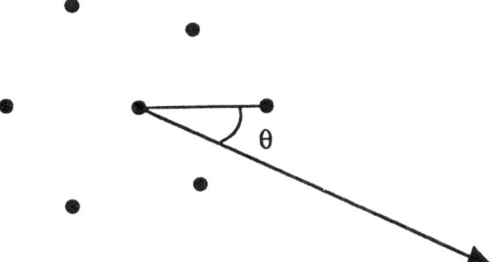

FIGURE 3.1. The definition of a bond angle $\theta$ between a laboratory axis and the imaginary bond joining a particle located at the origin to one of its nearest neighbors.

dependent correlation length which increases rapidly as the system approaches a critical point. In the ordered phase the function $P(\mathbf{r})$ remains a constant finite value as the distance $|\mathbf{r}|$ approaches infinity, namely,

$$\lim_{|\mathbf{r}|\to\infty} P(\mathbf{r}) = \text{constant} \neq 0. \tag{3.3}$$

This is called long-range order (LRO). In the transition between the ordinary liquid and crystalline phase, the order parameter is the translational order and the LRO implies a broken translational symmetry and shows up in the X-ray structure factor as delta-function Bragg peaks at the reciprocal lattice points. Between these two extreme cases, there exists a third situation in which $P(\mathbf{r})$ decays algebraically to zero,

$$P(\mathbf{r}) \sim |\mathbf{r}|^{2-d-\eta} \tag{3.4}$$

where $d$ is the spatial dimension of the system. This unique power-law decay can be found right at the second-order critical point. The parameter $\eta$ is a universal critical exponent and is expected to be dependent only on the dimensionality of the order parameter ($n$) and of the physical system ($d$) (Wilson, 1971; Wilson and Kogut, 1974), as well as the symmetry of the ordered state (Lee et al., 1986; Kawamura, 1987). To distinguish it from the previous two cases, it has been called quasi-long-range order (QLRO). This brief review of the principal ideas behind these three types of order is essential for the discussion of many different kinds of order existing in liquid crystal systems and is equally applicable to all other kinds of ordering related to various phase transitions. The central theme of this book is the hexatic phase found in many different physical systems. The order parameter associated with the hexatic phase is the bond-orientational order $[\Phi = \phi_6 * \exp(i6\theta_6)]$ in a system with sixfold symmetry. Figure 3.1 shows the definition of the bond angle ($\theta_6$) which is the orientation, relative to any fixed laboratory axis, of a bond between two nearest-neighbor molecules.

The balance of this chapter is arranged so that in Section 3.2, we describe the theoretical background of the hexatic phase and the experimental realization of the hexatic phase in liquid crystal mesophases and other physical systems, as well as provide general comments on the liquid crystal mesophases with special emphasis on those related to hexatic order. For reference, a list of liquid crystal compounds and mixtures exhibiting hexatic phases will be given. Because of space limitations, this is a fairly selective list and is not intended to be complete. Both thermodynamics and critical behavior related to the bulk liquid crystal hexatic phases will be described in Section 3.3. The majority of the experimental information presented here is obtained from heat capacity and thermal conductivity measurements. The experimental results from our recently developed free-standing liquid crystal film calorimeter will also be included in this section to demonstrate not only the feasibility but also the great capability of this new and unique experimental system. Theoretical attempts to explain the existing experimental data are discussed in Section 3.4. Finally we summarize this chapter in Section 3.5.

## 3.2    Fundamental Properties of the Hexatic Phase

### 3.2.1    Two-Dimensional Melting Theory

It has been demonstrated both experimentally and theoretically that the idea of universality works very well in describing the conventional continuous phase transitions (Wilson and Kogut, 1974). Namely, the nature of a continuous phase transition which can be characterized by a set of critical exponents is determined by the range of interaction related to the ordering, the system spatial dimension $(d)$, the order parameter dimension $(n)$, as well as the symmetry of the order parameter relevant to the given phase transition. For example, the simple cases of $n = 1, 2$, and 3 corresponding to the Ising, $XY$, and Heisenberg model are the most well-known occurring in nature. In $d = 3$ (three dimensions) all transitions with $n = 1, 2$, or 3 can be continuous transitions occurring at a nonzero temperature. On the other hand, in the case of $d = 2$, the Ising system $(n = 1)$ can still have a continuous phase transition at finite temperature; the Heisenberg system $(n = 3)$ cannot have a phase transition at any nonzero temperature; and the $XY$-system $(n = 2)$ is a very special case, namely, $d = 2$ is the lower-marginal dimension for the $XY$ system. Consequently, the two-dimensional melting transitions (i.e., hexatic-liquid and solid-hexatic) which are in the same universality as the $XY$ system are examples of a finite-temperature and lower-marginal-dimensionality phase transitions. Many review articles (Nelson, 1983; Minnhangen, 1987; Strandburg, 1988) have been published which give extensive discussions of the topic of two-dimensional melting phenomena. Here we will briefly summarize the relevant ideas which are pertinent to our discussion of hexatic order in liquid crystals.

Mermin and Wagner (1966) established that the continuous symmetry of the $XY$ and Heisenberg models in two dimensions could not be spontaneously broken at finite temperature. Namely, there exists no possibility of conventional long-range order, as discussed above, at nonzero temperature and any conventional order-disorder phase transition is excluded. In 1973, Kosterlitz and Thouless proposed a different mechanism for a phase transition by explicitly proving that the two-dimensional $XY$ model can have topological defects which mediate a new type of phase transition. In contrast to the conventional order-disorder phase transition, this novel transition involves unbinding of vortex-antivortex pairs. Such a phase transition can be continuous. Examples of topological defects include dislocations in the two-dimensional solid, vortices in the two-dimensional superconductor or superfluid, and vortices in the two-dimensional magnetic system with an $XY$ symmetry. In the low-temperature phase, only bound pairs of thermally generated defects are thermodynamically stable. Upon increasing temperature, the defects begin to unbind at the transition temperature $T_m$. For $T > T_m$, there exists a finite density of unbound defects causing an exponential decay of the relevant physical parameter, with a characteristic length being the average separation between unbound defects.

Defect-mediated transitions can be roughly divided into two generic categories. Transitions in the two-dimensional superfluid $^4$He films, superconductors, and roughening (solid-on-solid) model belong to one category. This kind of problem is characterized by a scalar interaction between defects. In general, these problems contain no cubic invariant in the Hamiltonian and the transition in three dimensions can be continuous. Much experimental evidence demonstrates that the Kosterlitz–Thouless (KT) mechanisms work very well for a variety of $^4$He film thickness, substrates, and concentrations of $^3$He impurities (Nelson and Kosterlitz, 1977; Rudnick, 1978; Bishop and Reppy, 1978; Webster et al., 1979; Roth et al., 1980).

The second category is characterized by a vector interaction between defects. An important example is melting in two dimensions which shows a strong discontinuous phase transition in three dimensions. The vector nature of the defect-defect interaction allows the melting transition to occur in two steps instead of the single step characteristic of the first category. The theory does not rule out the possibility of the two-step melting process being preempted by a single, discontinuous transition. Here we will not discuss the case with a first-order transition, even though this remains an important topic in the context of two-dimensional melting phenomena.

To describe systems belonging to the second category, Halperin and Nelson (1978, 1979) and Young (1979) have extended the ideas of the $KT$ mechanism to the two-dimensional melting problem. The basic idea of the theory is based on the separate unbinding of dislocations and disclinations. In this theoretical approach, they predicted the presence of a new two-

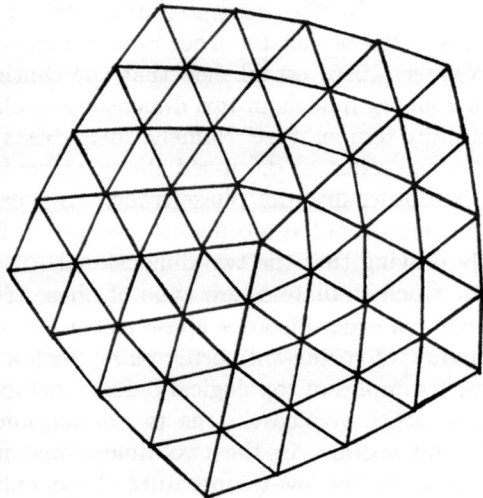

FIGURE 3.2. Schematic diagram of a disclination in a triangular lattice.

dimensional phase called the hexatic phase for systems with sixfold symmetry. Unlike the two-dimensional liquid, this phase would display quasi-long-range bond-orientational order (BOO) but not the quasi-long-range translation order indicative of the two-dimensional solid phase. If the melting process is characterized by an unbinding of dislocation pairs at a temperature $T_m$, then the increase in the density of free dislocations above $T_m$ will result in an exponential decay of the translational order parameter. However, the orientational ordering persists, in the sense that BOO decays only algebraically and displays QLRO. The elastic constant $K_A(T)$, which is the coefficient of the hexatic term in the free energy density expansion, is infinite just above $T_m$, but decreases with increasing temperature, until a temperature $T_i$, at which the dissociation of disclination pairs (see Figure 3.2) drives another transition and the system transforms into an isotropic phase in which both the orientational and translational order decay exponentially. At the transition temperature $T_i$, the elastic constant $K_A$ experiences a universal jump. Both the solid-hexatic and hexatic-liquid transition can be continuous. As mentioned above, the intermediate hexatic phase may be preempted by a single-step first-order melting transition of the two-dimensional solid directly to the two-dimensional liquid phase.

The $KT$ theory also offers a theoretical prediction for the heat-capacity behavior near a phase transition in a two-dimensional $XY$ system. Specifically, the temperature dependence of heat capacity contains only an essential singularity at the transition temperature. Due to the gradual dissociation of dislocation pairs above the transition temperature, the heat-capacity anomaly appears as a broad hump with the transition temperature located on the low-temperature side (Berker and Nelson, 1979; Solla and Riedel, 1981).

Two-dimensional melting without a substrate or on an ideally "smooth" substrate is very uncommon in nature. Furthermore, the experimental investigation of thermal properties of substrate-free systems to explore the nature of two-dimensional melting transitions is not an easy task. Consequently, the bulk of our knowledge on the thermal signature related to the substrate-free two-dimensional $XY$-like transition comes from computer simulations. Even in an ideal case, these computer simulations face many limitations. To name a few, the finite size effects and finite simulation times lead to many conflicting conclusions about the thermodynamic behavior of the two-dimensional $XY$-like transition, as well as the nature of other two-dimensional melting transitions (Strandburg, 1988). For example, by employing an isothermal-isobaric Monte-Carlo computer simulation of melting in a two-dimensional Lennard–Jones system, Abraham (1980, 1981) has concluded that melting in two dimensions is always first-order. A Monte-Carlo investigation (Saito, 1982) of a system of interacting dislocation vectors in two dimensions reveals the possibility of two types of phase transitions. The system with a large dislocation core energy has a continuous transition through unbinding of dislocation pairs. Reducing the core energy, the system favors a discontinuous transition by formation of grain boundaries. On the other hand, carrying out molecular dynamics simulation studies for a dense liquid structure in the microcanonical ensemble with periodic boundary conditions, Glaser and Clark (1990) recently concluded that melting in two dimensions is first-order, as indicated by a two-phase coexistence region in the equation of state. Moreover, while the bulk of theoretical arguments indicate that there exists no diverging heat-capacity singularity at the two-dimensional $XY$ transition, several computer simulations have challenged this theoretical prediction (Tobochnik and Chester, 1979; Van Himbergen and Chakravarty, 1981). Actually, the work by Van Himbergen and Chakravarty (1981) cannot rule out the possibility of a cusp in the heat capacity in the immediate vicinity of the transition temperature. On the other hand, similar computer simulations on the scalar Coulomb system (Saito, 1982) and computational-path-integral methods on the superfluid transition in a two-dimensional $^4$He system give a broad heat-capacity peak (Ceperley and Pollock, 1989).

Facing these conflicting simulation results and theoretical arguments for the thermal properties related to two-dimensional melting phenomena, we definitely need high-resolution experimental data from substrate-free systems. Recently we have developed a unique experimental technique, namely, differential quasiadiabatic ac calorimetry which enables us to measure the heat capacity of free-standing liquid crystal films with thickness ranging from three to a few thousand molecular layers (Geer et al., 1990, 1991). The experimental results provide us with important physical insight into the substrate-free two-dimensional system and may allow us to address the nature of two-dimensional melting transitions. While we are exploring many new physical phenomena related to substrate-free two-dimensional films, in particular, the effect of free surfaces, some preliminary and surprising results will be presented in the subsection entitled Calorimetric Studies on Free-Standing Films in Section 3.3.1.

## 3.2.2    Existence of Hexatic Order in Other Physical Systems

Besides the existence of hexatic order in some of the liquid crystal mesophases which will be introduced in Section 3.2.3 as the central part of this chapter, it has been speculated, suggested, or demonstrated that several other physical systems display hexatic order. First, a mechanical model consisting of a dense matrix of small ball bearings disturbed by a dilute concentration of larger ones has been studied (Nelson et al., 1982). A quantitative measure of translational and orientational order by the pair correlations clearly displays the existence of a quasi-long-range order in the bond orientation with short-range translational order.

The second example is that of the rare gases physisorbed on various substrates, in particular, xenon on graphite (Nagler et al., 1985; Gangwar et al., 1989; Jin et al., 1989) or xenon on a silver (111) surface (Greiser et al., 1987) near the mono-layer coverage. X-ray and vapor-pressure isotherm investigations indicate that, at near mono-layer coverage, the first-order melting of xenon on graphite or an Ag (111) becomes a continuous transition as the temperature becomes sufficiently large. This suggests the existence of the hexatic phase as an intermediate phase during the melting process. In addition to the interaction between the substrate and rare gas molecule, surface steps or defects have been suggested as promoting hexatic order, while the translational order remains short-range as observed in X-ray diffraction studies (Greiser et al., 1987). Recent detailed heat-capacity investigations (Jin et al., 1989) have ruled out the possibility of a continuous melting transition in the case of xenon on graphite. Consequently, it remains an open question as to the existence of the hexatic phase in the rare gases physisorbed on various substrates.

The third candidate for the hexatic phase is that of colloidal suspensions of highly charged submicron-sized spheres (polyballs) in water confined between two flat glass plates with a slight wedge (Murray and Van Winkle, 1987). The small angle between the two glass plates allows them to vary the density of polyballs across the sample in the direction of the wedge. By employing optical microscopy and digital imaging to locate the position of each polyball, both the orientational and translational correlation functions were calculated as a function of density. The corresponding correlation lengths suggest the existence of the hexatic phase between the solid and liquid phase. The reported hexatic phase exists only in a relatively narrow range of two-dimensional reduced density, however. The last but not least important example that will be mentioned here is the observation of hexatic order in the Abrikosov flux lattices in crystals of one of the high-$T_c$ superconductor materials (Murray et al., 1990) by in situ magnetic decoration of the flux lattice at 4.2 K. Computer imaging-analysis of the flux lattice leads to an exponential decay in positional order with a correlation length of only a few lattice constants, while the bond-orientational order exhibits a quasi-long-range order, indicative of the hexatic phase. Although the hexatic phase seems to occur relatively rarely in nature and it is therefore intriguing to identify it in various physical systems, it is also very important to experiment on systems that display a well-characterized hexatic phase. At present, liquid crystal systems seem to be the best candidate.

### 3.2.3  Hexatic Phases in the Liquid Crystal

Soon after the development of the two-dimensional melting theory by Halperin and Nelson (1978), Birgeneau and Litster (1978) applied this two-dimensional melting theory to three-dimensional liquid crystal phases consisting of stacked two-dimensional layers. They argued that the weak interlayer interaction could stabilize the quasi-long-range order found in two-dimensional systems. Then the BOO of the hexatic phase and the translational order of the solid phase in two dimensions become long-range order in the corresponding three-dimensional liquid crystal mesophase. In two-dimensional melting, one deals with the solid, hexatic, and liquid phases. Correspondingly, in the three-dimensional stacked system, one has the CryB, HexB, and SmA phases. The various types of orientational and translational order of these six phases in two-dimensional and three-dimensional systems, respectively, are listed in Table 3.1. Correspondingly, Figure 3.3 exhibits the schematic scattering profiles for a three-dimensional system with hexagonal symmetry in the liquid, hexatic, and crystal phases, respectively. Accordingly, the order parameter for the SmA–HexB transition is the bond-orientational order $\Phi = \phi_6 * \exp(i6\theta_6)$ (see Figure 3.1 for the definition of the bond angle $\theta_6$) and that for the HexB–CryB transition is the translational order. If we judge from the symmetry of these order parameters, the HexB–CryB transition should always be first-order, while the SmA–HexB transition can be continuous.

Nelson and Halperin (1980) generalized their two-dimensional melting theory by considering systems consisting of rigid rodlike molecules with an additional degree of freedom, namely, the rod-like molecules can tilt an angle $\psi$ away from the normal of the two-dimensional plane. In addition to the BOO and the translational order, this introduces another order parameter $\Psi = \psi * \exp(i\theta_t)$ (tilt orientation) which describes the projection of the long axis of the molecule (molecular director) in the two-dimensional plane. The possible couplings among those three order parameters give rise

TABLE 3.1. Orientational and translational order of six different phases in two-dimensional and three-dimensional systems.

| Two-dimensional System | Solid | Hexatic | Liquid |
|---|---|---|---|
| Positional order | QLRO | SRO | SRO |
| Bond-orientational order | LRO | QLRO | SRO |

| Three-dimensional System | CryB | HexB | SmA |
|---|---|---|---|
| Positional order | LRO | SRO | SRO |
| Bond-orientational order | LRO | LRO | SRO |

(a)                    (b)                    (c)

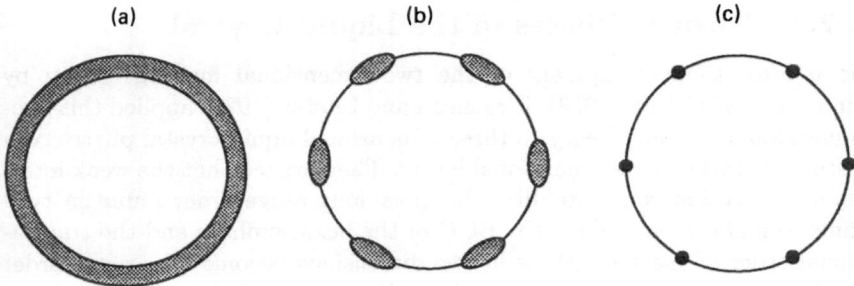

FIGURE 3.3. Schematic diagrams of the in-plane X-ray scattering profiles for the SmA (a), HexB (b), and CryB (c) phases in three dimensions.

to four possible fluid phases: isotropic liquid, hexatic, locked tilted hexatic, and unlocked tilted hexatic. One important result from this theory is that the molecular tilt degree of freedom can couple to the BOO and will induce a nonvanishing BOO in the tilted phase. Thus, thermodynamically, there exists no tilted two-dimensional liquid phase unless the coupling constant between those two degrees of freedom is smalller than a critical value.

Later, Bruinsma and Nelson (1981) solved a mean-field model with a coupling term between the BOO and the molecular tilt angle for bulk smectic liquid crystal phases. The resulting phase diagram is displayed in Figure 3.4. The fact that the molecular tilt will automatically induce the long-range bond-orientational order through the coupling term suggested that thermodynamically there exist only tilted hexatic phases [namely, the smectic-I (SmI) or the smectic-F (SmF)] and the smectic-C (SmC) phase will have a small long-range BOO. The dotted line in Figure 3.4 separates the region of the tilted hexatic phase into two parts. The region with a small value in bond-orientational stiffness constant ($K_6$) and correspondingly with a small value in the long-range BOO is the SmC phase. No sharp

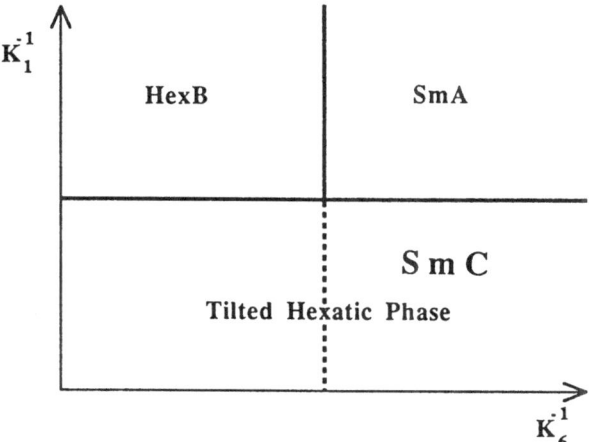

FIGURE 3.4. Schematic illustration of a phase diagram of a coupled system with bond orientation and molecular tilt order in the mean-field approximation. Here the dimensionless stiffness constants $K_1$ and $K_6$ characterize the interaction strength between the nearest-neighbor tilt angle and bond-orientational angle, respectively. The region of the tilted hexatic phase with a small value of $K_6$ is denoted as the SmC phase.

phase transition can be detected between the SmC and the tilted hexatic phase. Physically, this is equivalent to the paramagnetic-ferromagnetic transition under a small applied magnetic field. However, due to the observation of distinct microscopic textures between the SmC and the tilted hexatic phases and the fact that the coupling strength between the tilt and the BOO is not strong, the SmC phase may be treated as a separate phase. Another interesting point illustrated in Figure 3.4 is the triple point where the SmA, HexB, and tilted hexatic phases coexist. This unique thermodynamic point has been identified and investigated in one binary mixture system by Pitchford et al. (1986). Details will be discussed in Section 3.3.2.

Thus far, three different hexatic phases have been identified in those thermotropic liquid crystal compounds without chirality. The HexB phase has a layer structure with the long axis of the molecule (molecular director) parallel to the layer normal. The centers of mass of the molecules exhibit a three-dimensional long-range BOO but only a short-range translational order. Upon heating, the HexB phase can lose its long-range BOO and transform to the smectic-A (SmA) phase, in which within each smectic layer the centers of mass of the molecules are distributed randomly just like a two-dimensional liquid. Upon cooling, the molecules in the HexB phase can establish long-range positional order and exhibit the crystal-B (CryB) or the crystal-E (CryE) phase. In contrast to both the SmA and HexB phases, the CryB and CryE phases have long-range inter- and intra-layer translational order. Moreover, molecules can freely rotate around their long axes in the SmA, HexB, and CryB phases. They are optically uniaxial phases. In the CryE phase the free rotation around the molecular director is partially frozen and the molecules have herring-bone packing. It is an optically biaxial phase. Before the high-resolution X-ray diffraction work, on the basis of limited experimental information, several theoretical attempts were made (deGennes and Sarma, 1972; Meyer and McMillan, 1974; Huberman et al., 1975) to explain the physical properties of the conventional smectic-B phase as some kind of plastic crystals or a two-dimensional crystal with a weak interlayer interaction. Now it is clear that the so-called smectic-B materials can be classified as either CryB or HexB materials and therefore require very different theoretical interpretations.

The other two stacked hexatic phases are found in mesophases with molecular directors tilting away from the smectic layer normal. The tilted counterpart of the SmA phase is the SmC phase. Depending on the direction of the molecular tilt with respect to the orientation of the hexagonal packing, there are two groups of mesophases in thermotropic liquid crystals which are tilted counterparts of the HexB, CryB, and CryE phases. For the SmI, crystal-J (CryJ), and crystal-K (CryK) phases, the molecular director tilts toward the nearest-neighbor molecule. For the SmF, crystal-G (CryG), and crystal-H (CryH) phases, the molecular director tilts toward the midpoint of the line joining the nearest-neighbor molecules (the molecular bond). Thus, among the thermotropic liquid crystal molecules without chirality, the HexB, SmF, and SmI phases are the three known stacked hexatic phases which lack translational order but have long-range BOO In the case of lyotropic liquid crystal systems, detailed X-ray investigations by Smith et al. (1988) have revealed another unusual tilted hexatic phase, namely, smectic-L (SmL). In this new mesophase, the molecular long axis tilts toward an angle which is neither toward the neighbors nor toward the middle point between the neighbors. Consequently, the SmL phase has a lower symmetry than either the SmI or the SmF phase. With the same alphabetic nomenclature, the crystalline phases corresponding to the CryB and CryE phases are named CryM and CryN, respectively. Furthermore, the coupling between the tilt and bond-orientation order in the tilted hexatic phases distorts the short-range positional symmetry from being hexagonal to being rectangular. A schematic diagram of the molecular arrangements in these relevant liquid crystal phases is displayed in Figure 3.5.

Long before the term "hexatic phase" was invented in the two-dimensional melting theory (Halperin and Nelson, 1978), Demus and co-workers in 1971 first discovered the SmF phase in 2-(4-$n$-pentylphenyl)-5-(4''-$n$-pentyloxyphenyl) pyrimidine (see Section 2.4 for the molecular structure and typical transition sequence and transition temperatures). The molecular arrangement in this intriguing SmF phase was investigated by Leadbetter et al. (1979a) and by Benattar et al. (1979) on different liquid crystal compounds about eight years later employing X-ray diffraction. They reported pseu-

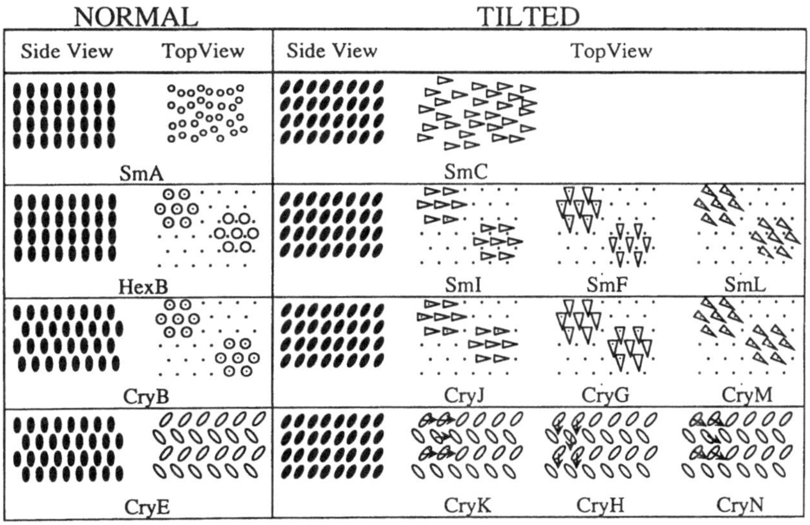

FIGURE 3.5. Schematic illustration of the molecular order in various liquid crystal phases with a layer structure. The small dots indicate the two-dimensional hexagonal lattices. The triangles are used to represent molecular tilt direction. Part of the figure is adapted from Gray and Goodby (1984).

dohexagonal molecular arrangement within the smectic layer of both the SmF and SmI phases. A satisfactory explanation for this observed pseudo-hexagonal molecular arrangement has required that the ideas of the hexatic order be developed in the two-dimensional melting theory.

At approximately the same time as theorists put forth these exciting hypotheses, experimentalists (Leadbetter et al., 1979b), employing X-ray diffraction studies on various liquid crystal compounds, demonstrated the existence of two types of smectic-B phases with different ranges of inter-layer correlation lengths. This observation was subsequently followed by several high-resolution X-ray diffraction investigations on free-standing liquid crystal films (Moncton and Pindak, 1979; Pindak et al., 1981; Pershan et al., 1981). First, the conventional SmB phase in 40.8 [$n$-(4-butyloxy-benzylidene)-4-$n$ octylaniline] was demonstrated to be a three-dimensional crystalline phase (Moncton and Pindak, 1979). The more interesting liquid crystal mesophase with long-range bond-orientational order and short-range positional order was subsequently found in 65OBC ($n$-hexyl-4'-$n$-pentyloxybiphenyl-4-carboxylate). In Figure 3.6, the X-ray scattering scan along one of the "crystalline" axes ($Q_{\parallel}$ scan) displays a short-range positional order, while the $Q_{\perp}$ scan exhibits a very weak interlayer coupling which is similar to the SmA phase. Figure 3.7 shows the development of six-fold modulation in a $\chi$ scan below the transition temperature $T_{AB} = 67.9$ °C which was determined by high-resolution heat-capacity measurements

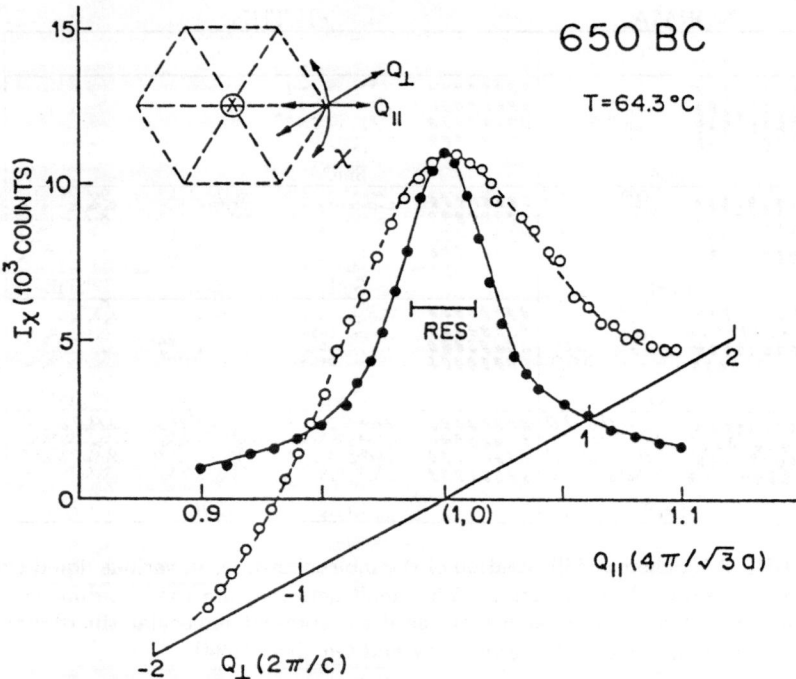

FIGURE 3.6. $\chi$-averaged X-ray scattering intensity for a $Q_{\shortparallel}$ scan (solid dots) and a $Q_\perp$ scan (open circle) in the HexB phase of 65OBC. Note that different scales are used for the $Q_{\shortparallel}$ and $Q_\perp$ axis. Three different scans, namely, the $\chi$, $Q_{\shortparallel}$, and $Q_\perp$ scans, are shown in the inset. (Adapted from Pindak et al., 1981.)

(Huang et al., 1981). The sixfold modulation indicates that the phase below $T_{AB}$ exhibits a three-dimensional long-range BOO. Recently, employing electron diffraction, Cheng et al. (1988) have obtained photos characterizing the HexB order from very thin free-standing liquid crystal films with thickness less than ten molecular layers. As a result of these experimental evidences, the traditional smectic-B phase has been reclassified into two phases, i.e., the HexB and CryB phases.

Because it is very difficult to prepare well-aligned bulk samples of many liquid crystal mesophases, the free-standing film technique (Young et al., 1978), which can give us high-quality single SmA domain samples, offers a unique approach for high-resolution X-ray investigations and optical studies on many liquid crystal mesophases. The fact that free-standing liquid crystal films are substrate-free and films as thin as two layers are stable provides us with an important and unique physical system to investigate the evolution from three-dimensional (thick films) to two-dimensional (thin films) behavior, the nature of two-dimensional phenomena and the effect of free surfaces. Furthermore, reducing the dimensionality of a system tends

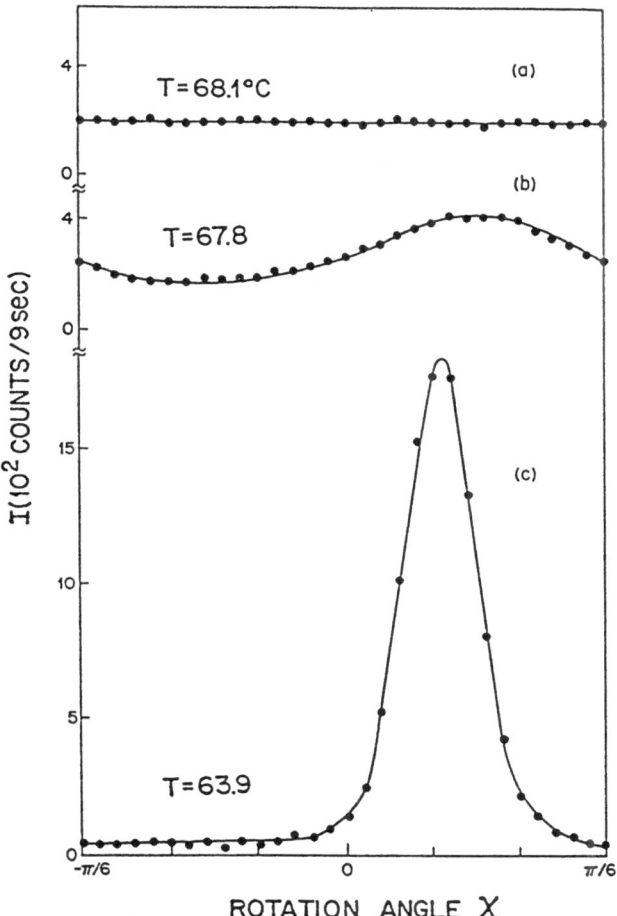

FIGURE 3.7. $\chi$ scans of a 65OBC sample at three different temperatures: (a) in the SmA phase; (b) just below the SmA-HexB transition; and (c) well into the HexB phase. Both (b) and (c) clearly exhibit the sixfold modulations characterizing the hexatic order. (Adapted from Pindak et al., 1981.)

to enhance the importance of thermal fluctuations. On the other hand, the surface tension on free-standing liquid crystal films will suppress the one-dimensional density fluctuations associated with the smectic-A ordering on the surfaces. This, in turn, may enhance the other types of order, e.g., the hexatic order (Pindak et al., 1980; Heinekamp et al., 1984; Ocko et al., 1986; Chen et al., 1989; Geer et al., 1989; Amador and Pershan, 1990). Thus, for some phase transitions free-standing films may not be an ideal system to investigate the simple crossover behavior due to finite sample thickness, but are unique systems to study the surface ordering phenomena, as well as to investigate the competition between thermal fluctuations and the surface ordering as the film thickness is reduced (Sirota et al., 1987; Geer et al., 1989; Geer et al., 1991a).

One of the surprising outcomes from the free-standing film technique is the existence of intermediate hexatic phases in some range of film thickness of liquid crystal compounds which have no such phases in bulk samples or thick films (Moncton et al., 1982; Collett et al., 1984; Sirota et al., 1987). For example, the free-standing films of 4-$n$-heptyloxy benzylidene-4-$n$-heptylaniline (7O.7) display the SmF and SmI phases only for films thinner than approximately 280 and 25 layers, respectively. However, in both thick films and bulk samples, neither the SmF nor the SmI phase exists in 7O.7 (Sirota et al., 1987).

The hexatic order in the HexB, SmI, and SmF phases of liquid crystals has been well-characterized by many experimental tools in both three-dimensional and two-dimensional systems. Thus far, liquid crystals have provided us with the most unambiguous hexatic ordering. However, numerous experimental questions remain to be answered. This chapter will address the subject of the thermal properties related to the hexatic order in liquid crystals. Before closing this section, we will provide a partial list of important liquid crystal compounds on which most of the experimental work on the hexatic order has been performed.

### 3.2.4    Pure Compounds and Mixtures Exhibiting Hexatic Phases

So far three different kinds of hexatic phases have been identified among thermotropic liquid crystal compounds without chirality. For future reference, here is the list of the most interesting compounds and mixtures which exhibit hexatic phases. Due to the limitations of space and the lack of detailed characterizations of many new liquid crystal compounds, our list is not intended to be complete.

#### Pure Compounds Exhibiting the HexB Phase

So far, we have found the following four homologous series that possess the HexB phase.

a) $n$-alkyl-4'-$n$-alkoxybiphenyl-4-carboxylate (nmOBC)

The general molecular structure for this homologous series is

$$C_mH_{2m+1}-O-\langle\!\langle\bigcirc\rangle\!\rangle-\langle\!\langle\bigcirc\rangle\!\rangle-COO-C_nH_{2n+1}$$

These compounds are very stable chemically and show a direct isotropic (I)-SmA transition without the nematic in between. So far all of the SmA-HexB transitions are found to be continuous among these compounds and some of the binary mixtures in this homologous series. The pioneering research work on the properties of the HexB phase was carried out on 65OBC which has the following transition sequence:

I(85 °C) SmA(67 °C) HexB(61 °C) CryE.

So far, we have studied nine different nmOBC compounds with the HexB phase. Upon cooling, none of them exhibits the HexB-CryB transition in bulk samples.

To explore the nature of the SmA-HexB transition, we have experimented on some binary mixtures in this homologous series. The temperature-composition phase diagram of one typical homologous series $n(10)$OBC with $n$ ranging from one to four is shown in Figure 3.8 (Nounesis, 1990). This phase diagram displays two special thermodynamic points, i.e., the SmA-HexB-CryE and SmA-HexB-SmI point where the relevant three phases coexist. Experimental investigations near these two thermodynamic points will be presented later.

b) 4-acetyl-4'-$n$-alkanoyloxyazo-benzene

The general chemical formula (Goodby, 1981) is

$$C_mH_{2m+1}-CO-\langle\!\langle\bigcirc\rangle\!\rangle-N=N-\langle\!\langle\bigcirc\rangle\!\rangle-COO-C_nH_{2n+1}$$

Because of the azo center linkage, these compounds show a bright orange color and are not very stable chemically at elevated temperatures. The transition sequence for the most well-characterized compound in this series, namely, 4-propionyl-4'-$n$-heptanoyloxyazo-benzene (PHOAB), is shown below

I(144 °C) N(142 °C) SmA(88 °C) HexB(84 °C) CryB.

In contrast to the I-SmA-HexB-CryE (or recrystallization) transition sequence found in the nmOBC compounds exhibiting the HexB phase, the

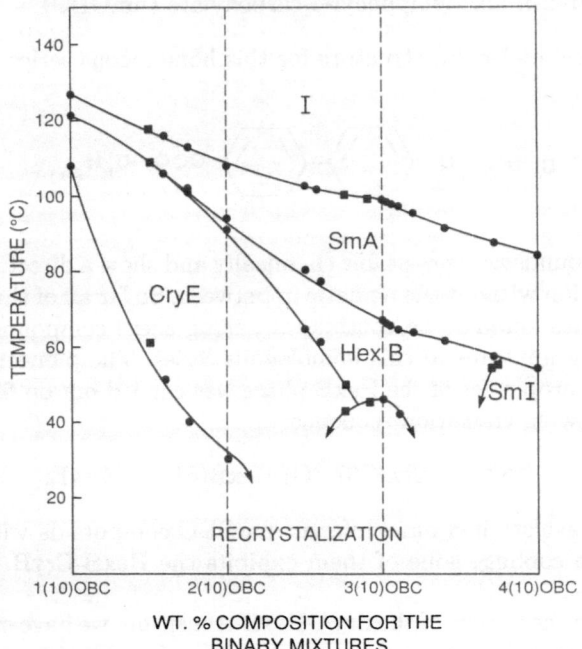

FIGURE 3.8. Temperature-concentration phase diagram for the 1(10)OBC-2(10)OBC, 2(10)OBC-3(10)OBC, and 3(10)OBC-4(10)OBC binary mixtures. The transition temperatures are determined by calorimetric measurements (circles) and by optical microscopy studies (squares).

I-N-SmA-HexB-CryB transition sequence is found in this compound. Unfortunately, both the HexB and CryB phases are monotropic phases. Samples of this compound prepared using the free-standing film technique will recrystallize once the temperature is below the SmA-HexB transition. Thus, X-ray diffraction measurements have to be carried out on the bulk sample (Albertini et al., 1984) with much less resolution. In contrast to the continuous SmA-HexB transition of nmOBC compounds, our heat-capacity studies on PHOAB show a first-order SmA-HexB transition (Huang et al., 1986).

c) $n$-alkyl-4′-$n$-alkoxybiphenyl-4-thiocarboxylate (nmOSBC)

The general molecular structure (Goodby, 1986) is

$$C_m H_{2m+1} - O - \langle\text{biphenyl}\rangle - COS - C_n H_{2n+1}$$

The phase sequence for one typical compound (65OSBC) is

$$I(150\ °C)\ SmA(121\ °C)\ HexB(91\ °C)\ Crystal.$$

Again there exists no nematic phase. Both the SmA and HexB phases stabilize at a higher temperature in comparison with the corresponding nmOBC compound. This group of compounds is also very stable chemically. Our calorimetric studies near the SmA-HexB transition of 65OSBC lead to broad heat-capacity anomalies with different magnitudes between successive heating and cooling runs. The data failed to be fitted to a power-law expression. Consequently, it appears to be a first-order transition.

d) $n$-alkyl-4′-$n$-pentanoyloxy-biphenyl-4-carboxylate (n4COOBC)

This homologous series of liquid crystal compounds has the following molecular structure:

$$C_4 H_9 - COO - \langle\text{biphenyl}\rangle - COO - C_n H_{2n+1}$$

which is very similar to that of nmOBC. The difference involves a change from alkoxy to acyloxy (Surrendranath et al., 1985). For $n = 3$ to 9, it exhibits the SmA-smectic-B (SmB) transition. Using heat-capacity studies near the first-order SmA-SmB transitions of two compounds ($n = 3$ and 6 homologs), Mahmood et al. (1988) concluded that the SmB phase of these compounds should be the HexB phase. Further confirmation of the hexatic-type order by either X-ray or electron diffraction studies is necessary.

## Mixtures Exhibiting the Hexatic-B Phase

In addition to the binary mixtures in the $n(10)$OBC compounds, the following three mixtures have been prepared and studied in the context of either the SmA-HexB or the HexB-CryB transition.

### a) 65OBC-4O.8 mixture

Here 4O.8 refers to $n$-(4-$n$-butyloxy-benzylidene)-4-$n$-octylaniline which has the following transition sequence:

$$I(78\ °C)\ N(64\ °C)\ SmA(50\ °C)\ CryB(38\ °C)\ Crystalline.$$

The compound 65OBC with the well-characterized HexB phase has a HexB-CryE transition instead of a HexB-CryB one. Employing optical microscopy studies, Goodby and Pindak (1981) first identified a SmA-HexB-CryB transition sequence in the mixture of 69.1% by wt. of 65OBC and 30.9% by wt. of 4O.8.

### b) 3(10)OBC-PHOAB mixture

In order to investigate the evolution of the first-order SmA-HexB transition in PHOAB to the continuous one in the nmOBC compound, Huang and co-workers (1989) have studied the nature of the SmA-HexB transition in this binary mixture system. The temperature-concentration phase diagram determined by both optical microscopic and calorimetric studies is shown in Figure 3.9. A detailed account of the heat capacity studies on this mixture system will be presented in the subsection entitled Calorimetric Studies on Bulk Samples with the Hexatic-B Phase in Section 3.3.1.

### c) 65OBC-PP5CC mixture

Employing a polarizing microscope and calorimeters, Mahmood and colleagues (1986) have carried out detailed studies of the SmA-HexB transition in this binary mixture. Here PP5CC refers to 4-propionylphenyl-$trans$-(4-$n$-pentyl) cyclohexane carboxylate with the following transition sequence:

$$I(107\ °C)\ N(75\ °C)\ SmA.$$

This pure compound does not have the HexB phase below the SmA phase. However, from the experimental results of the mixtures, the authors concluded that there might exist a hidden HexB phase below the crystalline phase of the pure PP5CC compound. In this binary mixture, the magnitude of the heat-capacity anomaly decreases very rapidly as the concentration of PP5CC increases. Furthermore, the associated critical exponents display an unusual evolution. More experimental work is needed to clarify the nature of the variation on these heat-capacity anomalies.

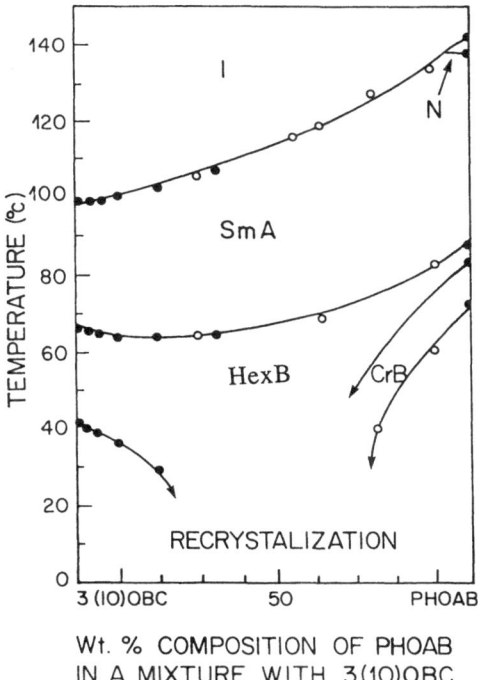

FIGURE 3.9. Temperature-concentration phase diagram of 3(10)OBC and PHOAB mixtures. The closed and open circles denote the results from calorimetric and optical microscopy studies, respectively.

## Compounds Exhibiting the Tilted Hexatic Phase

Unlike the HexB phase which seems to be relatively rare among the liquid crystal compounds, for some reason, both the SmI and the SmF phases are not that uncommon. Due to the limitation of space, we will list only a few homologous series which possess the SmI or SmF phase and have been studied to some extent.

   a) 2-(4-$n$-alkylphenyl)-5-(4-$n$-alkyloxyphenyl) pyrimidine

The general chemical formula is

This is the first liquid crystal compound in which the tilted hexatic phase was identified using the optical microscopy technique (Demus et al., 1971). Because the compound is relatively difficult to prepare, the first clear X-ray identification of the molecular order in the SmI or SmF phase was carried out in the compounds of the following two homologous series.

   b) Racemic     4-(2′-methylbutyl)-phenyl-4′-$n$-alkyloxybiphenyl-4-car-boxylate (2M4PnOBC or nOSI)

The general chemical formula is (Leadbetter et al., 1979a)

Some members of this homologous series have an extremeley rich variety of liquid crystal mesophases. For example, 2M4P8OBC has the following transition sequence:

   I(174 °C) N(171 °C) SmA(132 °C) SmC(79 °C) SmI(72 °C)
   CryJ(61 °C) CryK.

Although some mesophases appear at relatively high temperatures, chemically they are very stable compounds. This combined with the fact that the SmA-SmC transition is well-isolated from the other transitions enabled Huang and Viner (1982) to experimentally demonstrate the mean-field-like heat-capacity anomaly and propose an extended mean-field model to unambiguously characterize the nature of the SmA-SmC transition.

Detailed X-ray (Brock et al., 1986 and 1989a), light scattering (Sprunt and Litster, 1987), heat capacity (Hobbie et al., 1988; Garland et al., 1989), and thermal conductivity (Hobbie et al., 1989) studies have been carried out in the vicinity of the SmC-SmI transition of this compound. Within experimental resolution, the smectic layer spacing does not show any jump through the SmC-SmI transition of 2M4P9OBC (9OSI) (Leadbetter et al., 1979a). Soon after finding that the SmA-HexB transition in the nmOBC compounds was continuous with a sharp and symmetric heat-capacity anomaly, Viner and Huang (1983) first showed that the SmC-SmI transition of 2M4P9OBC displayed an intrinsically rounded and asymmetric heat-capacity peak. They attributed the difference in heat-capacity anomalies between the SmA-HexB transition of 65OBC and the SmC-SmI transition of 2M4P9OBC to the existence of a finite coupling between the BOO and molecular tilt ordering as proposed by Brunisma and Nelson (1981).

c) Terephthal-bis-(4-$n$)-alkylaniline (TBnA)

The general molecular structure is

$$C_nH_{2n+1} \diagdown \diagdown N{=}CH \diagdown \diagdown CH{=}N \diagdown \diagdown C_nH_{2n+1}$$

Because of the Schiff base bonding in the center part of the molecule, they are relatively unstable chemically at elevated temperatures. The compounds with $n$ less than 5 do not exhibit the SmF phase. The transition sequence of one of the compounds (TB5A or TBPA) is

I(233 °C) N(212 °C) SmA(179 °C) SmC(149 °C)
SmF(140 °C) CryG(61 °C) CryH.

Usually, the SmC-SmF transition of these compounds occurs in a relatively high-temperature range. This combined with the chemical instability makes high-resolution experimental work on the SmC-SmF transition fairly difficult. For some reason, the SmI phase is more commonly found than the SmF phase. This homologous series is well known to have the SmF phase below the SmC phase. Employing X-ray diffractions, Kumar (1981) found that the smectic layer spacing experienced a jump through the SmC-SmF transitions of this homologous series with $n$ ranging from 5 to 8. This implies a jump in the tilt angle at the transition temperature. Consequently, the transition should be first-order. Recently, Noh and co-workers (1989) have carried out much more detailed X-ray studies near the SmC-SmF transition of this homologous series with $n$ ranging from 5 to 7 to explore the variation of the BOO. In contrast to the noticeable BOO in the SmC phase of 8OSI (Brock et al., 1986), the induced BOO in the SmC phase by the molecular tilt angle is found to be immeasurably small in the TBnA compounds. This indicates that the coupling constant between the molecular tilt and bond-orientational order is smaller in the TBnA compounds than in 8OSI.

d) Racemic 4-(2′-methylbutyl)-phenyl-4′-$n$-octylbiphenyl-4-carboxylate (2M4PnBC or nSI)

The general chemical formula is

$$C_nH_{2n+1}\text{—}\underset{\phantom{a}}{\text{biphenyl}}\text{—}COO\text{—}\underset{\phantom{a}}{\text{phenyl}}\text{—}CH_2\text{—}\overset{\displaystyle H}{\underset{\displaystyle CH_3}{C}}\text{—}C_2H_5$$

In comparison with the second homologous series cited in this category, these compounds do not have the oxygen atom between the alkyl chain and the biphenyl group. Because of this, all of the transition temperatures are lower. For example, the transition sequence for 8SI is

I(141 °C) N(135 °C) SmA(84 °C) SmC(70 °C) SmI(64 °C) CryJ(62 °C) CryK.

Detailed X-ray characterization of the SmI, CryJ, and CryK phases was carried out by Budai et al. (1984). Many optical investigations of the hexatic order in the SmI phase of this compound were performed by Dierker and co-workers (1986 and 1987).

# 3.3    Thermal Properties

There exist two major obstacles in the experimental investigation of HexB materials. As a system approaches a continuous transition from the disordered phase, fluctuation-driven domains of the ordered phase will appear. The way in which these domains grow provides very important insight into the nature of the transition, as well as the properties of the ordered phase. This information is usually obtained through the correlation function of the order parameter which, in principle, can be studied by a proper scattering probe. In many phase transitions between liquid crystal mesophases, X-ray scattering is a very powerful and useful tool to serve this purpose. Unfortunately, in order to define a bond angle, one must consider the relative position of two molecules. Therefore, the correlation function related to BOO requires simultaneously probing the positions of four molecules. Currently, there exists no appropriate technique to determine a four-point correlation function from X-ray scattering. Second, no physical field is readily available to act as the conjugate field to the HexB order parameter. (*Note*: This is not the case for the tilted hexatic phases.) In this respect, the HexB order parameter behaves the same as the one that describes the normal to superfluid $^4$He transition.

Perhaps the easiest method of producing samples that display excellent alignment in the ordering related to the molecular director and smectic layering is to spread the compounds over a hole to produce free-standing films. In this films, the smectic ordering will display only a single domain, but in the HexB phase, the films will consist of many domains with different bond-orientational angles. The presence of small domains, coupled with the fact that since the smectic planes are parallel to the film, the experimental geometry is severely limited, makes optical investigations difficult. Fortunately, heat-capacity measurements probe the energy-energy correlations and do not require well-aligned samples. The existence of domains may, however, smear out the transition to such a degree that the critical exponent cannot be accurately obtained by the experiment. From both X-ray diffraction (R. Pindak, private communications) and electron-scattering studies (J.T. Ho, private communications), the domain size for a given bond orientation in the HexB phase is found to be approximately 1 mm in linear dimension. If we assume the bare correlation length to be the size of the molecule i.e., about 10 Å, the rounding in the measured heat capacity due to the existence of domains will occur around $|T - T_c| = 10^{-9}$ K, which is much smaller than our experimental resolution, and therefore, the accuracy with which we obtain the critical exponents is determined by experimental resolution rather than the domain size.

To get a better understanding of the stacked hexatic phase, we have carried out heat-capacity and thermal conductivity studies on bulk samples and heat-capacity measurements on free-standing films. Here is a brief summary of the experimental results. The heat-capacity data from free-standing liquid crystal films near the SmA-HexB transition of both 75OBC and 65OBC clearly show layer-by-layer transitions. Separate surface and interior transitions are detected for film thicknesses greater than two molecular layers. Our data clearly indicate that the two-dimensional thermal behavior has been achieved experimentally in the films of 750BC with two molecular layers in thickness. In the bulk sample, in the vicinity of the SmA-HexB transition of nmOBC compounds, three critical exponents determined from heat capacity ($\alpha$), thermal conductivity ($\eta$), and birefringence measurement ($\beta$), respectively, satisfy the scaling relation, even though the exponents do not agree with the three-dimensional $XY$ exponents. For example, the value of critical exponent $\alpha$ equals 0.60. The nature of the SmC-SmI (or -SmF) transition is enriched by the coupling between the molecular tilt order and the bond-orientational order. Consequently, the data analysis is more complicate than that associated with the SmA-HexB transition.

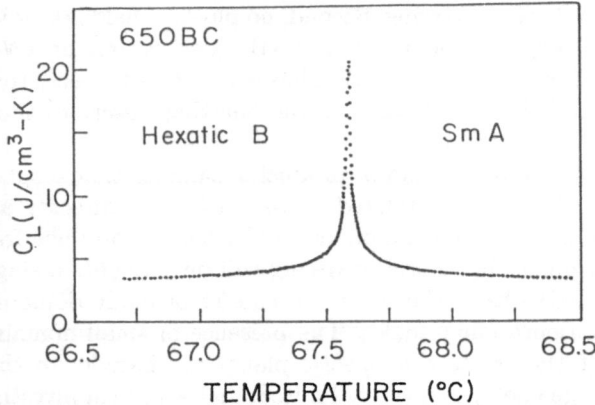

FIGURE 3.10. Temperature dependence of the heat capacity near the SmA-HexB transition of 65OBC.

In the following, we will present experimental results related to the HexB phase first and then present those related to the tilted hexatic phases.

### 3.3.1    Heat Capacity

#### Calorimetric Studies on Bulk Samples with the Hexatic-B Phase

According to our previous discussions, the symmetry-breaking field associated with the SmA-HexB transition is the bond-orientational order which can be described by a complex number $\Phi_6 = \phi_6 \exp(6i\theta_6)$ as the relevant order parameter. In the language of modern critical phenomena, this transition belongs to the three-dimensional $XY$ $(n = 2)$ universality class which is characterized by a unique set of critical exponents. For example, the critical exponent $(\alpha)$ associated with the heat-capacity anomaly near the transition should have a value of $-0.007$ (LeGuillon and Zinn-Justin, 1985) and the temperature dependence of the order parameter is described by another critical exponent, $\beta = 0.35$.

The first experimental evidence of the BOO in the HexB phase was obtained by Pindak and co-workers in 1981. The liquid crystal compound used for this pioneering X-ray diffraction work was 65OBC. Subsequently, employing an ac calorimetry technique (Sullivan and Seidel, 1968), Huang and co-workers (1981) performed high-resolution heat-capacity investigations in the vicinity of the SmA-HexB transition of 65OBC and the SmA-CryB transition of 4O.8. The heat-capacity anomalies associated with those two transitions are dramatically different (see Figures 3.10 and 3.11). The SmA-CryB transition shows a rounded and highly asymmetric heat-capacity peak with thermal hysteresis. This indicates that the SmA-CryB transition is first-order. In contrast, the SmA-HexB transition has a sharp and symmetric heat-capacity anomaly with no detectable hysteresis. Further-

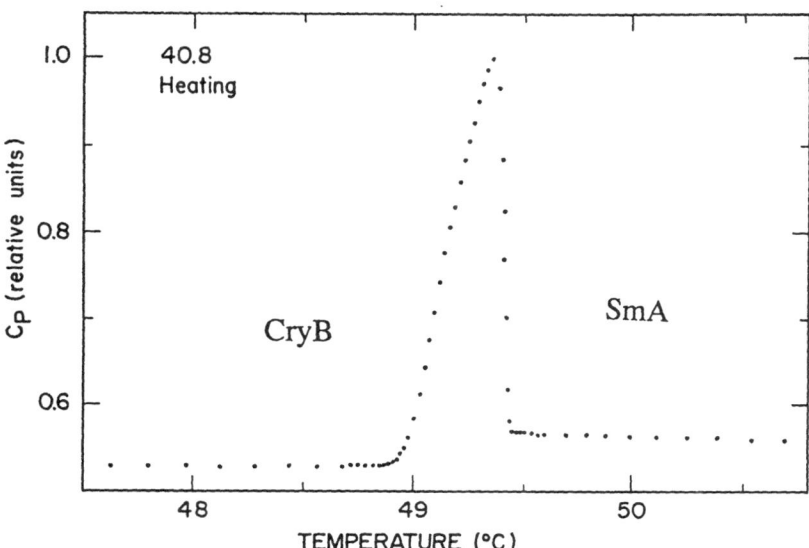

FIGURE 3.11. Temperature variation of the heat capacity in the vicinity of the SmA-CryB transition of 40.8. The data were taken in a heating run.

more, the experimental data can be fitted to a power-law expression with scaling correction terms, namely,

$$C = \begin{cases} A^+ |t|^{-\alpha}[1 + E^+ |t|^x] + B_1 + Dt, & T > T_c \\ A^- |t|^{-\alpha}[1 + E^- |t|^x] + B_1 + Dt, & T < T_c. \end{cases} \quad (3.5)$$

Here $A^+$, $A^-$, $B_1$, $D$, $E^+$, and $E^-$ are constants and $t = (T - T_c)/T_c$ is the reduced temperature. Without any theoretical information we chose the scaling correction exponent $x = 0.75$ which is larger than the value of $\alpha$ we have obtained. This guarantees a smooth variation of the background heat capacity through the transition temperature. Typical fitting results for one of the nmOBC compounds are displayed in Figure 3.12. The SmA-HexB transition of this compound can be described by the above power-law expression with a critical exponent $\alpha = 0.56 \pm 0.03$. Moreover, no thermal hysteresis can be detected within the experimental resolution, which indicates that the SmA-HexB transition seems to be continuous.

Although most of the theoretical models predict $\alpha \simeq 0$ in three dimensions, there are some situations in which large values of $\alpha$ are predicted. To the best of our knowledge, the Gaussian tricritical fixed point ($\alpha = 1/2$), "frustrated" spin systems of various type ($\alpha = 0.4$ for the Heisenberg antiferromagnetic on layered-triangular lattices) (Kawamura, 1987), and the three-state Potts model ($\alpha \simeq 0.5$) (Burkhardt et al., 1976; Herrmann, 1979) are the only systems predicting a large value in the critical exponent or "apparent" critical exponent $\alpha$. So far, most of the theoretical advances

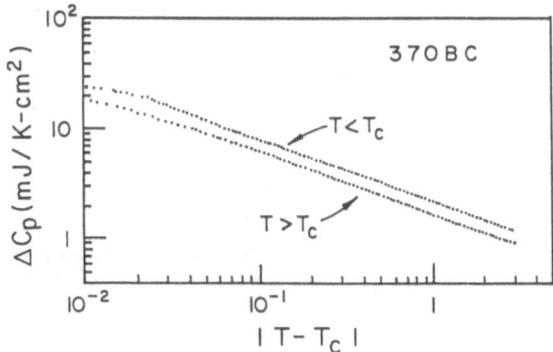

FIGURE 3.12. Anomalous part of the heat capacity $\Delta C = C - B_1 - Dt - A^\pm E^\pm |t|^{x-\alpha}$ vs. $|T - T_c|$ for 37OBC in the neighborhood of the SmA-HexB transition.

have been concerned with the idea of Gaussian tricriticality. We will discuss them in detail in Section 3.4. Some preliminary ideas related to the latter two scenarios will be addressed briefly here. After numerous theoretical investigations, it is believed, in general, that the three-state Potts model in three dimensions is weakly first-order with noticeable pretransitional phenomena due to fluctuations (Fukugita and Okawa, 1989). Many attempts have been made to determine the critical exponents characterizing the pretransitional effects before the system gets close enough to the transition temperature to exhibit first-order behavior. Monte Carlo simulations on the three-state Potts model in three dimensions by Herrmann (1979) lead to a relatively weak first order transition and give $\alpha = 0.52 \pm 0.16$ and $\beta = 0.17 \pm 0.01$. Incidentally, these values of critical exponents are in reasonably good agreement with $\alpha = 0.60 \pm 0.03$ and $\beta = 0.19 \pm 0.03$ (Rosenblatt and Ho, 1982) for the SmA-HexB transition of the bulk 65OBC sample. On the other hand, the simulation results exhibit a very asymmetric heat-capacity anomaly with amplitude ratio $A^+/A^- = 0.25$, which is definitely different from the symmetric heat-capacity peak associated with the SmA-HexB transition with $A^+/A^- = 0.85$ (Pitchford et al., 1985). Within our experimental resolution ($10^{-5}$ in the reduced temperature scale), the SmA-HexB transition among the nmOBC compounds is continuous. Furthermore, in the HexB phase both electron diffraction (from thin films with film thickness less than eight molecular layers) and X-ray studies reveal a hexatic order which definitely does not have three-state Potts symmetry. In addition, X-ray studies on thick free-standing films of nmOBC compounds reveal well-defined herring-bone peaks (see Figure 3.13). The herring-bone order (see the insert of Figure 3.13) can have a three-state Potts symmetry. Consequently, cooling through the SmA-HexB transition of nmOBC compounds may develop not only the long-range bond-orientational order but

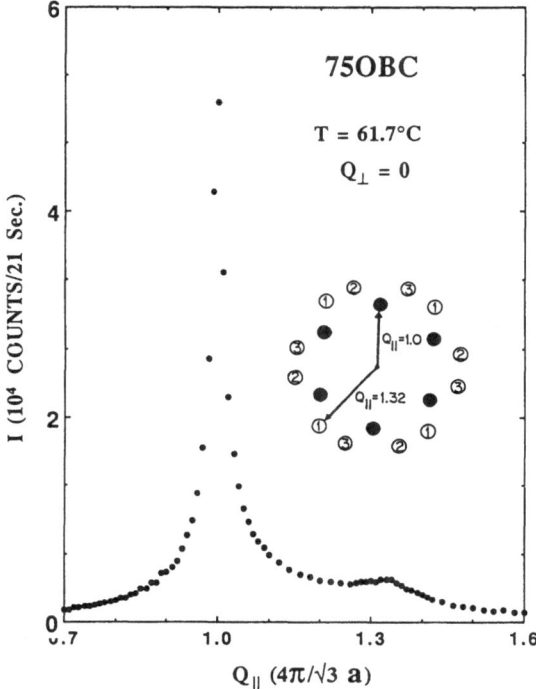

FIGURE 3.13. $Q_{\|}$ scan showing the additional scattering at $Q_{\|} = 1.32$ due to the existence of the herring-bone packing. The scattering from a herring-bone packing is shown schematically in the inset. The three-state Potts nature of the herring-bone packing is indicated by the labels 1, 2, and 3.

also the herring-bone order. Possibly, a mechanism similar to the system of Heisenberg antiferromagnets on layer-triangular lattices may be applicable to describe the nature of the intriguing SmA-HexB transitions in the nmOBC compounds. Currently, we are exploring this possibility. It is, however, too early to draw any conclusion. For the purposes of this chapter, we will assume that the BOO is the only symmetry-breaking field related to the SmA-HexB transition. With this restriction, the transition intrinsically belongs to the $XY$ universality class. Under these circumstances, only tricriticality can produce a large heat-capacity critical exponent. Several relevant experiments have been carried out to test this idea, the details of which will be presented in Section 3.4.

In describing thermodynamic behavior near a tricritical point, detailed theoretical calculations show the existence of logarithmic correction terms to the simple power law which has been demonstrated near the paramagnetic-ferromagnetic transition of LiTbF$_4$ (Ahlers et al., 1975; Griffin et al., 1977). The expression for the heat-capacity anomaly including the logarith-

mic correction term was obtained by Gorodetskii and Zaprudskii (1977) and
has the following expression:

$$
C = \begin{cases}
(T/T_t^2)(\pi k/4b^{1/2})\theta_0^2 t^{-1/2}[\ln(b/t)]^{-1/2}, & T > T_t \\
\\
(T/T_t^2)(\theta_0^2/b^{3/2})|t|^{-1/2}\{\frac{\sqrt{5}}{4}[\ln(b/|t|)]^{1/2} \\
\quad -(\frac{\sqrt{5}}{3} + \frac{3ka}{16})[\ln(b/|t|)]^{-1/2}\}, & T < T_t.
\end{cases}
\tag{3.6}
$$

Here the constant $a = 3.66$ and the reduced temperature $t = (T - T_t)/T_t$.
From the ratio of the heat-capacity anomaly near the $^3$He–$^4$He tricritical
point, the constant $k$ has been estimated to be about 0.2 (Gorodetskii and
Zaprudskii, 1977). Except for the common factor $T\theta_0^2/T_t^2$, the only free
parameter in Eq. (3.6) is $b$ which characterizes the amplitude of the diver-
gent behavior of order parameter susceptibility $\chi = b/t$. Obviously, from
Eq. (3.6), there exist different logarithmic correction terms to the heat-
capacity anomaly above and below the tricritical temperature $T_t$. Above
$T_t$, the singularity in the heat-capacity anomaly becomes weaker than a
square root, while below $T_t$, it becomes stronger. This indicates that the
temperature variation of heat capacity near a Gaussian tricritical point is
again highly asymmetric with respect to the transition temperature $(T_t)$. To
demonstrate this asymmetry, for two different values of $b$, Figures 3.14(a)
and (b) display the calculated heat capacity as a function of temperature
near $T_t$. Due to the effect of the logarithmic correction terms, the asymmetry is larger for a smaller value of $b$. It is obvious that this highly asymmetric heat-capacity anomaly is very different from our result displayed
in Figure 3.10. However, this is not the whole story of the heat-capacity
anomaly near a tricritical point. Experimentally, there exists at least one
exception. Namely, the heat-capacity anomaly near the tricritical point of
the nematic-SmA transition exhibits a fairly symmetric peak with critical
exponent $\alpha = 0.5$ and amplitude ratio $A^+/A^- = 1$ (Thoen et al., 1984).

To test the generality of the large heat-capacity critical exponent found
near the SmA-HexB transition of 65OBC, we have carried out similar heat-
capacity measurements on seven additional pure nmOBC compounds and
binary mixtures of $n(10)$OBC with $n$ ranging from 1 to 3. The primary idea
behind all these experimental studies is to see if the exponent $\alpha$ depends
on the temperature range of either the SmA phase or the HexB phase. All
of our heat-capacity results indicate that the critical exponent $\alpha$ remains
very large (approximately 0.6) for the nmOBC compounds with a wide
temperature range for either the SmA phase (14 to 32 K) (Pitchford et
al., 1985) or the HexB phase (0.8 to 22 K) (Nounesis et al., 1989). These
results are summarized in Tables 3.2 and 3.3. The implication of these
experimental results will be discussed in Section 3.4.

Employing optical microscopy studies, Goodby and Pindak (1981) identified the first SmA-HexB-CryB point in the binary mixtures of 65OBC and
4O.8. Subsequently, we performed heat-capacity studies on this binary mix-

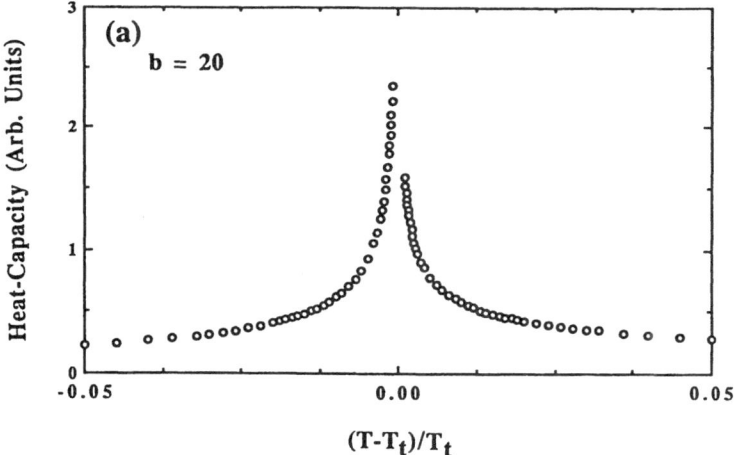

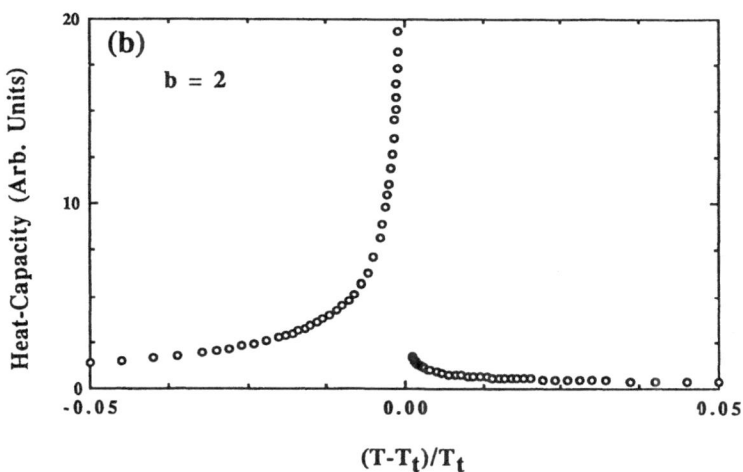

FIGURE 3.14. Calculated temperature dependence of the heat capacity from Eq. (3.6) near a tricritical point. The constant $b$ was set equal to (a) 20 and (b) 2 in our calculation.

TABLE 3.2. The critical parameters obtained from fitting heat-capacity anomalies near the SmA-HexB transition of various nmOBC compounds.

| Compound | $T_c$ (°C) | $\Delta T_A(K)$ | $\alpha$ | $A^+/A^-$ |
|---|---|---|---|---|
| 45OBC | 72.27 | 21.3 | 0.48±0.03 | 1.00±0.08 |
| 65OBC | 67.28 | 17.8 | 0.60±0.03 | 0.85±0.08 |
| 75OBC | 64.37 | 16.7 | 0.62±0.03 | 0.91±0.08 |
| 26OBC | 100.02 | 16.2 | 0.59±0.03 | 0.91±0.08 |
| 36OBC | 75.38 | 31.7 | 0.58±0.03 | 0.93±0.08 |
| 46OBC | 67.25 | 24.6 | 0.60±0.03 | 0.80±0.08 |
| 37OBC | 70.98 | 31.3 | 0.56±0.03 | 0.78±0.08 |
| 2(10)OBC | 94.67 | 14.1 | 0.64±0.03 | 0.91±0.08 |
| 3(10)OBC | 67.09 | 31.9 | 0.56±0.03 | 0.75±0.08 |

ture, especially near the SmA-HexB-CryB point and found only one broad and hysteretic heat-capacity peak, presumably associated with a first-order SmA-HexB transition (Viner et al., 1983). The supporting arguments will be elaborated in the following paragraph.

Later on, we carried out calorimetric studies (Huang et al., 1986) on the PHOAB compound which exhibits the SmA-HexB-CryB transition sequence. Upon cooling, both the SmA-HexB and HexB-CryB transition displays first-order transitions (see Figure 3.15). Because both the HexB and CryB phase are monotropic phases, the HexB-CryB transition peak disappears in the subsequent heating run. This is also the reason that this HexB phase cannot be prepared with free-standing film techniques, making high-resolution X-ray data on this HexB phase unavailable. Two salient features are clearly discernible from Figure 3.15. First, the heat-capacity anomaly associated with the SmA-HexB transition has a sharper drop in the SmA side than that in the HexB side and is much broader than the same phase transition found in the nmOBC compounds (see Figure 3.10). This heat-capacity anomaly cannot be fitted to a simple power law. In both the N-SmA and SmA-SmC (or SmC*) transitions, one general empirical trend has been found. The transition is continuous for compounds with a sufficiently large temperature range in the disordered phase (e.g., the nematic phase temperature range for the N-SmA transition). This trend seems to be inapplicable to this SmA-HexB transition which is first-order and has a fairly large temperature range for the SmA phase (54 K) compared to the nmOBC compounds. Second, the heat-capacity anomaly is much smaller for the HexB-CryB transition than that for the SmA-HexB transition. This means that most of the entropy difference between the liquidlike SmA phase and the crystalline phase is removed at the SmA-HexB transition. This is consistent with the fact that while the BOO becomes long-range in the HexB phase, its ordering significantly enhances the in-plane translational

TABLE 3.3. The critical exponent and amplitude ratio obtained from fitting of the heat-capacity data near the SmA-HexB transition for 3(10)OBC and 2(10)OBC, as well as various samples of 1(10)–2(10)OBC and 2(10)–3(10)OBC binary mixtures. Here $\Delta T_A$ and $\Delta T_{\text{Hex}}$ are the SmA and HexB phase temperature ranges, respectively.

| Sample | $T_c$ (°C) | $\Delta T_A(K)$ | $\Delta T_{\text{Hex}}(K)$ | $\alpha$ | $A^+/A^-$ |
|---|---|---|---|---|---|
| Pure 3(10)OBC | 67.086 | 31.9 | 20.2 | 0.56±0.03 | 0.75±0.08 |
| 2(10)–3(10)OBC 40–60% in wt. | 77.617 | 24.3 | 18.6 | 0.58±0.03 | 0.86±0.08 |
| 2(10)–3(10)OBC 75–25% in wt. | 87.495 | 18.0 | 9.0 | 0.58±0.03 | 0.96±0.08 |
| Pure 2(10)OBC | 94.670 | 14.1 | 3.3 | 0.64±0.03 | 0.91±0.08 |
| 1(10)–2(10)OBC 25–75% in wt. | 102.302 | 11.0 | 0.82 | 0.58±0.03 | 1.03±0.08 |

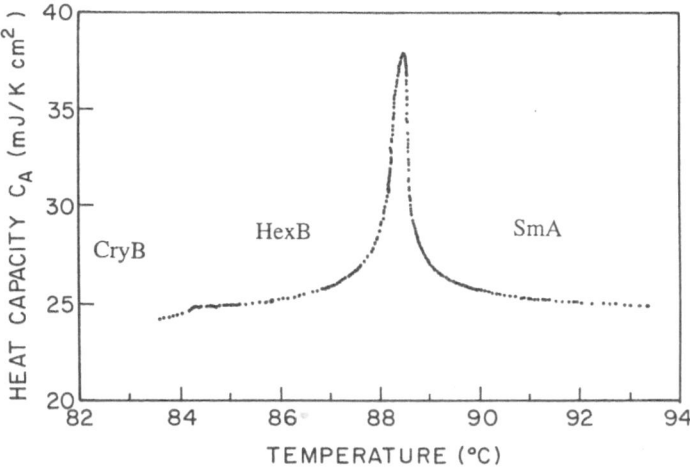

FIGURE 3.15. Temperature variation of the heat capacity in the vicinity of the SmA-HexB and HexB-CryB transition of PHOAB. The data were acquired in a cooling run.

TABLE 3.4. Critical parameters obtained from the fitting results for the 3(10)OBC-PHOAB binary mixture samples.

| wt. % of PHOAB | $T_c$ (°C) | $\Delta T_A(K)$ | $\Delta T_{Hex}(K)$ | $\alpha$ | $A^+/A^-$ |
|---|---|---|---|---|---|
| 0 | 67.09 | 31.9 | 20.2 | 0.59±0.03 | 0.71±0.08 |
| 2.5 | 65.84 | 33.0 | 24.9 | 0.64±0.03 | 0.65±0.08 |
| 5.0 | 65.16 | 34.5 | 25.9 | 0.58±0.04 | 0.64±0.09 |
| 10.0 | 64.71 | 36.3 | 27.7 | 0.60±0.04 | 0.64±0.09 |
| 20.0 | 64.46 | 39.0 | 34.5 | 0.65±0.05 | 0.54±0.10 |
| 35.0 | 65.56 | 41.5 | >50 | No critical | fitting |
| 100 | 88.44 | 49.6 | 4.1 | possible | |

order, and the correlation length increases from about 10 to about 60 Å (Pindak et al., 1981; Albertini et al., 1984). The in-plane translational order clearly remains short-range however. As a consequence, the development of long-range translational order through the HexB-CryB transition requires much less of an entropy change. This also gives a plausible explanation for our failure to detect the HexB-CryB transition near the SmA-HexB-CryB point of the 65OBC-4O.8 mixture. Surprisingly, the heat-capacity anomaly of the SmA-HexB transition in PHOAB mimics that of the SmC-SmI transition of nOSI (Viner and Huang, 1983; Hobbie et al., 1988; Garland et al., 1989) which will be presented in the subsection that follows.

To investigate the evolution from the second-order SmA-HexB transition in nmOBC compounds to the first-order SmA-HexB transition in PHOAB, we have carried out calorimetric studies on binary mixtures of 3(10)OBC and PHOAB. As the concentration of PHOAB increases, the sharp and symmetric heat-capacity peak associated with the SmA-HexB transition of 3(10)OBC gradually evolves into the broad and asymmetric heat-capacity peak of pure PHOAB. For the mixtures with PHOAB concentration larger than 20% in weight, power-law fittings to the heat-capacity data fail. A summary of the fitting results is shown in Table 3.4. The implication of these experimental results in the context of current theories will be discussed in Section 3.4.

## Calorimetric Studies on Bulk Samples with the Tilted Hexatic Phase

With a finite coupling constant between the tilt angle and BOO, the molecular tilt will induce a nonzero BOO. Unless this coupling constant is incidentally zero, the SmC phase is not a genuine thermodynamic phase, namely, there exists a very small but finite long-range BOO in the SmC phase. Under normal circumstances, the SmC-tilted hexatic phase transi-

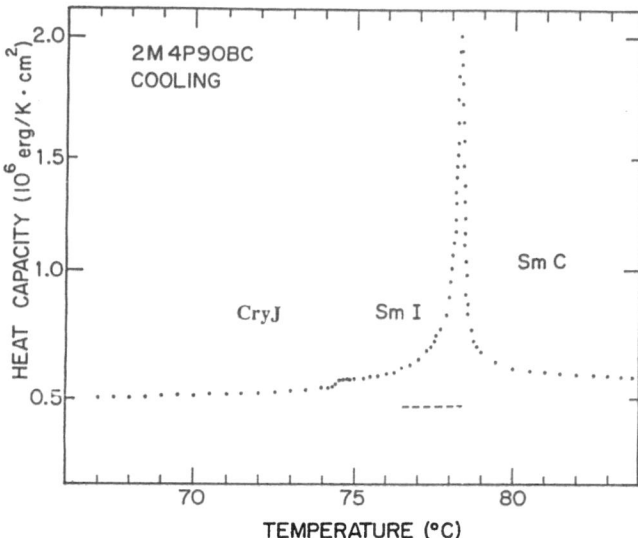

FIGURE 3.16. Temperature variation of the heat capacity near the SmC-SmI and SmI-CryJ transitions of 2M4P9OBC(9OSI). This is one of the cooling run results.

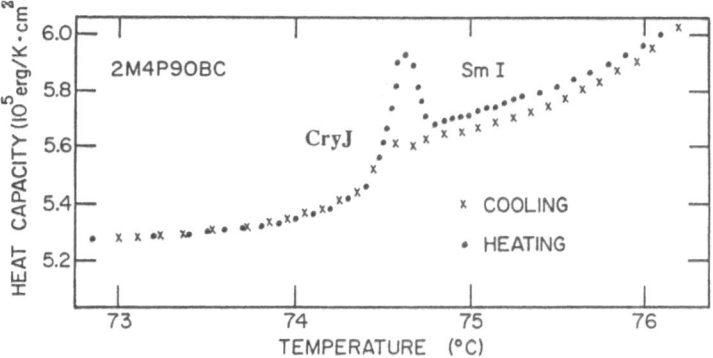

FIGURE 3.17. Temperature dependence of the heat capacity in the vicinity of the SmI-CryJ transition of 2M4P9OBC. To show thermal hysteresis, results from both heating and cooling runs are displayed.

tion will not be a sharp one. Viner and Huang (1983) carried out the first detailed heat-capacity studies near the SmC-tilted hexatic phase transition. The liquid crystal compound used for this study is the racemic 9OSI which in addition to other phase transitions has the SmC (78 °C) SmI (73 °C) CryJ transition sequence. The temperature variation of heat capacity near these two transitions is displayed in Figure 3.16. To show the nature of the first order SmI-CryJ transition, the heat-capacity data obtained from successive heating and cooling runs are shown in Figure 3.17. Similar to the SmA-HexB-CryB transition sequence of PHOAB, the SmC-SmI transition has a much larger heat-capacity peak than the SmI-CryJ transition. We believe that a large increase in the BOO through the SmC-SmI transition is also accompanied by an enhancement in the short-range in-plane translational order. Consequently, much more entropy is removed through the SmC-SmI transition than through the SmI-CryJ transition. In comparison with the results from the SmA-HexB transition of the nmOBC compounds, the SmC-SmI heat-capacity peaks of 9OSI or 8OSI not only display some asymmetry but also give fairly poor fitting to a power-law expression (Hobbie et al., 1988; Garland et al., 1989). These results are consistent with the recent X-ray results which clearly indicate the existence of the small but finite long-range BOO in the SmC phase of 8OSI (Brock et al., 1986; 1989a). Garland et al. (1989) have tried to fit simultaneously both heat-capacity and bond-orientational order data to a phenomenological equation of state (Schofield et al., 1969; Ho and Lister, 1970) with a temperature-dependent tilt field. The fitting results are displayed in Figure 3.18. The effective critical exponents ($\alpha = 0.47$ and $\beta = 0.077$) seem to indicate that the SmC-SmI transition of 8OSI would be weakly first-order in the limit that the tilt field approaches zero. To describe the heat-capacity anomaly, Brock et al. (1989a) have taken a different approach with the following three assumptions: the SmI phase has weakly coupled two-dimensional layers; the mean-field approximation is used for the coupling between the layers, and the Kosterlitz–Thouless expression for the two-dimensional susceptibility is employed. This model gives a qualitative but not quantitative description of the measured heat-capacity data. Further theoretical investigation beyond the ordinary mean-field approximation appears to be necessary.

In light of detailed X-ray investigations near the SmC-SmF transition of the TBnA compounds with $n$ ranging from 5 to 7, Stine and Garland (1990) have carried out calorimetric studies on these compounds. The X-ray results suggest that the SmC-SmF transition changes from a discontinuous jump in TB5A to a continuous transition in TB7A. In contrast to the X-ray results by Noh et al. (1989), the heat-capacity data demonstrated that the TBnA ($n = 5$, 6, and 7) compounds exhibit first-order SmC-SmF transitions.

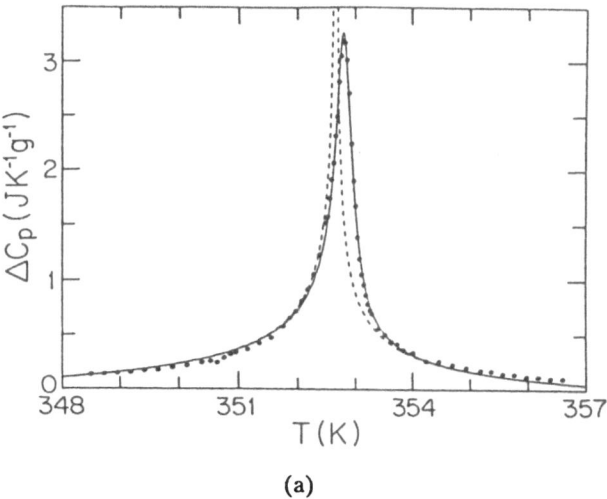

(a)

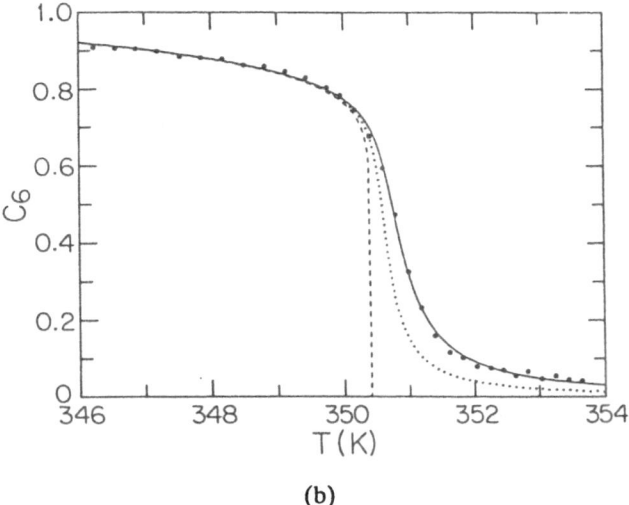

(b)

FIGURE 3.18. (a) Anomalous part of the heat capacity near the SmC-SmI transition of 2M4P8OBC (8OSI) and (b) temperature dependence of bond-orientational order near the SmC-SmI transition of 8OSI. The solid curves are the fitting results to the parametric equations and the dashed lines represent the predicted underlying critical behavior in the absence of a coupling field to the bond-orientational order. (Adapted from Garland et al., 1989.)

## Calorimetric Studies on Free-Standing Films

The free-standing liquid crystal film (Young et al., 1978) is one of the amazing physical systems which allows us to vary the sample thickness from two to a few thousand molecular layers fairly easily, provided appropriate liquid crystal compounds are chosen. So far many different experiments (e.g., light-scattering, X-ray diffraction, mechanical measurements, electron diffraction, optical observation, etc.) have been carried out on these types of liquid crystal systems to study the evolution of phase transitions as the system is taken from the thick-film to thin-film limit, as well as to investigate substrate-free two-dimensional transitions and the effect of free surfaces. The majority of these experimental techniques require extensive measurements for each discrete sample temperature. As a consequence, it is not an easy task to obtain high data density as a function of temperature necessary to probe the small signature related to surface ordering or to describe the diverging properties near a continuous transition. The fact that a quasiadiabatic ac calorimeter enables one to monitor heat capacity continuously as a function of temperature with high resolution makes an appropriately designed free-standing-film calorimeter a powerful and unique experimental tool to investigate the nature of phase transitions of free-standing films related to either surface ordering or two-dimensional transitions. In order to achieve sufficiently high resolution in measuring heat capacity of thin free-standing films, we have developed a diferential quasiadiabatic ac calorimeter in our laboratory (Geer et al., 1989; 1991b). This calorimeter enables us to measure the heat-capacity anomaly from films as thin as two molecular layers thick near the SmA-HexB transition of 75OBC (Geer et al., 1991a; Geer, 1991).

Once one learns how to make free-standing liquid crystal films, it would seem to be a relatively simple task to prepare them for different liquid crystal compounds in their smectic phases. In reality, for the purpose of high-resolution experimental work, we definitely need films with uniform thickness over a hole with a reasonable size. Among various nmOBC compounds, we have tried 2(10)OBC, 3(10)OBC, 37OBC, 26OBC, 36OBC, 46OBC, 65OBC, and 75OBC. We have been able to prepare uniform films over a hole (1 cm in diameter) with various thicknesses for 3(10)OBC, 46OBC, 65OBC, and 75OBC and have carried out heat-capacity measurements accordingly. The other nmOBC compounds either failed to produce films or gave us films with various thicknesses with very long relaxation times (> 1/2 day). One general guideline obtained from our experience is that those compounds with an alkyl chain length that is too short are not good candidates for preparing uniform free-standing films.

Undoubtedly, in addition to many other important contributions in physics, high-resolution heat-capacity investigations have revealed the importance of order parameter fluctuations near the majority of the phase transitions which have been studied. In the context of testing theoretical

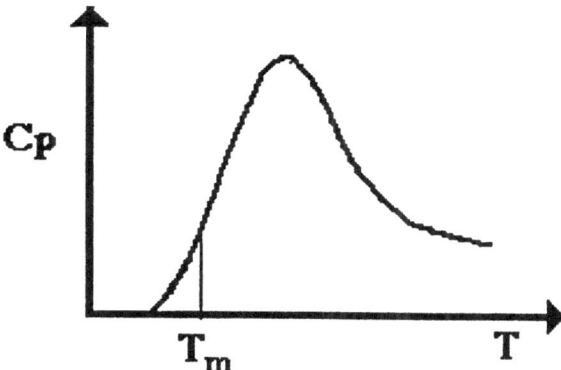

FIGURE 3.19. Vortex contribution to the heat capacity of an $XY$ model in two dimensions. $T_m$ is the melting temperature for a two-dimensional $XY$ system.

models and/or obtaining a better understanding of either two-dimensional transitions or related surface transitions, detailed heat-capacity studies, in conjunction with other measurements, will provide us with fruitful information concerning the important physical properties of free-standing films.

The $KT$-type theory for two-dimensional $XY$ phase transitions predicts only an essential singularity in the free energy which yields no diverging anomaly in the heat capacity at the transition temperature. The transition temperature is located on the low-temperature side of a broad heat-capacity hump (see Figure 3.19) (Berker and Nelson, 1979; Solla and Riedel, 1981). Other than free-standing films, substrate-free two-dimensional systems are not common in nature. In addition, the thermodynamic investigation of substrate-free two-dimensional systems is not an easy task. Thus, the bulk of our knowledge of thermal signatures associated with substrate-free two-dimensional $XY$-like transitions comes from computer simulations. As has been mentioned earlier, even in an ideal case, these simulations face many limitations. The "experimental" results from these simulations often conflict with each other. Consequently, the behavior of the heat capacity for two-dimensional $XY$ systems is theoretically unresolved and definitive experiments on well-characterized systems are extremely important.

So far we have carried out detailed heat-capacity studies of free-standing liquid crystal films near the SmA-HexB transition of 65OBC, 46OBC, and 75OBC. The thickness of each molecular layer is about 26 Å. Figures 3.20(a), (b), and (c) show our heat-capacity data near the SmA-HexB transitions of 65OBC with film nine, six, and and five molecular layers thick, respectively. Preliminary results on 65OBC free-standing films indicate that the surface liquid-hexatic transition occurs at about 72°C. Detailed investigations of the surface liquid-hexatic transition were carried out on 75OBC (Geer, 1991). The complete heat-capacity data from a four-layer film of 75OBC are shown in Figure 3.21. On the basis of our heat-capacity data from 75OBC free-standing films of various thickness and electron-diffraction studies, we have reached the following conclusions. The heat-capacity peak around 71°C is due to the liquid-hexatic transition from two outermost layers. The pronounced, sharp heat-capacity anomaly at about 64.8°C is due to the liquid-hexatic transition of two interior layers. In bulk samples, these three compounds display reasonably large HexB phase ranges (larger than 5 K). To our surprise, in the case of 75OBC with film thickness less than 150 layers, the HexB phase exists only in a narrow temperature range (less than 1 K) and is followed by a transition to a surface CryE phase (see Figure 3.21) (Geer et al., 1991a; Geer, 1991). This surface hexatic-CryE transition is characterized by an extremely sharp jump in heat capacity. In contrast to the surface liquid-hexatic transition, this surface hexatic-CryE transition possesses a large thermal hysteresis between heating and cooling runs. Ultimately, two-layer films of 75OBC exhibit a single heat-capacity anomaly associated with the liquid-hexatic transition around 71°C (Geer, 1991).

The evolution in the heat-capacity anomaly from nine- to four-layer films displays very distinct layer-by-layer transitions. It also demonstrates a unique path to the thermal properties of a two-dimensional substrate-free system, a two-layer film. Since both the SmA and HexB phases can be described by a one-dimensional density wave, the layer structure cannot be very rigid in three dimensions and experiences strong fluctuations. In free-standing films, the surface tension will suppress the layer fluctuations and enhance the SmA order on the surface. This, in turn, may promote the HexB order. We believe that this is the origin of separate surface peaks. Recently X-ray diffraction studies (Tweet et al., 1990) from free-standing films of the compound 70.7 near the SmC-SmI transition have clearly demonstrated the enhancement of the surface order due to the free-surface tension.

This newly developed ac free-standing-film calorimeter has opened many unique opportunities to investigate not only the nature of surface phase transitions but also phase transitions of substrate-free systems in the two-dimensional limit. One of the important theoretical predictions related to the heat capacity near a two-dimensional melting transition is that the peak position of the heat capacity is not the transition temperature. To give this theoretical prediction a crucial test, we plan to redesign our calorimeter system with some innovative ideas such that the physical properties related to the hexatic order and the heat capacity can be measured simultaneously.

## 3.3.2    Nature of the SmA-HexB-SmI Point

The fact that both the SmA-SmC and the CryB-CryG transitions have been observed in single-component liquid crystal compounds, but the HexB-SmI (or SmF) transition has not, makes the discovery of the HexB-SmI transition as well as the SmA-HexB-SmI triple point very exciting. In principle, the SmA-SmC, HexB-SmI (or SmF), and CryB-CryG (or CryJ) transitions

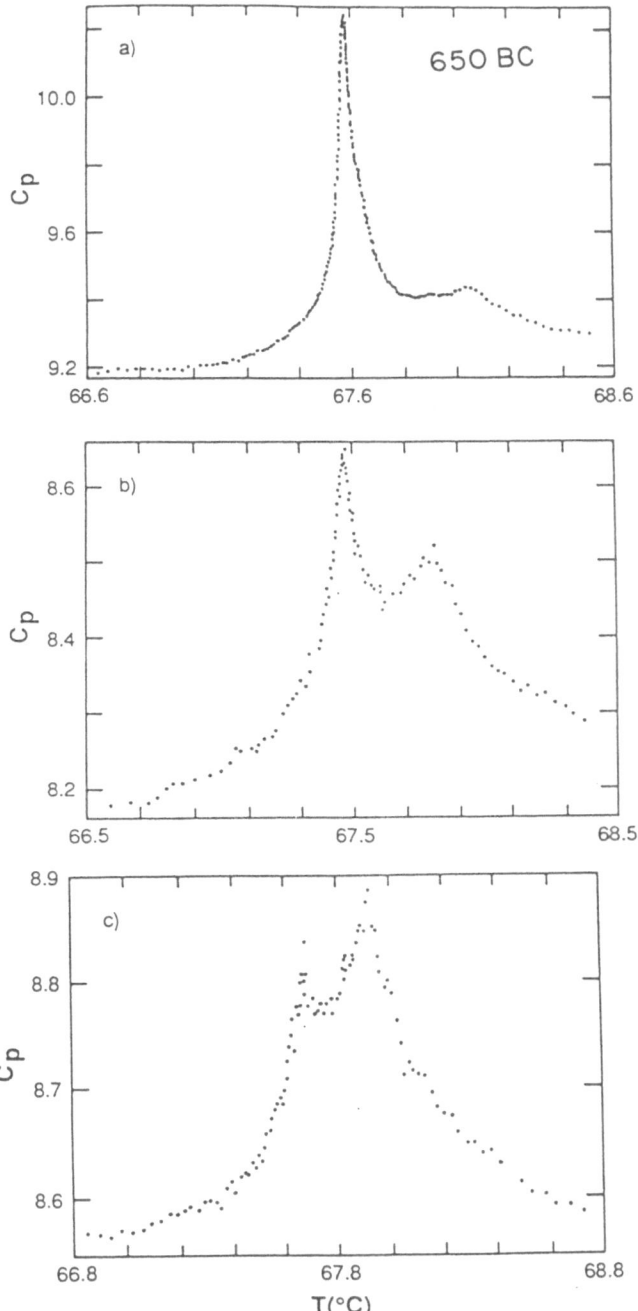

FIGURE 3.20. Evolution of heat capacity anomalies near the SmA-HexB transition of 65OBC obtained from a free-standing liquid crystal film calorimeter. (a) Nine-layer film; (b) six-layer film; (c) five-layer film. Note: The two outermost layers exhibit a liquid-hexatic transition at about 72°C.

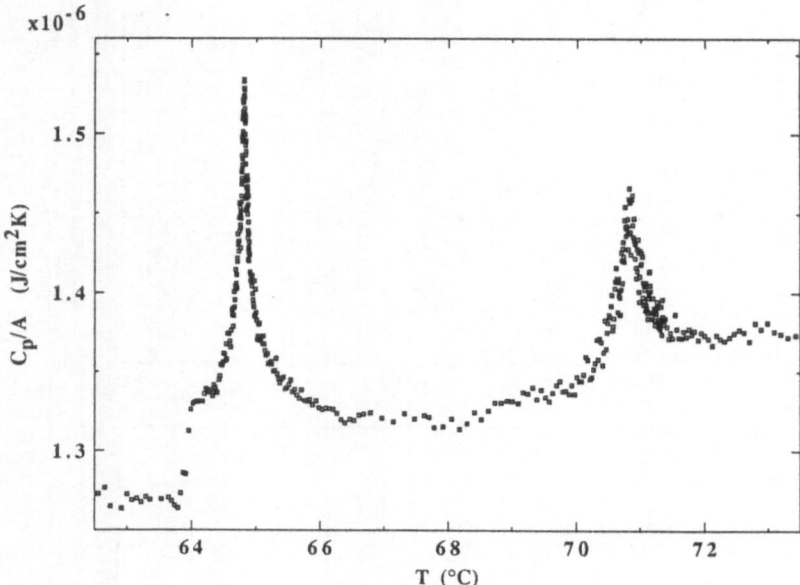

FIGURE 3.21. Heat capacity anomaly obtained from a four-layer film in the vicinity of the SmA-HexB transition of 75OBC. Both surface ($\approx 71°$C) and interior ($\approx 64.8°$C) transitions can be seen as separate heat capacity peaks. Here the sharp jump at about 64°C signals the surface-enhanced HexB-CryE transition.

can be simply driven by the molecular tilt away from the layer normal. Due to the symmetry of the simple tilt order parameter $[\Psi = \psi \exp(i\theta_t)]$, all three transitions can be continuous. Thus far, the majority of the SmA-SmC transitions, involving mesophases with liquidlike intraplanar molecular arrangement, have been found to be second-order and mean-field-like (Safinya et al., 1980). They have also been found to be well characterized by an extended mean-field theory proposed by Huang and Viner (1982) (Huang and Lien, 1985). On the other hand, the CryB-CryG transition, involving a transition between solidlike mesophases, is first-order (Leadbetter et al., 1979c). Since its range of positional correlations is intermediate relative to these two extreme cases, the HexB-SmI transition should be a good candidate for studying the effect of already established BOO and/or greatly enhanced positional order on the phase transitions which are driven mainly by molecular tilt. In principle, for a compound with a sufficiently large HexB phase temperature range, the BOO should be well established in the HexB phase and the HexB-SmI transition could be a purely molecular-tilt-driven transition which can be a second-order transition.

According to symmetry arguments, the SmA-HexB transition also can be continuous but the SmA-SmI transition should definitely be first-order. Thus, the SmA-HexB-SmI point could be a very interesting multicritical point with two continuous transition lines and one first-order transition line joining at one point which, in principle, is similar to the intriguing N-SmA-SmC multicritical point (Johnson et al., 1977; Chen and Lubensky, 1976). In at least one system, this is not the case, however, as Pitchford and coworkers (1986) have found that the SmA-HexB-SmI triple point in the binary mixture of 3(10)OBC–4(10)OBC is the intersection of three first-order phase transition lines (see Figure 3.8). In the vicinity of the SmA-HexB-SmI triple point, heat-capacity measurements clearly showed the anomalies of the SmA-HexB and SmA-SmI transition but failed to detect any peak associated with the HexB-SmI transition. With X-ray diffraction, the nature of the HexB-SmI transition was determined by the molecular tilt angle and found to be first-order. Furthermore, calorimetric studies in the region relatively far away from this triple point and along the SmA-HexB transition line of this binary mixture system seem to resemble the shape of the SmA-SmI transition (Nounesis, 1990). This suggests that the existence of the SmA-SmI transition has some effect on the SmA-HexB transition, even though these two phase transitions are separated by a triple point. A global model for the transition among the SmA, HexB, and tilted hexatic phase has to be developed before we can understand this unusual behavior. Experimental exploration of other SmA-HexB-tilted hexatic phase triple points is continuing. Unless there exists some unknown mechanism related to the ordering in both the HexB and the tilted hexatic phase, we believe that an interesting SmA-HexB-tilted hexatic phase triple point with two continuous phase transition lines joining at that point may exist in nature.

In an attempt to look for another SmA-HexB-SmI point, we have carried out optical microscopy studies of the phase diagram of the $[3(10)OBC]_{1-x}$ $- [8OSI]_x$ mixture. Here 3(10)OBC exhibits the HexB phase and 8OSI has the SmI phase. There exist some possibilities of finding a SmA-HexB-SmI point in this binary-mixture system. The results are shown in Figure 3.22. It looks likely that four phases, namely, the SmA, SmC, HexB, and SmI phases, may coexist near $x = 70$ wt. % of 8OSI. From optical microscopic textures, it is very difficult to distinguish these four phases near that concentration. Other measurements, e.g., heat capacity may allow us to resolve the sequence of phase transitions. So far, we have not found any transition from the HexB phase to the SmI phase. Thus, the phase boundary between these two phases seems to be very close to a vertical line as indicated in Figure 3.22.

### 3.3.3    Thermal Transport Studies

In addition to developing liquid crystal free-standing-film calorimetry in our laboratory, Huang and co-workers (1985) have established another important extension of the ac calorimetric technique to carry out high-resolution thermal conductivity measurements. The fact that it requires a very small temperature oscillation (approximately 3 mK rms) for the thermal conductivity measurement makes this technique extremely powerful for investigating dynamic critical phenomena. Furthermore, only a very small amount of sample (about 20 mg) is needed and a relative resolution better than 1% has been achieved. The fact that the thermal conductivity of the liquid crystal is about one-fifth that of glass makes our extension of the calorimetric approach fairly impressive. So far, many attempts have been made to study the thermal conductivity of various liquid crystal phases. They are the steady-state method (Pieranski et al., 1972), the thermal-lens effect (Koren, 1976), and the forced Rayleigh light-scattering technique (Rondelez et al., 1978). All of these techniques require a fairly large temperature gradient inside the sample in order to obtain a measurable signal. Furthermore, most of them cannot achieve 1% resolution. As a consequence, these methods are not suitable for studying critical dynamics near a continuous phase transition.

Most liquid crystal molecules are obviously very anisotropic and quite a number of mesophases display anisotropic effects. It would be important therefore to measure the thermal conductivity of aligned samples and study this anisotropy. Unfortunately, after a few attempts, we have not been able to achieve good alignment within our delicate heat-capacity sample cells for nmOBC compounds or the 8OSI compound. Our results are therefore obtained from multidomain samples. Figure 3.23 displays the temperature dependence of the thermal conductivity in the vicinity of the SmA-HexB transition of 37OBC (Geer et al., 1989). Both heat capacity and thermal conductivity can be measured in the same sample. The pronounced anomalies in both of these quantities are clear (Figures 3.10 and 3.23). The ratio

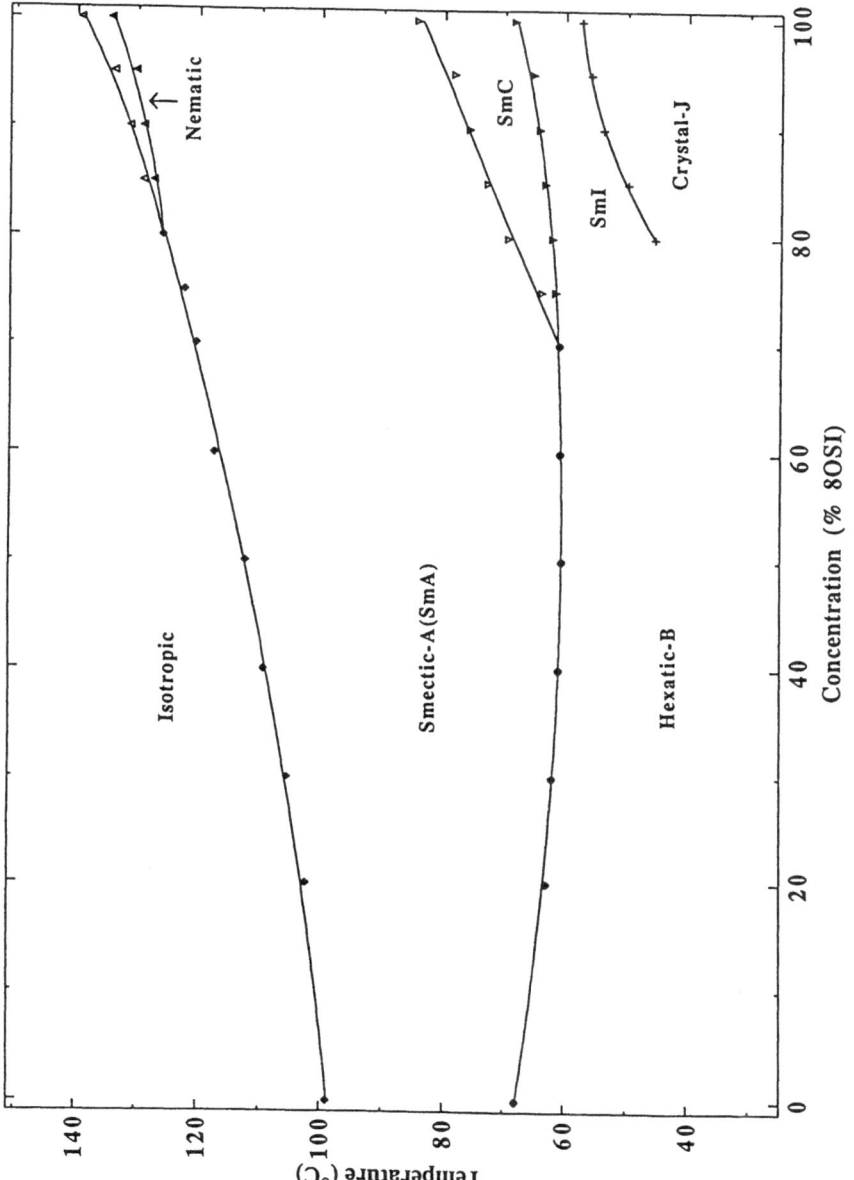

FIGURE 3.22. Temperature-concentration phase diagram of the binary mixture $[3(10)OBC]_{1-x} - [8OSI]_x$. Here $x$ is in wt. %.

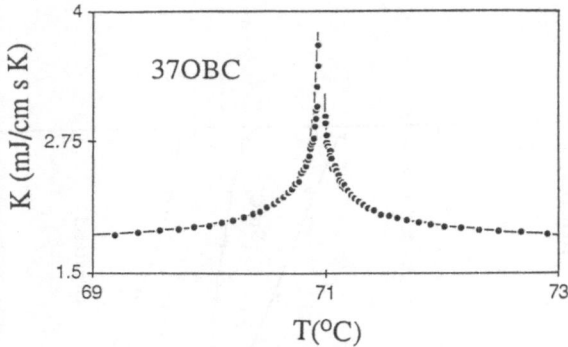

FIGURE 3.23. Temperature dependence of the thermal conductivity near the SmA-HexB transition of 37OBC.

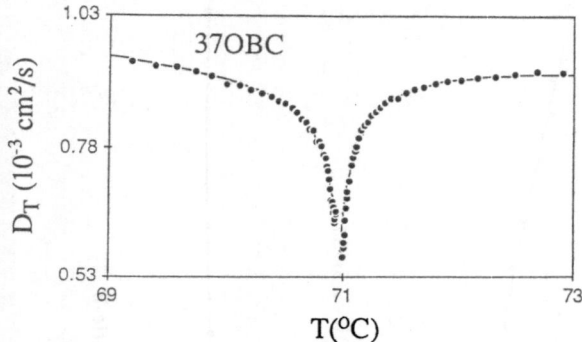

FIGURE 3.24. Temperature variation of the thermal diffusivity in the vicinity of the SmA-HexB transition of 37OBC.

of the thermal conductivity to the heat capacity is the thermal diffusivity $(D_T)$. The quantity $D_T$ has been calculated from our measured data and is shown in Figure 3.24. The drop in $D_T$ near $T_c$ clearly indicates the important phenomenon of the critical slowing down of thermal fluctuations near the critical temperature. To the best of our knowledge, this is the first set of data which clearly demonstrates the critical slowing down of thermal fluctuations for temperature regions both above and below the transition temperature.

The analysis of the thermal conductivity data was based on the assumption that the total conductivity is the sum of a regular ($K_{reg}$) and a singular ($K_{sing}$) contribution,

$$K = K_{reg} + K_{sing} = B_3 + D_3 t + \Lambda |t|^{-a}. \tag{3.7}$$

This assumption is an approximation which may well become outdated by future theoretical advances because it may be incorrect to simply add the contributions from two separate origins. Here $\Lambda^+$ and $\Lambda^-$ are the amplitudes of the singular term for $T > T_c$ (+) and $T < T_c$ (−), respectively, $a$ is the dynamic critical exponent, and $B_3$ and $D_3$ are the coefficients of the nonsingular term. Our thermal conductivity data of 65OBC are well described by Eq. (3.7) in the temperature range 20 mK $< |T - T_c| < 3$ K. The important parameters obtained are $a = 0.51 \pm 0.04$, and $\Lambda^+/\Lambda^- = 0.93 \pm 0.10$ (Nounesis et al., 1986).

Employing the Kubo expression for the thermal conductivity (Forster, 1975), Hobbie and Huang (1989) derived the following expression for the ratio $C/D_T$:

$$C/D_T = F_s^{-1} + B_4, \tag{3.8}$$

where the ratio has been written as a sum of two parts. The background $B_4$ is the nonsingular part; and $F_s^{-1}$ is the singular part given by

$$F_s^{-1} = (\Gamma_\psi \chi \xi^5)^{-1} \left(\frac{\partial \chi}{\partial t}\right)^2, \tag{3.9}$$

where the kinetic coefficient $\Gamma_\psi$ associated with the order parameter $\Psi$ is a nonsingular function of temperature. $\xi$ and $\chi$ are the correlation length and order parameter susceptibility, respectively. These two latter physical quantities have strong diverging behaviors near a continuous phase transition and are characterized by critical exponents $\nu$ and $\gamma$, namely, $\xi \simeq |t|^{-\nu}$ and $\chi \simeq |t|^{-\gamma}$. By employing scaling relations between the critical exponents, the singular part can be expressed again as

$$F_s^{-1} \simeq |t|^{-b}, \tag{3.10}$$

where $b = \alpha + (2 - \alpha)\eta/3$. Figure 3.25 displays our fitting results of the experimental data to Eqs. (3.9) and (3.10). The data can be fitted over more than two and a half decades in the reduced temperature scale. The critical exponent obtained from the fitting is $b = 0.69 \pm 0.05$. The fitting to the heat-capacity anomaly obtained from the same sample leads to $\alpha = 0.60 \pm 0.03$. This results in an unsually large and negative value for $\eta = -0.19 \pm 0.03$. Direct experimental measurements on the critical exponent $\eta$ are essential to confirm our interpretation of the ratio $C/D_T$. Assuming the validity of scaling relations, we find that $\beta = 0.19$, which is consistent with the value obtained by Rosenblatt and Ho (1982) for the SmA-HexB transition of 65OBC and $\gamma = 1.04$ as well as $\nu = 0.47$, which are only slightly removed from their mean-field values. Similar results have been obtained for the 65OBC near the SmA-HexB transition.

Recently thermal conductivity studies near the SmC-SmI transition of 8OSI have been carried out by Hobbie et al. (1989). As found in the SmA-HexB transition of 65OBC, an increase in the thermal conductivity and a drop in the thermal diffusivity in the neighborhood of the SmC-SmI transition have been observed. Even though these anomalies are much more

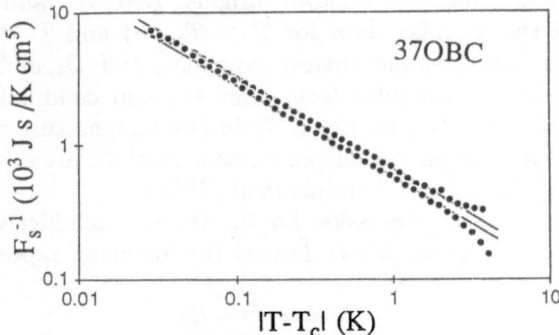

FIGURE 3.25. Log-log plot of anomalous part of $F_s^{-1}$ ($= C/D_T$) vs. $|T - T_c|$ for the 37OBC compound.

rounded than those of 65OBC, the data clearly demonstrate the critical slowing down of the thermal fluctuations near the SmC-SmI transition. Due to the finite coupling between the tilt angle and BOO, the explanation of the experimental data is much more complicated and less satisfactory than is the case for nmOBC (Hobbie, 1990).

## 3.4    Criticality of the Smectic-A-Hexatic-B Transition

Bond-orientational order was first clearly demonstrated in some temperature range below the SmA phase of 65OBC (Pindak et al., 1981). According to the symmetry of the BOO, one would expect to obtain $XY$-like critical exponents associated with the SmA-HexB transition. To our surprise, not only the first heat-capacity investigations on 65OBC (Huang et al., 1981; Viner et al., 1983) but also subsequent calorimetric studies on many other compounds in the homologous series of nmOBC (Pitchford et al., 1985) gave continuous SmA-HexB transitions with a very large value in the heat-capacity critical exponent ($\alpha = 0.60$). This value is drastically different from the $XY$-like exponent ($\alpha = -0.007$). As mentioned before, one of the plausible explanations for the nature of the SmA-HexB transitions of nmOBC with a large heat-capacity critical exponent is that this transition is in the vicinity of a tricritical point with the critical exponent being renormalized by the coupling of the BOO to some other degree of freedom with short-range order, e.g., translational, herring-bone, etc. Several theoretical attempts along this line have been proposed. We will discuss their relevance to the existing experimental data in this section.

In light of the existence of short-range (maybe) herring-bone fluctuations as detected by X-ray diffraction studies near the SmA-HexB transition of 65OBC (Pindak et al., 1981), Bruinsma and Aeppli (1982) formulated a theory that considered both the hexatic and herring-bone order $[\Phi_2 = \phi_2 \exp(2i\theta_2)]$. In this theoretical approach, both the hexatic and herring-bone order (assume that it enters as short-range fluctuations) belong to the $XY$ universality class which intrinsically has a negative heat-capacity critical exponent. In a phase diagram where two phase transition lines (responsible for the hexatic and herring-bone order, respectively) intersect, the order parameters are decoupled. In the mean-field approach, the SmA-HexB transition is continuous. By including the fluctuations and the coupling term $h * \mathrm{Re}(\Phi_6 * \Phi_2^3)$, the SmA-HexB transition can display a tricritical point separating second-order and first-order behavior. This argument may be applicable to the SmA-HexB transition of 65OBC, which upon cooling will undergo the HexB-CryE transition and establish long-range herring-bone order and positional order. However, the mixtures of 3(10)OBC and PHOAB have a very large temperature range for the HexB phase ($> 40$ K) with PHOAB weight concentration between 30 and 70%. If there exist herring-bone fluctuations near the SmA-HexB transition, one would expect that they would be very small due to the large HexB temperature range and the SmA-HexB transition should be continuous. To the contrary, the corresponding SmA-HexB transition is found to be first-order. Recently, we have carried out X-ray diffraction studies on 75OBC and found that the strength of the herring-bone peaks is weaker than those of 65OBC (Huang et al., 1990). In principle, if one assumes that 65OBC is near a tricritical point, 75OBC should be further removed from this point. The same heat-capacity critical exponent is obtained for these two compounds, however. This experimental result suggests that this theoretical model based on short-range herring-bone fluctuations may be irrelevant in explaining the nature of the SmA-HexB transition. Another crucial experimental test of this theoretical interpretation rests on detailed X-ray studies of the existence of herring-bone fluctuations near the SmA-HexB transition of other compounds. Also it is very important to determine experimentally whether the herring-bone order is truly short-range or not. Presumably, a long-range herring-bone order can exist in a system with long-range BOO without long-range translational order. If this is the case for nmOBC compounds, then our general picture of the HexB phase has to be changed completely.

In the other two well-characterized phase transitions among liquid crystal mesophases, namely, N-SmA and SmA-SmC transitions, it has been demonstrated that the temperature range of the disordered phase has a strong effect on the nature of the associated phase transition. For example, with decreasing nematic temperature range, the N-SmA transition changes from being second-order to first-order with a tricritical point in between (Thoen et al., 1984; Ocko et al., 1984). With this in mind, Pitchford et al. (1985) carried out extensive studies on the nature of the SmA-HexB transition among the nmOBC compounds with a relatively wide variety of SmA temperature ranges (from 14 to 32 K). The heat-capacity critical exponent remains fairly constant among these nmOBC compounds (see Table 3.2). In other words, the nature of the SmA-HexB transition in the nmOBC compounds remains unchanged with respect to different SmA temperature ranges. Moreover, both the N-SmA and the SmA-SmC transitions (Huang and Lien, 1985) favor continuous behavior for a large temperature range of the nematic and smectic-A phase, respectively. So far, the compound with the largest SmA temperature range (54 K for PHOAB) has exhibited a first order SmA-HexB transition (Huang et al., 1986). From these experimental facts, we conclude that unlike the N-SmA and SmA-SmC transition, the nature of the SmA-HexB transition does not depend on the temperature range of the disordered phase (namely, the SmA phase).

Because the layer ordering in both the SmA and HexB phase is one-dimensional, the layers are not planar but constantly undergo large undulations. Recently, Selinger (1988) constructed a Hamiltonian in terms of the hexatic order and the layer fluctuations $u$. Due to the geometrical frustration introduced by the curvature of the layers, the order parameter $\Phi_6$ and $u$ are coupled. After integrating out the parameter $u$, the effective Hamiltonian for the order parameter $\Phi_6$ gives a first-order SmA-HexB transition, provided the hexatic stiffness constants are sufficiently large. This means that the layer fluctuations will enhance the first-order SmA-HexB transition. Larger hexatic stiffness constants suggest that the HexB phase will exist in a higher temperature range, so that the SmA phase temperature range should be correspondingly smaller. Unfortunately, among the very limited number of liquid crystal compounds which exhibit the SmA-HexB transition, all of them have a fairly large SmA phase temperature range ($>$ 14 K). To give a critical test of the above two ideas, compounds with a much smaller SmA phase temperature range are essential. On the other hand, PHOAB again appears as a counterexample since it has a fairly large SmA phase temperature range (about 54 K), suggesting a small hexatic stiffness constant, but the SmA-HexB transition is first-order. Consequently, we feel that the coupling between $u$ and $\Phi_6$ alone cannot be responsible for the intriguing behavior in the SmA-HexB transition.

Both the SmA-crystal (X) and HexB-X transitions lead to long-range three-dimensional translational order and therefore should be first-order transitions. Here the crystal (X) phase refers to the CryB, CryE, or crystalline phase. Therefore, there are two possible scenarios near the ABX point along the SmA-HexB transition line. First, the ABX point may be the critical end point of the SmA-HexB transition line. The SmA-HexB transition would then be continuous along the entire transition line with an $XY$-like behavior, provided that $\Phi_6 = \phi_6 \exp(6i\theta_6)$ is the symmetry-breaking field at the phase transition. Alternatively, the second-order SmA-HexB transition may become first-order before joining the ABX point. Under these circumstances the ABX point is a simple triple point. In this case, there must be a tricritical point separating the first-order and second-order parts of the SmA-HexB transition line. The latter scenario is the generic phase diagram suggested by Aharony and co-workers (1986). As a consequence, sufficiently far away from the tricritical point, the SmA-HexB transition should resume three-dimensional $XY$ behavior.

Employing optical microscopy studies and heat-capacity measurements, Nounesis et al. (1989) have completed the temperature-concentration phase diagram of the n(10)OBC homologous series, as shown in Figure 3.8. Here $n$ has a value ranging from 1 to 4. One of the key features of this phase diagram is the existence of a special thermodynamic point at which the SmA, HexB, and CryE phases coexist. Here the CryE phase has both long-range translational order and herring-bone molecular order. Because in both the SmA-CryE and HexB-CryE transitions, long-range translational order is developed, such transitions should be first-order. Consistent with this hypothesis, large hysteresis has been observed in our heat-capacity measurements through these transitions. To explore the nature of this unique ABX point, we have carried out detailed calorimetric studies along the SmA-HexB transition line.

The measured heat capacity in the vicinity of the SmA-HexB transition of pure 2(10)OBC, 3(10)OBC, and three binary mixtures are shown in Figure 3.26. Table 3.3 summarizes the fitted exponent $\alpha$ and the amplitude ratio $A^+/A^-$ along with the standard deviations. All the heat-capacity critical exponents we have obtained along the SmA-HexB transition line are the same and approximately equal to 0.60. The salient feature of this result is that the $[2(10)OBC]_{0.75} - [1(10)OBC]_{0.25}$ mixture has a very small temperature range (approximately 0.8 K) for the HexB phase and still displays a continuous SmA-HexB transition with $\alpha = 0.58$ (Nounesis et al., 1989). This suggests that if there exists a tricritical point separating the first-order SmA-HexB transition from the continuous one, such a point must be extremely close to the SmA-HexB-CryE triple point.

On the basis of the two scenarios we presented before, let us discuss the implications of these heat-capacity results. First, if the triple point is also the critical end point for the continuous SmA-HexB phase transition line, then the large value in $\alpha$ we have obtained along this SmA-HexB transition line is intrinsic to the SmA-HexB transition, but not due to the closeness of the given SmA-HexB transition to a tricritical point. Second, if there exists a tricritical point along the SmA-HexB transition, then this point should be located between the triple point and $[2(10)OBC]_{0.75} - [1(10)OBC]_{0.25}$ mixture point. Far away from this tricritical point one should observe either crossover behavior from the three-dimensional $XY$ exponent ($\alpha = -0.007$) to a tricritical exponent ($\alpha = 0.5$) or entirely three-dimensional $XY$-like behavior. The consistency of the exponent $\alpha$ along the SmA-HexB transition line with the HexB phase temperature range varying from 0.8 to 20 K seems to rule out the general belief that the large value in the heat-capacity exponent associated with the SmA-HexB transition implies that this transition is close to a tricritical point. It also seems odd that the whole

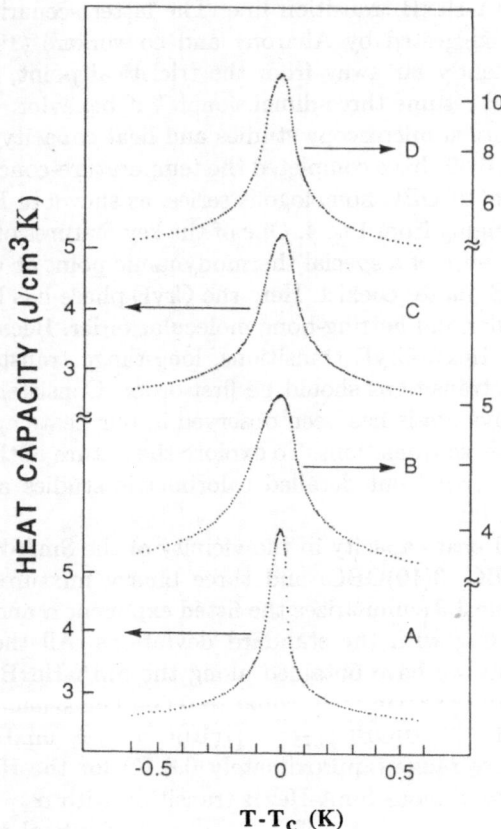

FIGURE 3.26. Temperature dependence of the heat capacity anomalies near the SmA-HexB transitions of various pure compounds and binary mixtures. Curve A, pure 3(10)OBC compound; B, [2(10)OBC].75 [3(10)OBC].25 mixture; C, pure 2(10)OBC compound; D, [1(10)OBC].25 [2(10)OBC].75 mixture.

SmA-HexB phase transition line is in the vicinity of a very special thermodynamical point, namely, a tricritical point. In conclusion, we have yet to detect any indication of the three-dimensional $XY$ heat-capacity critical exponent near the continuous SmA-HexB transition and the existing theories cannot give satisfactory explanations for all of the results related to the SmA-HexB transition.

## 3.5     Conclusions

After approximately one decade of extensive experimental and theoretical investigations on traditional phase transitions, Wilson introduced the celebrated renormalization group theory to provide excellent explanations of many important aspects of critical phenomena (Wilson and Kogut, 1974). Here in liquid crystals, we have found some unusual phase transitions. Among many other possible phase transitions in liquid crystal systems, the nematic-SmA, SmA-SmC [or -chiral-smectic-C (SmC*)], and SmA-HexB transitions have been studied extensively and were found to be continuous in many compounds. Due to the symmetry of the order parameters, the nematic-SmA, SmA-SmC (or -SmC*), and SmA-HexB transition, are all expected to belong to the three-dimensional $XY$ universality class. So far, $XY$-like fluctuations have not been found in any of these three phase transitions. The SmA-SmC (or -SmC*) transition has been convincingly demonstrated to be mean-field-like with an unusually large sixth-order term in the Landau free energy expansion (Huang and Viner, 1982). After about 20 years of theoretical and experimental effort, the nature of the nematic-SmA transition remains an unsolved problem in the investigation of critical phenomena. Consequently, it seems that sophisticated theories in addition to the renormalization group theory are required in order to explain the phase transitions in liquid crystals.

Although the idea of hexatic order introduced in the context of two-dimensional melting theories provides great insight into the molecular ordering in the three liquid crystal stacked hexatic phases, the mechanism behind the hexatic ordering in these liquid crystal mesophases may not be the same as the one suggested by the two-dimensional melting theory. The building block for the theory is a spherical or cylindrical object with no internal structure. The main assumption is that only the long-range part of the defect interaction has to be considered. Thermally created dislocations develop in heating from the corresponding crystalline phases and are responsible for destroying the positional order. These ideas may not hold for the liquid crystal systems. This viewpoint is supported by two experimental facts. First, the observed position for the diffuse peaks in the SmF phase and the Bragg peaks in the CryG phase through high-resolution X-ray diffraction studies are essentially coincident. This indicates that the CryG-SmF transition does not change the short-range structure and establishes that the SmF phase is a faulted version of the CryG phase (Benattar et al., 1979; Sirota et al., 1985). Furthermore, investigating the correlation between aliphatic chain length and the shape of the X-ray powder spectrum, Moussa et al. (1983) pointed out that aromatic cores and aliphatic chains behave quite differently in the SmF phase of the homologues of TBnA. The aliphatic chains perform disordered motions, giving rise to liquidlike correlations between them while the aromatic cores are responsible for the molecular order. This two-sublattice model gains further support from recent NMR studies of the SmF phase of 4-$n$-nonyloxybenzylidene-4'-butylaniline (9O.4) (Figueirinhas et al., 1987). Combining these two experimental results, one can imagine that the evolution from sharp Bragg peaks in the CryG to the diffusive peaks in the SmF phase is probably due to the melting of the aliphatic chains, but not the core part of the molecule because the X-ray is scattered by all the electrons in each molecule.

Finally, we would like to end this chapter by noting that it seems to be relatively difficult to find the theoretically predicted hexatic phase in nature. So far, the idea of BOO is extremely helpful in understanding many liquid crystal mesophases but is insufficient to explain the phase transitions related to the most well-characterized hexatic phases in liquid crystals. Consequently, further theoretical and experimental advances are necessary before we can clearly address the nature of the hexatic phases in liquid crystals.

*Acknowledgments.* The author acknowledges stimulating and helpful discussions with R. Birgeneau, R. Bruinsma, C. Dasgupta, J.W. Goodby, J.T. Ho, P. Hohenberg, D. Huse, D. Johnson, D. Litster, D. Nelson, R. Pindak, and M. Schick. The major part of this chapter is based on the Ph.D. thesis work of J. Viner, T. Pitchford, G. Nounesis, E. Hobbie, and R. Geer. In particular, Prof. J.W. Goodby has provided the author with many nmOBC compounds. Without those compounds, this work would have been practically impossible. I would like to thank R. Geer and T. Stoebe for reading this manuscript and offering useful comments. This work was supported by the Graduate School and the Center for Microelectronic and Information Sciences of the University of Minnesota, the Donors of the Petroleum Research Fund, administered by the American Chemical Society, and the National Science Foundation, Solid State Chemistry, Grant No.'s DMR-84-04945, DMR-85-03419, and DMR-89-19334. This is a revised version of the paper that had been submitted to *Advances in Physics*.

# References

F.F. Abraham, *Phys. Rev. Lett.* **44**, 463 (1980).

F.F. Abraham, *Phys. Rep.* **80**, 339 and references found therein (1981).

A. Aharony, R.J. Birgeneau, J.D. Brock, and J.D. Litster, *Phys. Rev. Lett.* **57**, 1012 ) (1986).

A. Aharony and M. Kardar, *Phys. Rev. Lett.* **61**, 2855 (1988).

G. Ahlers, A. Kornblitt, and H.J. Guggenheim, *Phys. Rev. Lett.* **34**, 1227 (1975).

G. Albertini, S. Melone, G. Poeti, F. Rustichelli, and G. Torquati, *Mol. Cryst. Liq. Cryst.* **104**, 121 (1984).

S.M. Amador and P.S. Pershan, *Phys. Rev. A* **41**, 4326 (1990).

L.J. Bellamy, *The Infra-Red Spectra of Complex Molecules*, Wiley, New York, 1957, p. 14.

J.J. Benattar, J. Doucet, M. Lambert, and A.M. Levelut, *Phys. Rev. A* **20**, 2505 (1979).

A.N. Berker and D.R. Nelson, *Phys. Rev. B* **19**, 2488 (1979).

R.J. Birgeneau and J.D. Litster, *J. Phys. Lett.* (Paris) **39**, L399 (1978).

D.J. Bishop and J. Reppy, *Phys. Rev.* **40**, 1727 (1978).

J.D. Brock, A. Aharony, R.J. Birgeneau, K.W. Evans-Lutterodt, J.D. Litster, P.M. Horn, G.B. Stephenson, and A.R. Tajbakhsh, *Phys. Rev. Lett.* **57**, 98 (1986).

J.D. Brock, D.Y. Noh, B.R. McClain, J.D. Litster, R.J. Birgeneau, A. Aharony, P.M. Horn, and J.C. Liang, *Z. Phys. B* **74**, 197 (1989a).

J.D. Brock, R.J. Birgeneau, J.D. Litster, and A. Aharony, *Contemporary Phys.* **30**, 321 (1989b).

R. Bruinsma and D.R. Nelson, *Phys. Rev. B* **23**, 402 (1981).

R. Bruinsma and G. Aeppli, *Phys. Rev. Lett.* **48**, 1625 (1982).

J. Budai, R. Pindak, S.C. Davey, and J.W. Goodby, *J. Phys. Lett.* (Paris) **45**, L1053 (1984).

T.W. Burkhardt, H.J.F. Knops, and M. Nijs, *J. Phys. A* **9**, L179 (1976).

D.M. Ceperley and E.L. Pollock, *Phys. Rev. B* **39**, 2084 (1989).

S. Chandrasekhar, *Liquid Crystals*, Cambridge University Press, New York, 1977.

J.H. Chen and T.C. Lubensky, *Phys. Rev. A* **14**, 1202 (1976).

W. Chen, L.J. Martinez-Miranda, H. Hsiung, and Y.R. Shen, *Phys. Rev. Lett.* **62**, 1860 (1989).

M. Cheng, J.T. Ho, S.W. Hui, and R. Pindak, *Phys. Rev. Lett.* **61**, 550 (1988).

J. Collett, P.S. Pershan, E.B. Sirota, and L.B. Sorensen, *Phys. Rev. Lett.* **52**, 356 (1984).

S.C. Davey, J. Budai, J.W. Goodby, R. Pindak, and D.E. Moncton, *Phys. Rev. Lett.* **53**, 2129 (1984).

P.G. DeGennes, *Mol. Cryst. Liq. Cryst.* **21**, 49 (1973).

P.G. DeGennes, *The Physics of Liquid Crystals*, Oxford University Press, London, 1974.

P.G. DeGennes and G. Sarma, *Phys. Lett.* **38A**, 219 (1972).

D. Demus, S. Diele, M. Klapperstuck, V. Link, and H. Zaschke, *Mol. Cryst. Liq. Cryst.* **15**, 161 (1971).

S.B. Dierker, R. Pindak, and R.B. Meyer, *Phys. Rev. Lett.* **56**, 1819 (1986).

S.B. Dierker and R. Pindak, *Phys. Rev. Lett.* **59**, 1002 (1986).

J. Figueirinhas, S. Zumer, and J.W. Doane, *Phys. Rev. A* **35**, 4389 (1987).

D. Frenkel and J.P. McTague, *Phys. Rev. Lett.* **42**, 1632 (1979).

D. Forster, *Hydrodynamic Fluctuations, Broken Symmetry, and Correlation Functions*, Benjamin/Cummmings, New York, 1975.

M. Fukugida and M. Okawa, *Phys. Rev. Lett.* **63**, 13 (1989).

R. Gangwar, N.J. Collela, and R.M. Suter, *Phys. Rev. B* **39**, 2459 (1989).

C.W. Garland, J.D. Litster, and K.J. Stine, *Mol. Cryst. Liq. Cryst.* **170**, 71 (1989).

R. Geer, C.C. Huang, R. Pindak, and J.W. Goodby, *Phys. Rev. Lett.* **63**, 540 (1989).

R. Geer, H.Y. Liu, E.K. Hobbie, G. Nounesis, C.C. Huang, and J.W. Goodby, *J. Phys. France* **50**, 3167 (1989).

R. Geer, PhD thesis, University of Minnesota (1991).

R. Geer, T. Stoebe, C.C. Huang, R. Pindak, G. Srajer, J.W. Goodby, M. Cheng, J.T. Ho, and S.W. Hui, *Phys. Rev. Lett.* **66**, 1322 (1991a).

R. Geer, T. Stoebe, T. Pitchford, and C.C. Huang, *Rev. Sci. Intrum.* **62**, 415 (1991b).

M.A. Glaser and N.A. Clark, *Phys. Rev. A* **41**, 4585 (1990).

J.W. Goodby, *Mol. Cryst. Liq. Cryst. Lett.* **72**, 95 (1981).

J.W. Goodby and G. Gray, *J. de Phys.* (Paris) **40**, C3–27 (1979).

J.W. Goodby and R. Pindak, *Mol. Cryst. Liq. Cryst.* **75**, 233 (1981).

E.E. Gorodetskii and V.M. Zaprudskii, *Sov. Phys. JETP* **45**, 251 (1977).

G.W. Gray and J.W. Goodby, *Smectic Liquid Crystals: Textures and Structures*, Leonard Hill, Glasgow and London, 1984.

N. Greiser, G.A. Held, R. Frahm, R.L. Greene, P.M. Horn, and R.M. Suter, *Phys. Rev. Lett.* **59**, 1706 (1987).

J.A. Griffin, J.D. Litster, and A. Linz, *Phys. Rev. Lett.* **38**, 251 (1977).

B.I. Halperin and D.R. Nelson, *Phys. Rev. Lett.* **41**, 121 (1978).

S. Heinekamp, R.A. Pelcovits, E. Fontes, E.Y. Chen, R. Pindak, and R.B. Meyer, *Phys. Rev. Lett.* **52**, 1017 (1984).

H.J. Herrmann, *Z. Physik B* **35**, 171 (1979).

J.T. Ho and J.D. Litster, *Phys. Rev. B* **2**, 4523 (1970).

E.K. Hobbie, H.Y. Liu, C.C. Huang, and J. Liang, *Phys. Rev. A* **37**, 3963 (1988).

E.K. Hobbie and C.C. Huang, *Phys. Rev. A* **39**, 4154 (1989).

E.K. Hobbie, H.Y. Liu, C.C. Huang, and J.C. Liang, *Phys. Rev. A* **39**, 4159 (1989).

E.K. Hobbie, PhD thesis, University of Minnesota, 1990.

C.C. Huang, J.M. Viner, R. Pindak, and J.W. Goodby, *Phys. Rev. Lett.* **46**, 1289 (1981).

C.C. Huang and J.M. Viner, *Phys. Rev. A* **25**, 3385 (1982).

C.C. Huang and S.C. Lien, *Phys. Rev. A* **31**, 2621 (1985).

C.C. Huang, J.M. Viner, and J.C. Novack, *Rev. Sci. Instrum.* **56**, 1390 (1985).

C.C. Huang, G. Nounesis, and D. Guillon, *Phys. Rev. A* **33**, 2602 (1986).

C.C. Huang, G. Nounesis, R. Geer, J.W. Goodby, and D. Guillon, *Phys. Rev. A* **39**, 3741 (1989).

C.C. Huang, R. Pindak, and G. Srajer, 1990 (unpublished).

B.A. Huberman, D.M. Lublin, and S. Doniach, *Solid State Commun.* **17**, 485 (1975).

A.J. Jin, M.R. Bjurstrom, and M.H.W. Chan, *Phys. Rev. Lett.* **62**, 1372 (1989).

D. Johnson, D. Allender, R. DeHoff, C. Maze, E. Oppenheim, and R. Reynolds, *Phys. Rev. B* **16**, 470 (1977).

H. Kawamura, *J. Appl. Phys.* **61**, 3590 and references found therein (1987).

G. Koren, *Phys. Rev. A* **13**, 1177 (1976).

J.M. Kosterlitz and D.J. Thouless, *J. Phys. C* **6**, 1181 (1973).

S. Kumar, *Phys. Rev. A* **23**, 3207 (1981).

A.J. Leadbetter, J.P. Gaughan, B. Kelly, G.W. Gray, and J.W. Goodby, *J. Phys.* (Paris) **40**, C3–178 (1979a).

A.J. Leadbetter, J.C. Frost, and M.A. Mazid, *J. de Phys. Lett.* (Paris) **40**, L325 (1979b).

A.J. Leadbetter, M.A. Mazid, B.A. Kelly, J.W. Goodby, and G.W. Gray, *Phys. Rev. Lett.* **43**, 630 (1979c).

D.H. Lee, J.D. Joannopoulos, J.W. Negele, and D.P. Landau, *Phys. Rev. B* **33**, 450 (1986).

J.C. LeGuillon, J.C J. Zinn-Justin, *J. Physique Lett.* **46**, L137 (1985).

G.R. Luckhurst and G.W. Gray, ed., *The Molecular Physics of Liquid Crystals*, Academic Press, London, 1979.

R. Mahmood, M. Lewis, R. Biggers, V. Surrendranath, D. Johnson, and M.E. Neubert, *Phys. Rev.* **33**, 519 (1986).

R. Mahmood, M. Lewis, D. Johnson, and V. Surrendranath, *Phys. Rev.* **38**, 4299 (1988).

N.D. Mermin and H. Wagner, *Phys. Rev. Lett.* **22**, 1133 (1966).

R.J. Meyer and W.L. McMillan, *Phys. Rev. A* **9**, 899 (1974).

P. Minnhangen, *Rev. Mod. Phys.* **59**, 1001 (1987).

D.E. Moncton and R. Pindak, *Phys. Rev. Lett.* **43**, 701 (1979).

D.E. Moncton, R. Pindak, S.C. Davey, and G.S. Brown, *Phys. Rev. Lett.* **49**, 1865 (1982).

F. Moussa, J.J. Benattar, and C. Williams, *Mol. Cryst. Liq. Cryst.* **99**, 145 (1983).

C.A. Murray and D.H. Van Winkle, *Phys. Rev. Lett.* **58**, 1200 (1987).

C.A. Murray, P.L. Gammel, D.J. Bishop, D.B. Mitzi, and A. Kapitulnik, *Phys. Rev. Lett.* **64**, 2312 (1990).

S.E. Nagler, P.M. Horn, T.F. Rosenbaum, R.J. Birgeneau, M. Sutton, S.G.J. Mochrie, D.E. Moncton, and R. Clarke, *Phys. Rev. B* **32**, 7373 (1985).

D.R. Nelson and J.M. Kosterlitz, *Phys. Rev. Lett.* **39**, 1201 (1977).

D.R. Nelson and B.I. Halperin, *Phys. Rev. B* **19**, 2457 (1979).

D.R. Nelson and B.I. Halperin, *Phys. Rev. B* **21**, 5312 (1980).

D.R. Nelson, M. Rubenstein, and F. Spaepen, *Phil. Mag. A* **46**, 105 (1982).

D.R. Nelson, *Phase Transitions and Critical Phenomena*, Vol. 7, edited by C. Domb and J.L. Lebowitz, Academic Press, New York, 1983, p. 1.

D.Y. Noh, J.D. Brock, J.D. Litster, R.J. Birgeneau, and J.W. Goodby, *Phys. Rev. B* **40**, 4920 (1989).

G. Nounesis, C.C. Huang, and J.W. Goodby, *Phys. Rev. Lett.* **56**, 1712 (1986).

G. Nounesis, R. Geer, H.Y. Liu, C.C. Huang, and J.W. Goodby, *Phys. Rev.* **40**, 5468 (1989).

G. Nounesis, PhD thesis, University of Minnesota, 1990.

B.M. Ocko, R.J. Birgeneau, J.D. Litster, and M.E. Nuebert, *Phys. Rev. Lett.* **52**, 208 (1984).

B.M. Ocko, A. Braslau, P.S. Pershan, J. Als-Nielsen, and M. Deutsch, *Phys. Rev. Lett.* **57**, 94 (1986).

P.S. Pershan, G. Aeppli, J.D. Litster, and R.J. Birgeneau, *Mol. Cryst. Liq. Cryst.* **70**, 861 (1981).

P.S. Pershan, *Structure of Liquid Crystal Phases*, World Scientific, Singapore, 1988.

P. Pieranski, F. Brochard, and E. Guyon, *J. Phys.* (Paris) **33**, 681 (1972).

R. Pindak, D.J. Bishop, and W.O. Springer, *Phys. Rev. Lett.* **44**, 1461 (1980).

R. Pindak, D.E. Moncton, S.D. Davey, and J.W. Goodby, *Phys. Rev. Lett.* **46**, 1135 (1981).

T. Pitchford, G. Nounesis, S. Dumrongrattana, J.M. Viner, C.C. Huang, and J.W. Goodby, *Phys. Rev. A* **32**, 1938 (1985).

T. Pitchford, C.C. Huang, J.D. Budai, S.C. Davey, R. Pindak, and J.W. Goodby, *Phys. Rev. A* **34**, 2422 (1986).

J. Prost, *Advances in Physics* **33**, 1 (1984).

F. Rondelez, W. Urbach, and H. Hervet, *Phys. Rev. Lett.* **41**, 1058 (1978).

C. Rosenblatt and J.T. Ho, *Phys. Rev. A* **26**, 2293 (1982).

J.A. Roth, G.J. Jelatis, and J.D. Maynard, *Phys. Rev. Lett.* **44**, 333 (1980).

I. Rudnick, *Phys. Rev. Lett.* **40**, 1454 (1978).

C.R. Safinya, M. Kaplan, J. Als-Nielsen, R.J. Birgeneau, D. Davidov, J.D. Litster, D.L. Johnson, and M.E. Neubert, *Phys. Rev. B* **21**, 4149 (1980).

Y. Saito, *Phys. Rev. Lett.* **48**, 1114 (1982).

136     C.C. Huang

P. Schofield, J.D. Litster, and J.T. Ho, *Phys. Rev. Lett.* **25**, 1098 (1969).

J.V. Selinger, *J. Phys.* (France) **49**, 1387 (1988).

E.B. Sirota, P.S. Pershan, L.B. Sorensen, and J. Collett, *Phys. Rev. Lett.* **55**, 2039 (1985).

E.B. Sirota, P.S. Pershan, L.B. Sorensen, and J. Collett, *Phys. Rev. A* **36**, 2890 (1987).

G.S. Smith, E.B. Sirota, C.R. Safinya, and N.A. Clark, *Phys. Rev. Lett.* **60**, 813 (1988).

S.A. Solla and E.K. Riedel, *Phys. Rev. B* **23**, 6008 (1981).

S. Sprunt and J.D. Litster, *Phys. Rev. Lett.* **59**, 2682 (1987).

K.J. Stine and C.W. Garland, *Mol. Cryst. Liq. Cryst.* **188**, 91 (1990).

K.J. Strandburg, *Rev. Mod. Phys.* **60**, 161 and references found therein (1988).

P.F. Sullivan and G. Seidel, *Phys. Rev.* **173**, 679 (1968).

V. Surrendranath, D.L. Fishel, A. deVries, R. Mahmood, and D.L. Johnson, *Mol. Cryst. Liq. Cryst.* **131**, 1 (1985).

J. Thoen, H. Marynissen, and W. vanDael, *Phys. Rev. Lett.* **52**, 204 (1984).

J. Tobochnik and G.W. Chester, *Phys. Rev. B* **20**, 3761 (1979).

D.J. Tweet, R. Holyst, B.D. Swanson, H. Stragier, and L.B. Sorensen, *Phys. Rev. Lett.* **65**, 2157 (1990).

J. Van Himbergen and S. Chakravarty, *Phys. Rev. B* **23**, 359 (1981).

J.M. Viner, D. Lamey, C.C. Huang, R. Pindak, and J.W. Goodby, *Phys. Rev. A* **28**, 2433 (1983).

J.M. Viner and C.C. Huang, *Phys. Rev. A* **27**, 2763 (1983).

E. Webster, G. Webster, and M. Chester, *Phys. Rev. Lett.* **42**, 243 (1979).

K.G. Wilson, *Phys. Rev. B* **4**, 3174, 3184 (1971).

K.G. Wilson and J. Kogut, *Phys. Repts.*, **12**, 75 (1974).

A.P. Young, *Phys. Rev. B* **19**, 1855 (1979).

C.Y. Young, R. Pindak, N.A. Clark, and R.B. Meyer, *Phys. Rev. Lett.* **40**, 773 (1978).

# 4

# Experimental Studies of Melting and Hexatic Order in Two-Dimensional Colloidal Suspensions

*Cherry A. Murray*

## 4.1   Introduction

Suspensions of monodisperse colloidal particles are a rich model experimental system for understanding the fundamental mechanism for melting of two-dimensional crystals, and for searching for the elusive hexatic phase, with order intermediate between that of a perfect crystal and disordered fluid. By performing imaging experiments on colloidal systems, one can surmount a number of problems that have plagued either past experiments on atomic scale systems or computer simulations. The colloid experiments share some useful features with computer simulations: In particular, the colloids can be made very rigidly two-dimensional, so that promotion to a second layer may not occur. The substrate potential can be made atomically smooth so that the coupling to underlying lattice symmetry is nonexistent or, on the other hand, modulated with any strength and symmetry. The particle-particle and particle-wall interactions can be adjusted from short- to long-range. The interparticle separation, collision time, and particle diameter can be chosen for optimum video imaging with visible light so that one can obtain direct visual evidence of lattice defects, their dynamics, and of bond-orientational order in real space in the system on relevant timescales. Compared to simulations, however, the colloid imaging experiments have some very important advantages: the system size can be very large, so that boundary conditions and sample size can be effectively eliminated as problems, and the system can be monitored for easily $10^7$ collisions, so that time for attainment of equilibrium may be reached. Brownian motion of the suspension assures Langevin dynamics for the spheres and a true thermodynamic temperature. Recently experiments on colloidal systems have shown convincing evidence for defect-mediated melting and hexatic order. I will summarize the theoretical predictions of melting of two-dimensional crystals in Section 4.1, review the experimental studies to date of two-dimensional melting using direct visualization of colloidal suspensions in Section 4.2, and point out some needs and exciting possibilities for future work in Section 4.3.

## 4.1.1   Theoretical Predictions of Two-Dimensional Melting

Nearly 20 years ago much interest was generated in the melting of two-dimensional crystals by the prediction of Kosterlitz and Thouless [1] that, unlike in the third dimension, melting in the second dimension can be continuous, driven by the breakup of thermally generated topological defects in the lattice. Kosterlitz and Thouless pointed out that a two-dimensional crystal is in the same universality class as the $xy$ model (Heisenberg spins in the plane), two-dimensional superfluids, and two-dimensional superconductors. An excellent review of the theory of defect-mediated phase transitions has been provided by Nelson [2] and can be found elsewhere in this book. I will concentrate here only on specific predictions of their theory and their experimental consequences relevant to two-dimensional crystals.

The two-dimensional crystal translational order parameter is a Fourier component of the density $\rho_G(r) \equiv \exp(i\mathbf{G} \cdot \mathbf{r})$, where $\mathbf{G}$ is a reciprocal lattice vector of the crystal. This order parameter is complex, continuous, and Abelian. The two-dimensional crystal is at a borderline dimensionality below which there is no perfect long-range order at finite temperature $T$, because for $T > 0$, long wavelength phonon fluctuations cause logarithmic corrections, perturbing the infinite long-range order of the crystal sufficiently to cause algebraic decay of the translational order of the system [2,3]. Although the mean translational order parameter of the infinite crystal is zero, the correlation function $g_G(r)$ of translational order decays algebraically in space with a very weak temperature-dependent power law:

$$g_G(r) \equiv \langle \rho_G(0)\rho_G(r) \rangle \sim r^{-\eta_G(T)}. \qquad (4.1)$$

This has been dubbed "quasi-long-range" order [2] as the system looks ordered on length scales that are typically probed by most experiments or simulations. The phonons of a crystal are thermal fluctuations of the phase of the order parameter $\rho_G$. Such a crystal is also subject to thermal fluctuations in the amplitude $\rho_G$ such as energetically low-lying topological defects in the lattice.

In the second dimension, there are two classes of topological defects that are low in energy and relevant to the order-disorder transition, both of which are point defects. The defect that Kosterlitz and Thouless considered is a dislocation, or lattice row end, illustrated for a triangular two-dimensional crystal in Figure 4.1. A dislocation is characterized in elasticity theory by its Burgers vector $\mathbf{b}$, a vector which closes a complete lattice circuit around the core of the dislocation as shown in the figure [4]. Bound dislocation pairs are thermally generated in a crystal near its melting temperature. These pairs have equal and opposite Burgers vectors and can be

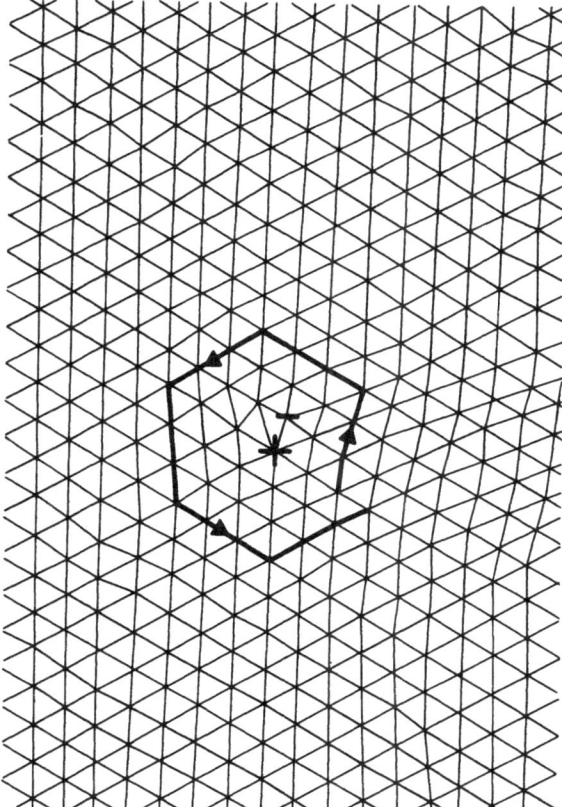

FIGURE 4.1. A dislocation in a two-dimensional triangular lattice, showing a Burgers circuit around the core. The local Burgers vector for the disclination is the vector required to close the circuit. A dislocation is composed of a fivefold disclination separated by one lattice spacing from a sevenfold disclination, as marked by the − and + symbols, respectively.

generated by a thermal fluctuation in which only one particle in the network is disturbed. On a smooth substrate with no underlying orientation field, the free energy of a pair of equal and opposite dislocations with the smallest magnitude Burgers vector $|\mathbf{b}| = a$, the nearest-neighbor distance in the lattice, is given in elasticity theory by

$$\frac{F}{k_B T} = -K(T)\left[\ln\frac{R}{a_c} - \tfrac{1}{2}\cos^2(\phi)\right] + 2E_c, \qquad (4.2)$$

where $k_B$ is Boltzmann's constant, $R \equiv |\mathbf{R}|$ is the distance between dislocation cores, $a_c$ is the radius of each core, $E_c k_B T$ is the energy needed to create a single dislocation core, $\phi$ is the angle between $\mathbf{b}$ and $\mathbf{R}$, and the dislocation pair interaction coefficient $K$ is given by the expression

$$K(T) = \frac{4\mu(T)[\lambda(T) + \mu(T)]a^2}{[\lambda(T) + 2\mu(T)]k_B T}, \qquad (4.3)$$

with $\lambda$ and $\mu$ the standard Lamé coefficients of the lattice. The pair interaction is basically logarithmic with core separation $R$.

The Kosterlitz–Thouless (KT) melting transition corresponds to the breakup of dislocation pairs into free dislocations as their density increases sufficiently for intervening pairs to screen (and renormalize) the interaction coefficient $K$. Some experimentally important consequences of the scaling of the problem using linear elasticity theory and renormalization group are the following: dislocation pairs begin to unbind above a temperature $T_m$. For $T > T_m$, they are separated by a diverging correlation length $\xi(T) \sim \exp[c/(T - T_m)^\nu]$, with $\nu = 0.369\dots$. Only essential singularities exist in the specific heat and compressibility at the transition. The specific heat should exhibit a weak peak above $T_m$ due to the gradual unbinding of dislocation pairs above the transition. The renormalized shear elastic constant in the crystal has a cusplike singularity: $\mu(T) \sim \mu(T_m)[1 + b|T - T_m|^\nu]$, and drops abruptly to zero above $T_m$. The dislocation pair interaction energy $K$ attains a universal limiting value from below $16\pi$ at $T_m$ and immediately jumps to zero above $T_m$ as the shear modulus jumps to zero. Finally the behavior of the structure factor at the reciprocal lattice points is singular as $T$ approaches $T_m$ from above:

$$S(G) \sim \xi^{2-\eta_G}, \qquad (4.4)$$

where $\eta_G$, the exponent of $g_G(r)$, the correlation function of the translation order parameter, is given by the expression

$$\eta_G(T) = \frac{|G|^2 k_B T[3\mu(T) + \lambda(T)]}{4\pi\mu(T)[2\mu(T) + \lambda(T)]}. \qquad (4.5)$$

A finite density of unbound dislocations will destroy the translational order of the crystal. How this physically comes about is demonstrated in

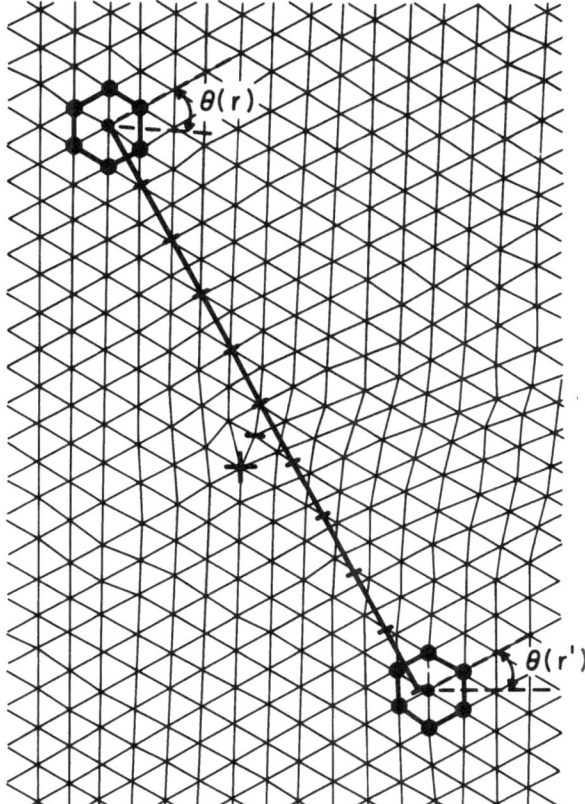

FIGURE 4.2. The disruption of translational order by a free dislocation in a triangular lattice. The line shown traversing the core of the dislocation starting from the lattice site at upper left has length $20a\sqrt{3}/2$ but misses falling onto a lattice point at lower right. On the other hand, the bond angles $\theta(r)$ and $\theta(r')$ are affected very little by the presence of the dislocation.

Figure 4.2. For $T > T_m$, the correlation function for translational order will decay exponentially in space with a correlation length $\xi(T)$ approximately equal to the mean spacing between free dislocations. One important difference between the KT transition in the second dimension and the case of ordinary melting in the third dimension is that the breakup of dislocation pairs and therefore the loss of translational order are continuous in KT melting, rather than first-order.

About five years after Kosterlitz and Thouless proposed the breakup of dislocation pairs as a possible two-dimensional melting mechanism, Halperin and Nelson [5] pointed out that there was another relevant order parameter for the two-dimensional crystal that was not addressed in the original Kosterlitz–Thouless treatment. The new order parameter comes about for the crystal because the dislocations must be described by vectors, rather than scalers as in the other members of the same universality class mentioned earlier. This bond-orientational order parameter can be expressed as $\psi_6(r) = \sum_{i=1}^{N} \exp[i\theta_i(r)]/N$, where $N$ is the number of nearest neighbors of a particle at $r$, and $\theta_i$ is the angle that the $i$th nearest-neighbor bond makes with an external direction. It measures the local mean orientation, modulo $\pi/3$, of all nearest-neighbor bonds in the lattice.

Halperin and Nelson (HN) showed that for $T > T_m$, the crystal melts into a phase that possesses quasi-long-range bond-orientational order, not into an isotropic liquid [2,5]. This intermediate "hexatic" phase is an oriented liquid, possessing no finite shear modulus and having short-range translational order, yet weak algebraic decay of bond-orientational order. One can get a physical picture of how this could come about by noting that this intermediate phase would consist of a low density sea of free renormalized dislocations, as well as dislocation pairs not yet broken up. In Figure 4.2 is a simple demonstration of the relatively weak perturbation of the bond angle field of the lattice by a free dislocation, compared to its strong disruption of translational order.

A free dislocation is composed of disclinations, the second class of relevant point topological defects in the lattice [4,5]. Disclinations are particles or lattice points with imperfect coordination: for a triangular two-dimensional lattice, anything but a coordination of six. Nearest-neighbor bonds are uniquely determined by a Voronoi polyhedron construction [6], so that disclinations are always uniquely identified. A dislocation in a triangular lattice is a tightly bound pair of seven- and fivefold disclinations, separated by one lattice spacing as shown in Figure 4.1. According to HN [2,5], the interaction free energy of the hexatic is described by

$$\frac{F}{K_B T} = \tfrac{1}{2} K_A(T) \int d^2 r [\nabla \theta(r)]^2, \qquad (4.6)$$

where $\theta(r)$ is the phase of the bond-orientational order parameter at $r$, and $K_A$ is an effective stiffness of the bond orientation field, called the Frank constant in analogy with that of liquid crystals. A renormalization group treatment of this problem leads to a second KT transition at $T_i > T_m$, above which the disclination pairs unbind to form an isotropic liquid.

The bond-orientational correlation function in the intermediate hexatic phase for $T_m < T < T_i$ is predicted to decay algebraically

$$g_6(r) \equiv \langle \psi_6(0)\psi_6(r) \rangle \sim r^{-\eta_6(T)}, \tag{4.7}$$

with the bond-orientational exponent given by

$$\eta_6(T) = \frac{18 k_B T}{\pi K_A(T)}. \tag{4.8}$$

The bond-orientational stiffness $K_A(T)$ diverges near $T_m$ and, analogous to the universal jump in the dislocation pair interaction strength $K$ at $T_m$, has a universal jump discontinuity at $T_i$ from $72\pi$ to zero above the transition. Therefore at $T_i$, the value of the bond-orientational exponent $\eta_6$ is predicted to be $1/4$, and it decays to zero at $T = T_m$. In the isotropic liquid phase at temperatures above $T_i$, the orientational correlation length $\xi_6$, essentially the mean separation of free disclinations, diverges as $\xi_6 \sim \exp[c'/|T - T_i|^{1/2}]$. As in the dislocation unbinding transition at $T_m$, the specific heat and energy will display only essential singularities at $T_i$, but the specific heat should exhibit a weak peak above $T_i$ due to the gradual unbinding of disclination pairs.

If we summarize the experimental consequences of the KTHN predictions for two-dimensional melting on a smooth substrate, the crystal melts through two continuous phase transitions at $T_m$ and $T_i$, with only essential singularities in thermodynamic quantities, and universal jump discontinuities of the dislocation pair interaction strength $K$ and orientational stiffness $K_A$, respectively, at the two transitions. In the crystal phase, the bond-orientational order is infinite: $g_6(r)$ is constant, while the translational order is quasi-long-range: $g_G(r) \sim r^{-\eta_G(T)}$. The limits on $\eta_G$ are set by the range of stability given by the two-dimensional Poisson's ratio $\sigma$ of the crystal, defined in terms of the renormalized Lamé coefficients of the lattice $\sigma(T) \equiv \lambda(T)/[\lambda(T) + \mu(T)]$. The expression Eq. (4.5) can also be written [5] as

$$\eta_G(T \to T_m^-) = |G|^2 \frac{a^2}{64\pi^2} [1 + \sigma(T)][3 - \sigma(T)]. \tag{4.9}$$

Since $\sigma$ cannot go negative or exceed unity for a stable lattice, the exponent $\eta_{G_0}(T_m)$ for the first Bragg point for a triangular lattice is bounded by the values $1/4$ and $1/3$.

For the intermediate hexatic phase, the translational correlation function $g_G(r)$ is predicted to fall off exponentially with a correlation length $\xi$ given by the mean distance between free dislocations which decreases singularly with $T$ [see discussion immediately following Eq. (4.3)]; while the orientational correlation function $g_6(r)$ decays algebraically with a temperature-dependent exponent $0 \leq \eta_6(T) \leq 1/4$ as the temperature is raised from

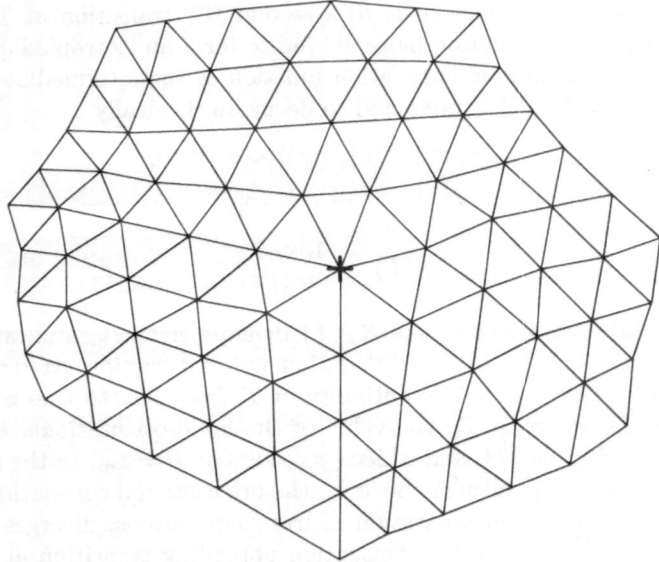

FIGURE 4.3. Sevenfold disclination in a two-dimensional triangular lattice. A free disclination completely disrupts both the translational and bond-orientational order.

$T_m$ to $T_i$. In the liquid for $T > T_i$, both the translational and bond-orientational correlations fall off exponentially with the same correlation length $\xi$ roughly given by the mean distance between free disclinations. The liquid is isotropic. The strong disruption of both translational and orientational order of a lattice by a free unbound disclination is illustrated in Figure 4.3.

If the two-dimensional system experiences an externally imposed orientational field, such as that caused by an underlying modulated substrate, then the phase diagram and the nature of the KT phase transitions can be altered dramatically, as shown by Halperin and Nelson and Young (Y) [2,5,7]. For $T > T_m$, the effect of a substrate potential will force the orientational order to be long-ranged. The disclination unbinding transition at $T_i$ will be washed out if the substrate has sixfold symmetry and be replaced by an Ising-like transition for a substrate with fourfold symmetry. This situation is relevant for all atomic scale experiments of two-dimensional melting, as all atomic substrates have some degree of modulation on the atomic length scale. There has been much controversy about the effect on the melting of a rare gas overlayer of the ordering field of graphite [8].

Of course, the possibility exists that a lattice instability or a spontaneous generation of classes of lattice defects not considered in the KTHNY scenario will preempt the Kosterlitz–Thouless phase transition. In particular, in the KTHNY theory an expansion in the fugacity of the system is made

[2,5] in the renormalization procedure. This amounts to a dilute gas approximation for the defects responsible for both phase transitions. If at the transitions the density of relevant topological defects is higher than a few percent, one would expect the approximation to break down due to strong multidefect interactions in addition to the pair interactions considered in the KTHNY theory. In this case, many have proposed that the melting transition will be first-order and directly into an isotropic liquid phase.

The density of dislocations will depend on the dislocation core energy which is usually not at all known and difficult to measure in experiments. Chui [9] has argued that the spontaneous generation of grain boundaries, which are arrays of dislocations, will always drive a first-order phase transition directly into an isotropic liquid before the dislocation unbinding transition can take place to the hexatic. He found a limit for the ratio at $T_m$ of dislocation core energy to dislocation pair interaction coupling energy $E_c/K < 0.71/4\pi$, (or $E_c < 2.84$), below which very strongly first-order grain boundary generated melting should be observed as the thermal density of dislocations is sufficiently high to form spontaneously small-angle grain boundaries which will disrupt the lattice abruptly into an isotropic liquid. However, for a larger core energy at $T_m$, $E_c/K > 0.71/4\pi$ (or $E_c > 2.84$), he predicts that the transition should be weakly first-order. His assumption that the dislocations comprising a grain boundary are less widely spaced than the grain boundaries themselves, however, breaks down in the large core energy regime so that the KTHNY dilute gas approximation may be more correct. I note that simulations of dislocation vector systems by Saito [10] in which the Hamiltonian of the system consists solely of dislocation pair interactions show evidence for a continuous KTHNY-like transition for $Ec < 2.84$ and for abrupt spontaneous grain boundary loop formation for $Ec > 2.84$. Also, in simulations of the Laplacian roughening model, which is dual to the two-dimensional crystal melting problem, Strandburg et al. [11] find a transition between a single first-order and two continuous transitions as an analogous parameter to the ratio between $Ec$ and $K$ is varied.

Others have also proposed first-order melting for two-dimensional crystals, often driven by other sorts of topological defects. Kleinert [12] proposed that disclination unbinding could occur simultaneously with dislocation unbinding if the screening of the disclination pairs by free dislocations were sufficiently strong. His calculations were, however, in the mean field limit, so that fluctuations, crucial in two-dimensional systems, were neglected. Simulations by Joos and Duesbery [13] on energies of localized vacancy and interstitial defects in Lennard–Jones systems show that the interaction energy of the dislocation pairs comprising these defects is much lower than the elastic limit. For vacancies stretched into lines, the elasticity limit is reached only at separations of $\sim 30a$, and for interstitials, only at $\sim 7a$. They question the validity of the KTHNY elastic description for real systems and propose that the low energy of localized vacancies may drive the melting transition, at least in systems with Lennard–Jones potential interactions. Fisher, Halperin, and Morf [14] find that the elastic limit is reached for $1/r$ potentials in $\sim 3a$, and Ladd and Horn [15] have similar findings for piecewise linear interaction potentials, so that the size of a dislocation core may be very system-dependent and is something to be aware of in experiments and simulations. It may be that the topological defects in the renormalized system at infinite separations which cause the transitions in the theory bear little resemblance to the microscopic defects observed in the real system on a length scale of a few lattice constants, a point that has been widely overlooked both by experimenters and simulators.

Ramakrishnan [16] has proposed a density wave theory of structurally driven two-dimensional freezing, in which the order parameters are $\rho_G$ with the $G$'s being the first two sets of reciprocal lattice vectors for the hexagonal crystal. In his theory, the fluid is placed into the lattice periodic potentials with amplitudes depending on $\rho_G$, and the transition takes place when the fluid free energy becomes larger than that of the crystal. He finds the two-dimensional system undergoes a very weakly first-order phase transition directly from the fluid to the crystal. His theory gives a specific prediction for the value near 5 of the height of the first peak in the two-dimensional structure factor at freezing, as well as the magnitude of the entropy change, fractional density change, and Debye Waller factor at freezing. This is a mean field theory in which fluctuations and topological defects are not explicitly considered.

Of course, there is no a priori reason not to have a first-order transition from the crystal to a hexatic, although this is not predicted by theory. The HN prediction of an intermediate hexatic phase possible in a system of isotropic particles without a substrate orientational field, that retains remnants of the crystalline long-range orientational order but has liquidlike short-range translational order, has generated not only theoretical activity, but also much experimental search and many computer simulations over more than a decade. An extensive review of the experimental and simulation literature is given by Strandburg [17] and also in other chapters in this book.

In experiments on atomic scale systems and simulations, little or no consensus exists in the theoretical, simulation, or experimental literature at present for the necessary conditions for KTHNY melting or even the existence of the hexatic phase in systems of spherically symmetric particles, much less the nature of its microscopic structure or particle motions. The hexatic phase is well established, however, in three-dimensional [18] and recently very thin layers [19] of liquid crystals composed of rodlike molecules, as discussed elsewhere in this book. The liquid crystal systems have an additional molecular tilt degree of freedom, which couples to the bond-orientational order much like an orienting field of a modulated substrate [20]. The transition from the crystal to the hexatic phase is first-order in the third dimension and grows more weakly first-order as the number of layers decreases.

A new approach to the search for a possible KTHNY phase transition and the hexatic phase in a spherically symmetric system began about four years ago, when a number of groups performed imaging experiments on the melting of two-dimensional submicron colloidal sphere suspensions. The experiments that have been performed so far are on spheres with either isotropic or dipolar interactions on smooth substrates. As will be detailed in the following section, since these experiments allow the study of lattice defects in real space and time on relevant timescales, they allow great insight into the complicated theoretical situation, into understanding defect melting mechanisms and their regimes of application. In some of these experiments there is the first good evidence of a smooth phase transition of a spherically symmetric system from crystal to fluid through a phase with spatial and bond-orientational correlations consistent with a hexatic.

## 4.1.2    Advantages and Disadvantages of Colloid Direct Imaging Experiments

Monodisperse colloidal spheres are a fascinating experimental system [21]. They have been used as a model system for condensed matter physics for more than a decade [22]. Most of the two-dimensional melting experiments done to date have used as samples monodisperse charge-stabilized colloidal polystyrene latex spheres which range from $\sim 0.1 - 10\,\mu m$ in diameter. These are formed by emulsion polymerization techniques [21,23] in which polar species, most often carboxylate or sulfate groups, are appended to the polymer ends on the surface of each sphere. When immersed into a polar solvent such as water, these polar groups dissociate, leaving one electron surface charge and one oppositely charged screening ion in solution. The surface charge density can be very high — as large as one electronic charge per $100\,\text{Å}^2$ on the sphere surface. Polystyrene latex particles and many other colloidal polymer particles are commercially available [24]. The latex spheres can be obtained with diameter and sphericity dispersions as small as 1% and corresponding surface charge uniformities, so that they are essentially identical particles with repulsive screened Coulomb interactions when in colloidal suspension in water electrolyte. For high surface charge density spheres the total surface charge for a $0.1\,\mu m$ diameter sphere can be typically $10^3$ to $10^4$ electronic charges.

Sterically stabilized colloidal polymer spheres also are available [25] with diameter dispersions of roughly 5%, nearly as good as those of the charged stabilized latex. These have spherical cores onto which another polymer layer has been adsorbed on the surface. The adsorbed layer can be a comb or graft polymer and is typically 10–50 Å thick. The polymer-coated spheres repel each other because of entropic steric hindrance from their polymer coatings [21] when they come sufficiently close for direct contact. These spheres can be suspended in hexane or other organic nonpolar solvents and exhibit "soft sphere" interparticle interactions [26], but to date have not yet been used for two-dimensional melting experiments. In addition, Skjeltorp [27] has demonstrated that by suspending latex spheres in ferrofluid (a suspension of 100–1000 Å size ferromagnetic particles in a solvent) in an applied magnetic field, one can create magnetic dipole "holes" in the liquid which can be easily imaged in the second dimension.

For a diameter range between 0.1 and 1 $\mu$m, the colloidal spheres in solvent suspension exhibit Brownian motion due to the collisions with their suspension medium, so that their dynamics can essentially be described by Langevin's equations and the system has a true thermodynamic temperature. For spheres of diameter 0.3 $\mu$m in water suspension, the sphere-sphere collision time is approximately 30 ms [28]. The mean sphere separation is roughly 1 $\mu$m. Thus, both the microscopic space and time scales of the suspension are readily accessible with video microscopy. Like computer simulations, the colloid experiments allow direct observation of microscopic particle and defect dynamics. This is particularly important in the study of order-disorder transitions and disordered systems in which diffraction is often difficult to interpret.

The study of melting in a layer of submicron spheres confined to two dimensions has several other advantages over experiments on atomic scale systems. It is possible to create a substrate that is absolutely smooth on the colloidal scale, unlike the case in atomic experiments where the modulation of a graphite substrate, e.g., is unavoidable and can have very large effects on the orientational order and phase diagram of a rare gas layer as mentioned earlier. Translational correlation lengths of $\sim 90a$ have been observed [29] for two-dimensional colloidal crystals, far in excess of the $\sim 30a$ correlation length near melting. Thus, it is experimentally possible to create a clean system unaffected by extraneous dirt or defects that could induce extrinsic translational disorder due to the nucleation of dislocations [30] or small random quenched displacements from lattice sites [31]. It is also possible to confine the spheres rigidly into a plane so that out-of-plane motions and second-layer promotion, which can have a dramatic effect on the melting transition due to easy vacancy creation in the first layer [32], are entirely avoided near melting. In addition, when desired, modulations of any symmetry and interaction strength can relatively easily be created as a model for substrate orientation fields to study their effects on the phase transition. To date, the effects of modulated substrates in the melting of two-dimensional colloidal crystals have not been carefully studied, however.

In experiments on colloidal crystals, one can avoid many of the serious problems with computer simulations associated with insufficient equilibration times or periodic boundary conditions by studying a small portion in the center of a much larger system, and allowing sufficient time to elapse for the entire system to come to statistical equilibration. It is easy to form liters of monodisperse suspension of particles and, more important, regions of two-dimensional layers that are uniform over $10^4$ or more particles. One can, in principle, mix up the suspension, fill an experimental cell, and wait for months to years for the system to equilibrate before beginning an experiment.

By adjusting the parameters of the specific colloidal suspension, one can create a model system of interacting macroparticles with forces of interaction that vary between short- and long-range. The charged sphere-sphere interactions are, in general, quite complex, as they include contributions from van der Waals, hard sphere, screened Coulomb, and many-body hydrodynamic contributions [33]. At the sphere densities and charges relevant to the two-dimensional melting experiments reviewed in this chapter, the repulsive screened Coulomb and indirect hydrodynamic interactions are the most important. For most experiments the system is, like electrons confined on a helium surface, dominated by repulsive interparticle interactions and thus must be confined by repulsive walls in order to attain sufficiently high densities for ordering.

It is not clear from the complexity of the interparticle interactions in this system even whether they can be modeled as pairwise additive interparticle potentials. However, the experimental three-dimensional phase diagram of the charged colloids [34,35] is qualitatively similar to that found [36] for a molecular dynamics simulation of hard spheres interacting with pairwise Yukawa (or Debye–Huckel) potentials with an effective sphere charge much smaller than the actual titratable charge. In this highly simplified model of colloid interactions, the sphere-sphere interaction potential for two spheres of diameter $d$ separated by a distance $r$ is given by

$$V(r) = f(a_s, d)\frac{Z}{\varepsilon r}\, e^{-(r-d/2)/\lambda_D}, \qquad (4.10)$$

where $Z$ is the effective charge of the sphere in the solution, $a_s \equiv (1/n)^{1/3}$ is the mean distance between sphere centers, $n$ is the sphere density, $\varepsilon = 80$ is the low-frequency dielectric constant of water, $\lambda_D$ is the effective Debye screening length in the solution. The geometrical correction factor $f(a_s, d)$ takes into account the volume lost to the screening ions due to the presence of the hard sphere core of diameter $d$, and can be incorporated into the effective charge, which will vary with sphere volume fraction. This interaction is often called the DLVO (Derjaguin, Langmuir, Verwey, Overbeek) potential when combined with dispersion interactions [37].

We must at all times keep in mind that the Yukawa potential is only a crude model for real charge-stabilized colloid interactions. Of course, the van der Waals attractive dispersion interaction has been ignored, and all of the complications of direct and indirect hydrodynamic interactions, solvation, and other steric interactions [21]. The Debye–Huckel or Yukawa interaction potential arises in calculations from a linearization of the mean field Boltzman equation in the solution of the electric potential around a charged sphere in an electrolyte solution with a certain density of screening charges. The surface charge densities of the latex spheres used in the experiments are typically several orders of magnitude too large for linearization to work. Alexander et al. proposed [38] that the Boltzman equation for the interaction of a point test charge with a highly charged macro-ion sphere in a spherically symmetric Wigner–Seitz cell in the suspension can be fitted to a linear solution near the zone boundary, and thus an effective charge and Debye screening length obtained. This is a reasonable method of getting an estimate for the effective charge felt by a test particle within the unit cell, but does not measure the macro-ion–macro-ion interaction in different unit cells, nor take into account the symmetry of the problem, the possible nonadditivity of the potentials, or fluctuation effects. More work is needed to obtain a more accurate measure of the actual macro-ion interactions, or to model the effective charge as a function of solution parameters in the real colloidal system.

The Yukawa system has two parameters which determine its entire equilibrium phase diagram [36]. The Coulomb screening parameter $\Lambda \equiv a_s/\lambda_D$ measures the ratio of the mean sphere separation to the effective Debye screening length in the electrolyte solution. The second parameter is the analog of the one-component plasma correlation parameter $\Gamma$, which measures the ratio of mean potential energy to kinetic energy of the system. For the Yukawa system, it is given by

$$\Gamma = \frac{Z(a_s)}{\varepsilon a_s k_B T} e^{-\Lambda}, \qquad (4.11)$$

where I have explicitly indicated a sphere density dependence of the effective charge $Z$ on each sphere. There are three distinct phases for particles interacting with the potential given by Eq. (4.10) above: liquid, BCC, and FCC crystals. The liquid phase exists in a region of the phase diagram for which $\Gamma$ is small. The crystalline phases are stable in the opposite limit of large $\Gamma$. The BCC phase is stable for $\Gamma \gg 1$ and $\Lambda \gg 1$ or large screening length compared to interparticle separation, basically the limit of the one-component plasma [39], which consists of point charges in a background screening dielectric. The FCC phase is stable for $\Gamma \gg 1$ and $\Lambda \ll 1$, basically the limit of hard spheres [40]. By altering the sphere diameter, the sphere volume fraction, and the Coulomb screening length in the charge stabilized suspension, one can adjust the experimental sphere-sphere interactions between the two limits of one-component plasma and hard spheres, though of course never experimentally reach either. The three-dimensional equilibrium phase diagram and properties of the two systems of hard spheres [40] and the one-component plasma [39] are well known from computer simulations.

The sterically stabilized colloidal systems exhibit a three-dimensional phase diagram which resembles that of hard spheres with an effective hard sphere diameter slightly larger than the experimental particle diameter [26].

Like hard spheres, this system is completely parametrized by sphere density or volume fraction. In experiments, both the charge [34] and sterically [26] stabilized spheres also exhibit three-dimensional glassy phases at high-volume fractions due to the small nucleation core sizes and extremely long time scales needed to reach equilibrium.

A disadvantage of working with colloids is that the suspension temperature is not a particularly useful experimental variable in this system. In general, raising the system temperature results in an unknown and uncontrolled influx of additional screening charges into the solution due to the activated nature of the ionic dissociation constants of the walls of the container, the spheres themselves, or anything else in contact with the solution. This can also be the case even in the sterically stabilized colloids, as it is difficult to maintain absolute charge neutrality in the suspensions. In addition, the dielectric constant and viscosity of water are strongly temperature-dependent. For this reason, most of the charge-stabilized two-dimensional melting experiments in water suspension have been carried out at room temperature with the colloid in direct contact with mixed bed H and OH ion-exchange resin in order to minimize ion gradients and thus large unknown changes in effective Coulomb screening lengths. Sphere density is generally used as the experimental variable, as it is intrinsically more controllable.

The ion-exchange resin acts as a getter pump, exchanging stray ions in solution that leach out of or leak into the container with H and OH ions. In equilibrium, the number of H screening ions per sphere is equal to the sphere charge density on average, and the number of OH ions in solution is determined by the rate constant for water dissociation. Obviously, the container for a charge-stablized colloid experiment can have no air leaks, which would be a continuing source of carbonate ions, or contain any substance which dissolves or dissociates at the natural pH of the suspension, which is generally near 4. This is a stringent experimental constraint. And of course, the colloid experiments are always more difficult in practice than in principle, as dust on the micrometer scale and stray salt and sphere density gradients are very hard to eliminate.

An additional inconvenience for colloid experiments in comparison with atomic scale systems is that the time scale needed for equilibration of the system, although workable, is very long on a human scale. It can be as long as roughly $10^7$ collisions [41], tens of hours or more, needed for a defect to move across a correlation length in the system. However, I must point out that the fact that this equilibration timescale is attainable in colloid imaging experiments is a great advantage compared to computer simulations which typically run for times that are three orders of magnitude shorter. I believe that this, along with the capability of studying a very large system in real space with at least $10^4$ particles and no periodic boundary conditions, is crucially important in the study of two-dimensional melting. I will touch on this later in the chapter.

# 4.2    Experimental Results—Melting of Two-Dimensional Colloidal Systems

In this section, I will review the experiments in which melting of two-dimensional colloidal crystals has been studied in sufficient detail to gain some insight into the topological defects of the system and the melting mechanism. For convenience, I will discuss the experiments in chronological order of publication and spend more time presenting the data-acquisition and analysis techniques for the first experiment discussed in Section 4.2.1, making comparisons when available in reviewing the later experiments. I will mention instances where data are lacking or further study is needed for each experiment as I go along. In Section 4.3, I will discuss potentially interesting new colloidal systems and experiments and areas for future work.

## 4.2.1    Charged, Confined between Flat Plates, Wedge Geometry

### Experimental Geometry

A straightforward method of creating a two-dimensional layer of colloid is to enclose a colloidal suspension between two smooth parallel walls that repel the colloidal particles, spaced apart by a distance of a few particle diameters. This method has been used by a number of groups to study thin suspensions of charge-stablized colloidal spheres in water [42–45]. Two groups [42,43] have mapped out the complex phase diagram of these spheres between two flat, parallel, smooth repulsive glass walls at colloid volume fractions sufficiently high for ordering of layers to occur. The colloidal spheres are repelled by positive image charges on the walls, as well as a density of charged OH$^-$ groups on the glass surface that can be as high as 0.01 Å$^{-2}$ [46], depending on the glass-cleaning procedures used.

As the spacing $t$ between the two walls becomes comparable to a few sphere diameters $d$, a single hexagonal two-dimensional crystalline layer can form. For smaller $t$, the spheres are repelled by the high external potential caused by the walls, their density (and therefore $\Gamma$) is lowered in equilibrium, and they melt. The sphere-sphere Coulomb interactions in this geometry are closer to three-dimensional than two-dimensional for $t \gg d$ as the screening counterion motion is nearly unrestricted in the third dimension. In addition, the sphere-sphere interactions may take on some higher multipole nature due to the image charges on the walls [47]. At any rate, the Coulomb interactions between spheres are presently less accurately modeled in this geometry than in bulk.

To study the equilibrium two-dimensional phase diagram of the system in detail, the trick is to create a small wedge angle between two smooth glass plates in order to produce a controlled gradual density change along the wedge and to achieve simultaneous equilibration of the two-dimensional fluid, two-dimensional crystal, and a surrounding three-dimensional crystalline reservoir directly in contact with each other. One is then able to sample equilibrated phases of colloid at various average densities. As the system is in direct contact with both thermal and particle reservoirs, constant chemical potential and temperature are maintained along the wedge in equilibrium. The thermodynamics of this system are analogous to that of an atomic fluid in a slit pore as discussed by Evans [48]. Changes in the grand free energy $\Omega$ of a fluid between two identical plates of area $A$ can be expressed as

$$d\Omega = -p\,dV - S\,dT - N d\mu + 2\gamma\,dA - Af\,dt, \qquad (4.12)$$

where $p$ is the bulk pressure, $V$ the volume, $S$ the entropy, $T$ the temperature, $N$ the number of particles, $\mu$ the chemical potential, $\gamma$ the surface tension, and $f$ the surface excess pressure. The wedge creates a small gradient in the plate separation $t$ across the system, keeping $\mu$, $T$, and $A$ fixed. If this gradient is sufficiently small, and one waits sufficiently long for the system to equilibrate, one can observe different states of the system along the wedge, at different mean values of $t$, and all in mutual equilibrium at the same $\mu$ and $T$. I will discuss equilibration times at length in the following section. One can obtain a gross estimate of the largest gradient in density for which one can observe a particular thermodynamic phase of the system by requiring that the density width of the phase should span more than a correlation length in the direction parallel to the gradient.

A melting transition in a single layer of spheres was noted in some very early experiments on spheres between plates [42–44], but the wedge angle was far too steep and the density gradient in the wedge too large to get meaningful results on the melting transition. In 1987, Murray and Van Winkle performed a series of experiments [29,41,49] on wedges that were more than an order of magnitude less steep, in which the two-dimensional melting transition was studied in detail. In these experiments we saw evidence for two-step melting.

Murray and Van Winkle used polystyrene sulfate spheres of 0.305-$\mu$m diameter ($\pm 2\%$) with a titratable surface charge of $\sim 2 \times 10^4$ electrons. The mean separation of the spheres in the wedge near the two-dimensional melting transition was $a_s \sim 2.5$, very close to that at the density at which the three-dimensional BCC–FCC transition occurs so that $\Lambda \sim 1$. In Figure 4.4 is shown a blow-up of the experimental wedge geometry. Melting of the two-dimensional layer was observed considerably before any out-of-plane motion ($\sim \pm 0.2d$) was measurable, so that to good approximation the spheres are rigidly confined to the plane through the two-dimensional melting transition. The imposed density gradient due to the wedge in the $y$ direction in the figure corresponds to a change of sphere density of roughly $\pm 1\%$ per image (of size $\sim 30$ sphere separations) along $y$. The wedge angle along $x$ was about a factor of 5 smaller than that along $y$.

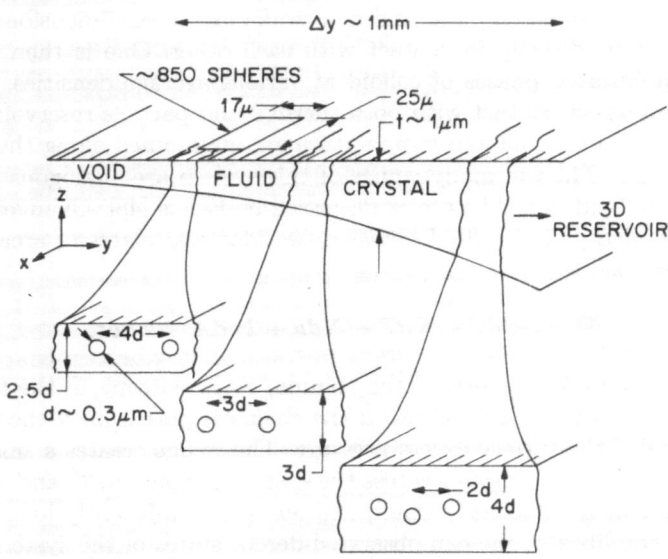

FIGURE 4.4. Experimental geometry for the wedge experiment (not to scale). The imposed density gradient in the $y$ direction is induced by a small wedge angle of $\approx 5 \times 10^{-4}$ rad. The wedge angle in the $x$ direction is at least five times smaller. The imaging volume in the $x - y$ plane is shown for the two time-resolved runs. For the seven static snapshot runs, the imaging volume was somewhat larger, $25 \times 38$-$\mu$m squared, up to 2000 particles.

During each run, which took approximately one half-hour, the system temperature changed by less than 0.1 K. This is important to insure that thermal expansion of the sample cell during the course of an experimental run did not greatly affect the wedge. In nine runs, the wedge did not change during this time to within experimental accuracy of $\sim 1 \times 10^{-5}$ radians. Surrounding the thin central region of the cell was a 100-cc reservoir of three-dimensional crystalline colloid in intimate contact with H and OH ion exchange resin and the two-dimensional phases. Once the colloidal phases were equilibrated in the wedge, a series of small ($\sim 35 \times 55$ spheres) areas along the wedge in the center of the two-dimensional region were then imaged with the 140X, numerical aperture 1.3, the oil immersion objective of an optical microscope. The extent of the sample with a two-dimensional density $n$ within $\pm 1\%$ accuracy was roughly at least $\sim 35 \times 350$ spheres.

The major disadvantage of a wedge geometry is the imposed density gradient in the direction of the wedge, which could smear out a density jump associated with a first-order melting transition. The two-dimensional crystal nearby would also serve as an orientation boundary condition for a possible hexatic phase in direct contact. A density gradient of 1% per image produces on average one-third extra free dislocation per image with a Burgers vector perpendicular to the gradient. This amounts to a background concentration of $\sim 3 \times 10^{-4}$ free dislocations. The major advantage of the wedge geometry is that all phases can be equilibrated in parallel in the experiment. The time required for equilibration in a statistical sense is extremely long, as will be discussed below.

In these experiments a charge-coupled device camera (resolution $576 \times 384$ pixels$^2$) was used to image a region of size $25 \times 38 -\mu m^2$ with between 1000–2000 particles. Digital imaging was used to locate particle centers in a series of one-time snapshots of the spheres at various locations along the wedge. Seven separate runs on different wedges were analyzed. These runs served to locate the transitions through the statistics of the instantaneous positions of topological defects and the behavior of the static correlation functions as a function of the mean density along the wedge. Subsequently at $\sim 10$ positions along two different wedges, a standard video camera and frame grabber (resolution $512 \times 480$ pixels$^2$) were used to digitize several snapshots of 700–850 particles separated by a time interval of 0.05 sec, sufficiently short that the movement of individual spheres was easily resolved. These time-resolved runs used colloid from the same batch, and identical geometry and bulk reservoir density to the earlier static runs. They allowed the determination of particle and defect motions, time-resolved correlation functions, and correlation times for the system at various average densities.

## Data Analysis Methods

The particle center locations (determined to 1 pixel accuracy) in each snapshot along the wedge were used to determine two-dimensional sphere density, instantaneous and time-resolved pair correlation functions, structure factors, and orientational correlation functions. The time-resolved functions are computed by cross correlations of the center locations in successive snapshots. For limiting thermodynamic behavior of these quantities and thus limiting spatial and temporal behavior, one would need to average the correlation functions over many ($10^3$) images displaced either in space by a distance larger than the correlation length or in time by an interval longer than the correlation time. For this reason, detailed fits of power laws to the calculated correlation functions for a single image are not definitive due to statistical fluctuations in defect concentrations. Nevertheless, the trends should be meaningful for images with > 1000 particles. This applies to all simulations, as well as to all of the other imaging experiments described in this chapter.

The topological defects in the images were determined by performing a Voronoi polyhedron analysis [6] on the sphere centers. This procedure uniquely locates the nearest-neighbor bonds in the network and thus the disclinations. Once the particle centers were located in a series of snapshots spaced apart by an interval $\delta t$, a straightforward algorithm was used to link each particle with itself in successive frames [51]. This linked list was then used to plot individual particle trajectories for the series.

## Static Correlation Functions

The two-dimensional density of spheres in reduced units of diameter$^{-2}$ versus position along the wedge for one of the runs is shown in Figure 4.5. The spread in density (the extent of the error bars) in each image is determined from the gradient imposed by the wedge. In Figure 4.6 are shown correlation functions for each of the snapshots labeled by the letters a–g in Figure 4.5. The static structure factor $S(K)$ and the pair correlation function $g(r)$ are standard [50], and the orientational correlation function $g'_6(r)$ is the correlation of the bond-orientational order parameter $\psi'_6(r) = \exp(i6\theta)$ at the midpoint of each bond, divided by the pair correlation function of bond centers. The long-range behavior of this correlation function is identical to $g_6(r)$ defined by Eq. (4.7).

The strongest evidence from these experiments for a two-stage melting process is the behavior of the correlation functions. A large change in orientational order is observed at density $n_1$ between snapshots c and d (rightmost column) with a much smaller increase in the translational order (center column). In Figure 4.7 are plotted the orientational and translational correlation lengths for the run, $\xi$ and $\xi_6$, defined by fits to an exponentially decaying pair correlation function for a perfect hexagonal lattice and pure exponential decay, respectively, to the correlation functions for single images displayed in Figure 4.6. As noted earlier, the correlation functions $g_G(r)$ and $g_6(r)$ should obey power-law decay in the hexatic and crystal in the KTHNY theory; however, $g(r)$ and $g'_6(r)$ were simply fitted to expo-

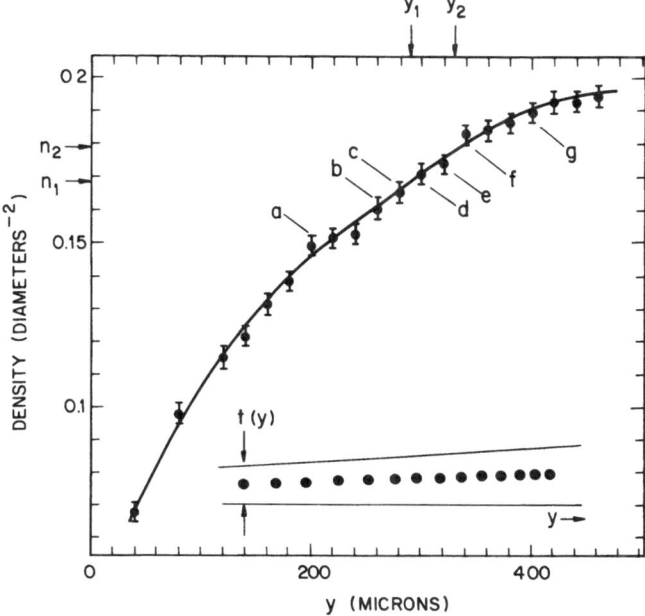

FIGURE 4.5. Two-dimensional density in units of diameters$^{-2}$ vs. position in the wedge $y$ for one of the experimental runs. Error bars are determined by counting accuracy. The letters $a$ through $g$ mark several densities of interest. The two densities of bond-orientational and translational ordering are marked by $n_1$ and $n_2$, respectively. Inset: schematic of the wedge (not to scale).

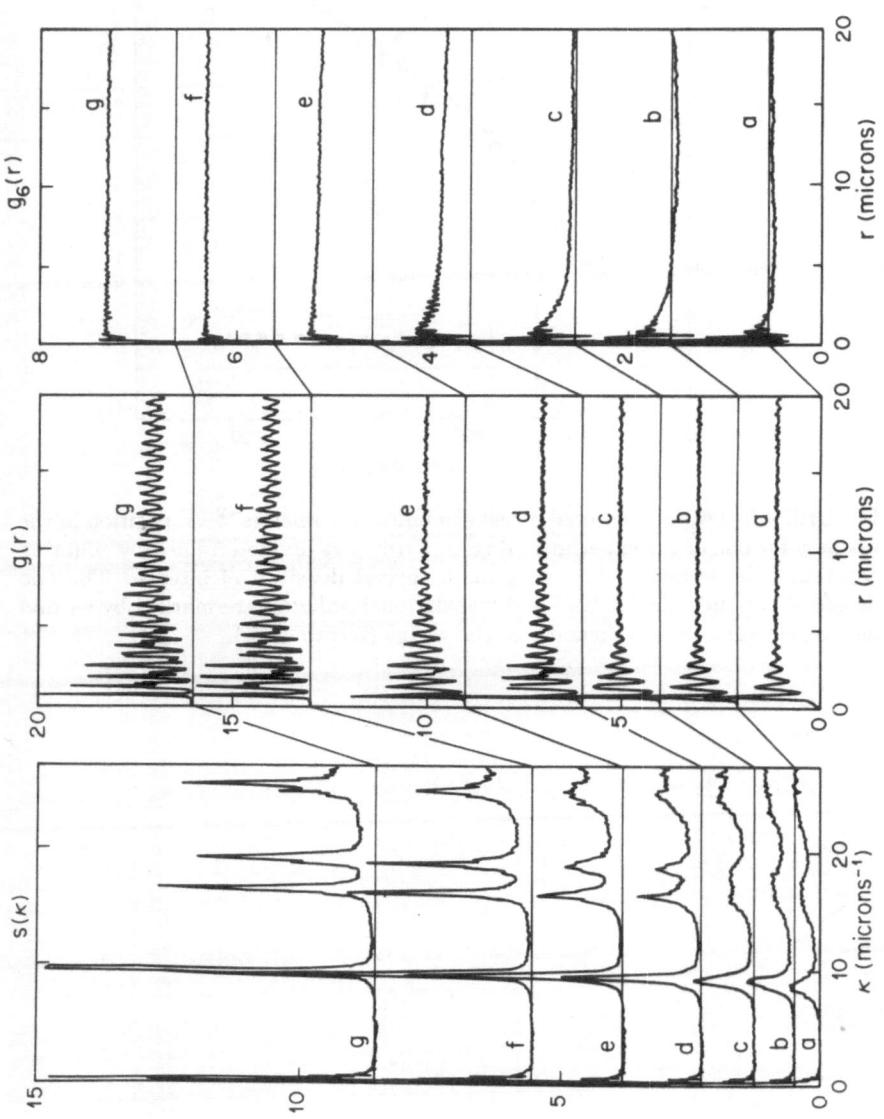

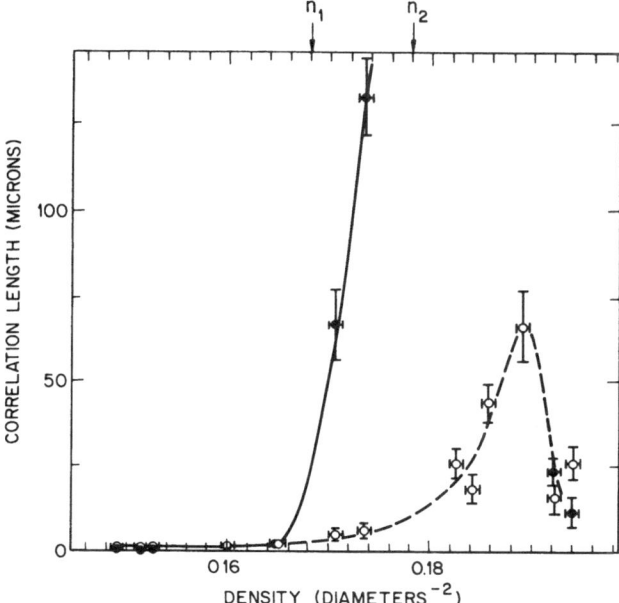

FIGURE 4.7. Translational (open circles and dashed line) and orientational (filled circles and solid line) correlation lengths corresponding to the correlation functions in Figure 4.6. The densities of bond-orientational and translational ordering transitions are marked by $n_1$ and $n_2$, respectively. Lines are guides to the eye. The fall off of both types of order beyond $n = 0.19$ is due to the beginning of out-of-plane motion of the spheres.

nential decays to note qualitative changes.

At $n = n_2$, about 4% higher than $n_1$, a large increase is observed in the translational correlation length (Figure 4.7), while the orientational correlation function flattens out to a constant on the distance scale of an image (between curves $e$ and $f$ in Figure 4.6). The height of the first peak of $S(K)$ reaches 5, the predicted limit at two-dimensional freezing by Ramakrishnan [16], consistent with freezing in simulations. This density can then be associated with the freezing transition. The translational correlation length

---

FIGURE 4.6. Static correlation functions for the densities marked by $a$ through $g$ in Figure 4.5 for the run. Left column: the structure factor $S(K)$; center column: the pair distribution function $g(r)$; right column: the bond-orientational correlation function $g_6'(r)$. There is a large change in orientational order between curves $c$ and $d$, with a slight increase of translational order there. Between curves $e$ and $f$, the translational order exhibits a dramatic increase, while $g_6'(r)$ becomes flat.

is about $30a$ at $n_2$, grows to a maximum of $90a$, and then decreases as $n$ increases. At $n \gg n_2$ in the dense crystal, above $n \sim 0.19$ for this run, the orientational and translational correlation lengths become comparable and roughly equal in value to $\sim 30a$. This is a result of the beginning of out-of-plane motion of the spheres, along with the insertion of a series of obvious grain boundaries separated by roughly $30a$, directly observed in the images.

The observed behavior of the translational and orientational correlation functions for $n_1 < n < n_2$ is remarkably similar to that predicted by the KTHNY theory for the hexatic phase, $\xi \ll \xi_6$. For a well-defined network of grain boundaries as predicted by Chui, one would expect similar behavior to that observed in the case of real grain boundaries for $n > 0.19$, where $\xi \sim \xi_6$. In all of the seven equilibrated runs the limiting $\eta_6$ obtained from power-law fits to $g_6'(r)$ at $n_1$ were consistent with the KTHNY prediction of $1/4$, but the imposed density gradient and accompanying lack of density resolution made the determination of exponents imprecise. Note that the intermediate "hexatic" phase occurs for an extent along the wedge $\Delta y > 4\xi$, so that we should be able to identify this phase [see discussion after Eq. (4.12)], but barely.

There is insufficient density resolution in the wedge experiment to determine whether the transitions are continuous or first-order; however, the observed $4 \pm 1\%$ extent of the intermediate region between $n_1$ and $n_2$ is several times larger than the thermodynamically allowed density change of $\sim 1\%$ for a two-dimensional system interacting with power-law potentials [57], and the 2.5% density jump predicted by Ramakrishnan [16] for a first-order density wave transition.

## Equilibrium Times

There are two time scales for equilibration: initial equilibration of gradients in the screening ion density with the ion-exchange resin takes place in roughly a month. Afterwards, any small adjustment made in the wedge will drive the spheres out of statistical equilibrium. Experimentally, consistency was observed for the two transitions after about 20 hr of equilibration after an adjustment was made to the wedge. In Figure 4.8 are plotted the densities at the two observed transitions for all seven of the static runs and for two video runs, plotted as a function of equilibration time after adjustment to the thickness of each wedge. The runs all used colloid from the same batch, but were taken with separate windows, wedges, and from one day to several months apart with identical three-dimensional reservoir density, and in direct contact with ion-exchange resin for periods of more than one month prior to each run to achieve equilibrium-screening ion densities. Twenty hr is long compared to the effective hard sphere collision or "rattling time" of $\sim 0.02$ sec, the time it would take for a sphere to diffuse in the solvent a distance $\sim 0.1a$ [28] in the crystal at a density

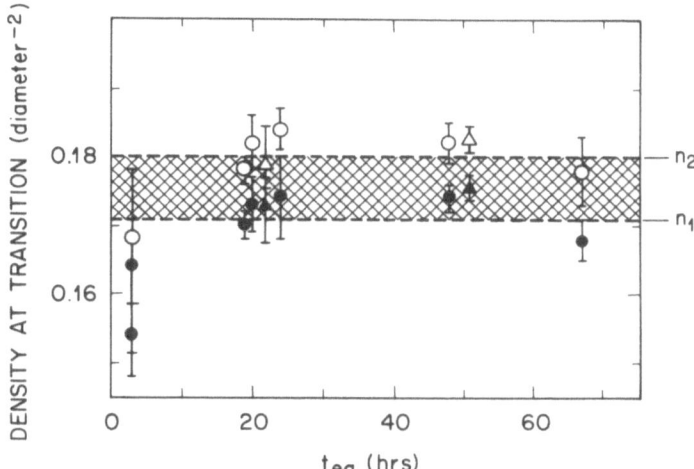

FIGURE 4.8. The densities of the bond-orientational and translational transitions $n_1$ and $n_2$ for nine experimental runs on different wedges with colloid from the same batch vs. equilibration time after a mechanical perturbation of the wedge. For all the runs with longer than 20 hr of mechanical equilibration, the transition densities are consistent to $\pm 2\%$. The triangles represent data for time-resolved runs and the circles the larger imaging field static runs. The cross-hatching delimits the mean hexatic region for the equilibrated runs.

just before melting. However, 20 hr is close to the time required for a dislocation to climb (or a vacancy or interstatial to diffuse) a translational correlation length just before melting. This is a minimum criterion for statistical equilibration [2]. Climb is the slowest movement of a dislocation. It is in the direction perpendicular to its Burgers vector and thus requires the presence of either a vacancy or interstitial [58]. Murray and Wenk [49] were able to observe dislocation glide in the crystal near melting. The dislocation climb diffusion coefficient was estimated to be the product of the coefficient for dislocation glide and the observed concentration of vacancies and interstitials in the lattice.

## Time Resolved Correlation Functions

In Figure 4.9 are shown the time-dependent van Hove correlation functions, $g(x, y, t)$, for times ranging between 0 and 0.2 sec (between 0 and roughly 10 collisions) from one of the time-resolved runs. This is shown for four time intervals and three different densities, in the fluid, the hexatic, and the crystalline regions. The van Hove correlation function measures the probability of finding a particle anywhere in space at time $t$, given a particle at the origin at $t = 0$. The particle at the origin is represented by a delta function at $t = 0$. In the fluid, this peak broadens into a Gaussian due to diffusion

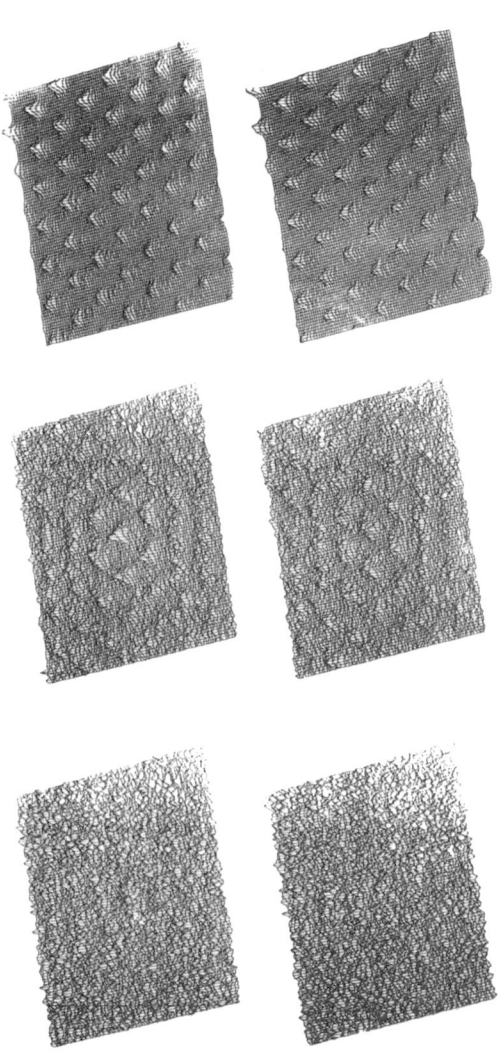

FIGURE 4.9. The van Hove correlation function $g(\mathbf{r}, t)$ plotted in the $xy$ plane for three different densities: left column fluid at $n = 0.96\, n_1$; center column hexatic at $n = n_1 + 0.33(n_2 - n_1)$; and crystal at $n = 1.003\, n_2$; at four different times: top row $t = 0$; first row from the top $t = 0.05$ sec or $\approx 2.5$ collisions; second row $t = 0.1$ sec (5 collisions); third row $t = 0.15$ sec (7.5 collisions); fourth row $t = 0.2$ sec (10 collisions). The sharp peak in the top row for each column represents the delta function in space for the particle at the origin $\mathbf{r} = 0(\hat{\mathbf{x}} + \hat{\mathbf{y}})$.

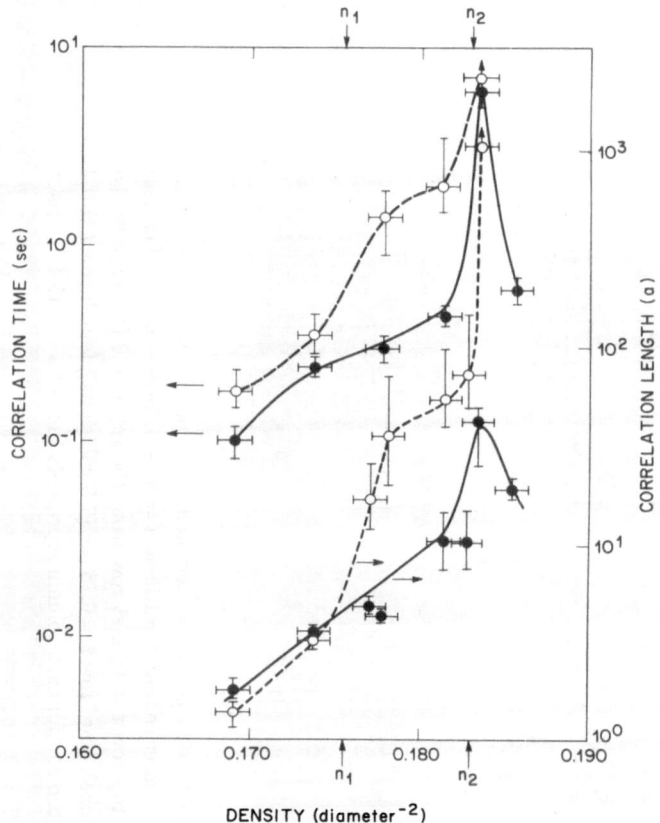

FIGURE 4.10. Correlation times (top curves) and lengths (bottom curves) for translational order (solid circles) and bond-orientational order (open circles) for a time-resolved run. Vertical arrows on data points imply lower bounds.

of the particle at the origin, and in the crystal, it broadens due to vibration about its lattice position. The translational correlations decay in time for the fluid rapidly, and for the crystal, not at all on this time scale. For the intermediate hexatic region, the behavior is intermediate. However, the orientational order decays much more slowly than the translational order. This is shown more precisely in Figure 4.10, which shows a measurement of the correlation times as well as lengths for translation and bond orientation for this particular run. The translational correlation time was defined as $[d \ln F(G_0, t)/dt]^{-1}$, the inverse of the best-fit exponential decay constant with time of the first peak of the intermediate scattering function $F(K, t)$, a measure of the lifetime of the correlations with spacing $a$. The orientational correlation time was defined as $[-d \ln g_6'(0, t)/dt]^{-1}$, the inverse of the exponential decay constant of $g_6'(r, t)$ with time at $r = 0$, a measure of the time decay of the average magnitude of the bond-orientational order at each point. The correlation times for bond-orientation are as long as $10^{2-3}$ collisions in the hexatic region, while the translational correlation times are an order of magnitude smaller.

## Defect Statistics

In Figures 4.11 and 4.12 are shown the instantaneous positions of particle centers for various average densities very close to melting. The Delaunay triangulations [6] of bonds (dual to the Voronoi diagram) are shown. Here, a sphere center is represented by a vertex of its nearest-neighbor bonds. The area between a sphere center and its nearest neighbors is shaded in the figures if it is not sixfold coordinated. These shaded areas mark the disclinations in the underlying topological lattice. Plots of the concentration of $n$-fold coordinated spheres for two runs are shown in Figure 4.13, from which it is observed that the two–dimensional fluid is over 80% sixfold coordinated very close to freezing. This has also been observed in simulations: e.g., two-dimensional Lennard–Jones particles [52] and hard and soft disks [53]. Basically, the two-dimensional fluid is considerably more ordered than its three-dimensional counterpart, as a result of the fact that the average number of nearest neighbors of the two-dimensional fluid is the same as that of the crystal [54]. While the densities at the transitions varied $\sim 2\%$ between run to run, the concentrations of four-, five-, seven-, and eightfold disclinations, the correlation functions, and the defect topologies were remarkably consistent to within 0.5% for all seven equilibrated runs at both transitions. It is clear from this data that both the translational and orientational ordering transitions are associated with the total defect concentrations.

## Microscopic Particle and Defect Motions

In order to probe microscopic models of melting, one needs to distinguish between disclinations that are bound tightly into dislocation cores and

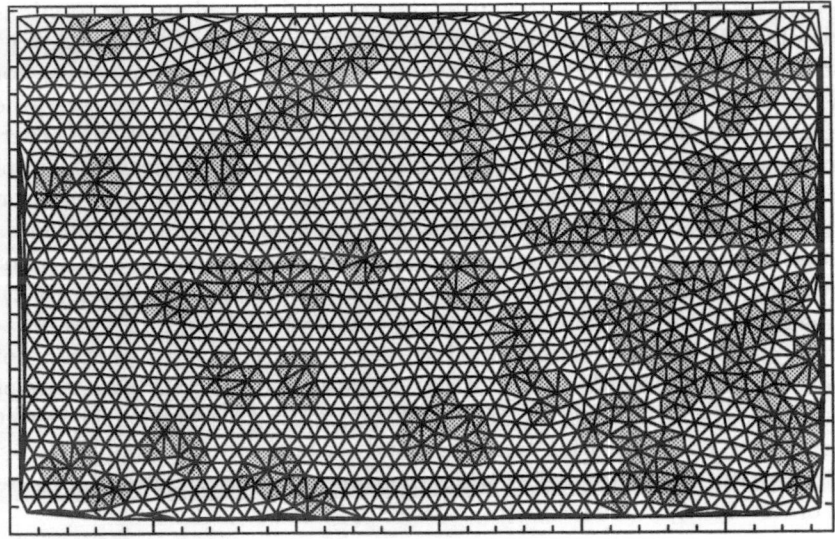

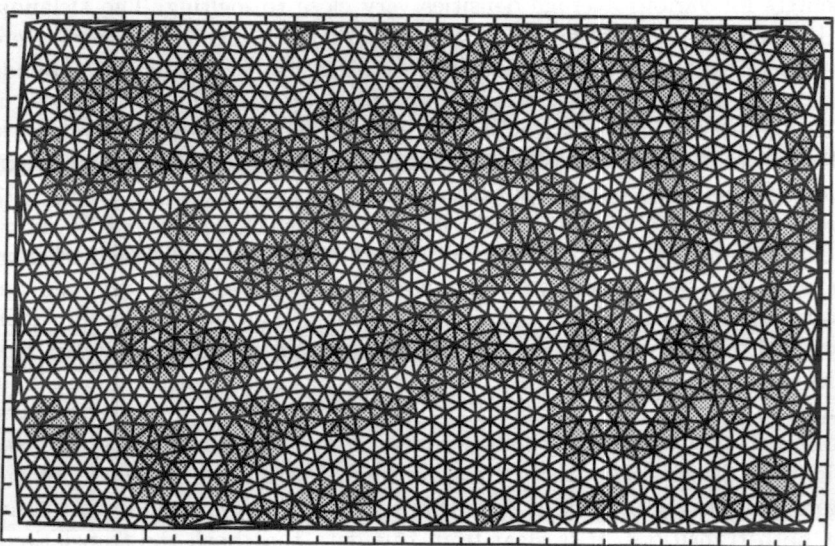

FIGURE 4.11. Delaunay triangulations with the nearest neighborhoods of disclinations in the lattice darkened. Centers of spheres are at the vertices of the triangulation. Top: crystal at density $n = 1.03n_1$; bottom: hexatic at density $n = n_1 + 0.6(n_2 - n_1)$. Dislocations are tight 5–7 disclination pairs, and dislocation pairs are quadruplets of 5–7–7–5 disclinations. Vacancies are usually eightfold disclinations with two fivefold neighbors.

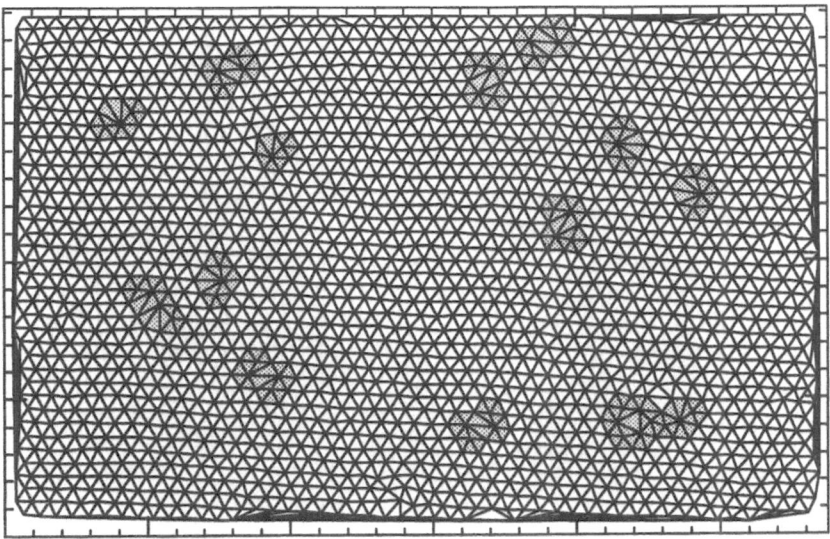

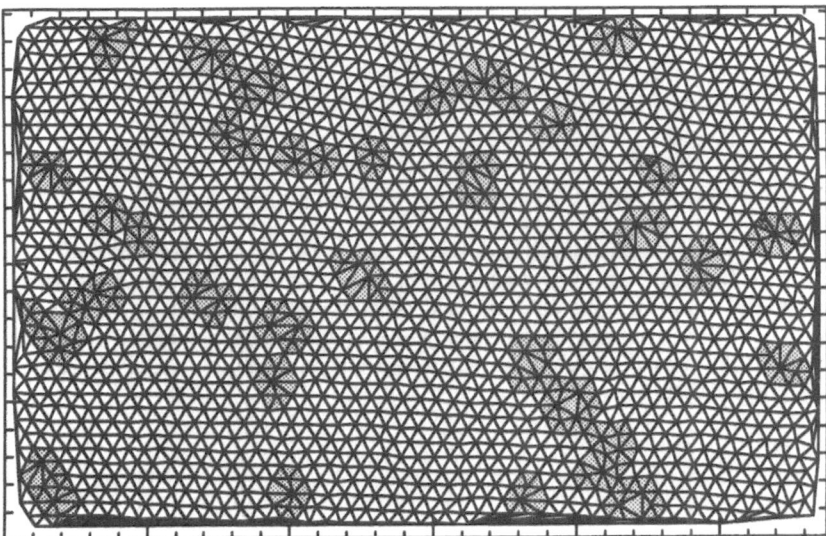

FIGURE 4.12. Delaunay triangulations with the nearest neighborhoods of disclinations in the lattice darkened. Centers of spheres are at the vertices of the triangulation. Top: hexatic at density $n = n_1 + 0.33(n_2 - n_1)$; bottom: fluid at density $n = 0.98n_1$. Dislocations are tight 5–7 disclination pairs, and dislocation pairs are quadruplets of 5–7–7–5 disclinations. Vacancies are usually eightfold disclinations with two fivefold neighbors.

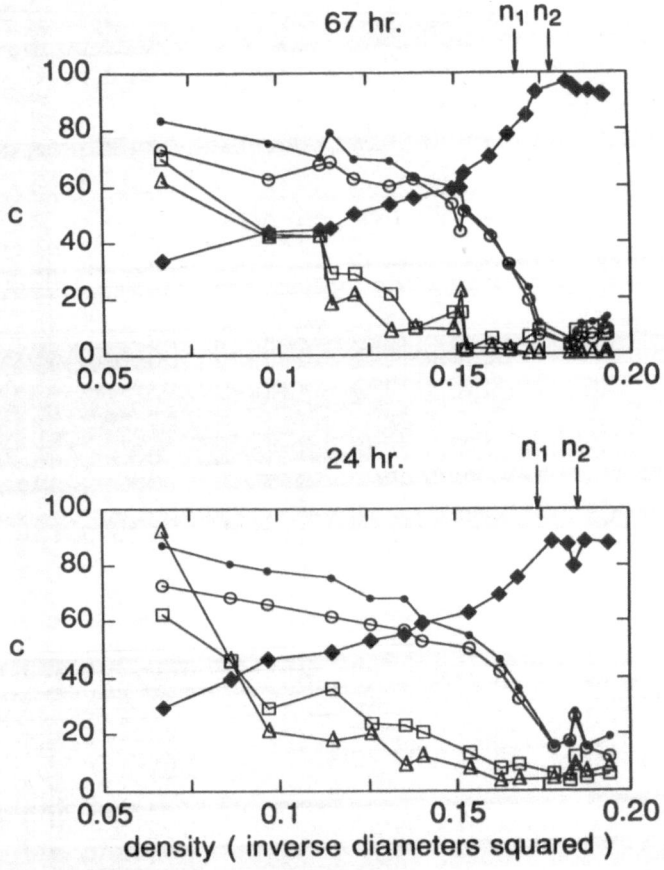

FIGURE 4.13. Disclination concentrations $c$ in percent of all spheres in the central 84% of the image field for two separate runs near melting. Solid diamonds: sixfold coordinated spheres; solid dots: fivefold coordinated spheres $\times$ 3; open circles: sevenfold $\times$ 3; open squares: eightfold $\times$ 10; open triangles: fourfold $\times$ 10. The equilibration times after disturbing the wedge mechanically for each run are listed on the top of each plot.

those that are either more loosely bound or unbound. Similarly, one would like to distinguish bound and unbound dislocation pairs. This is not a unique procedure at present [49], but is only possible by direct observation of the particle (and defect) locations and trajectories. Individual particle trajectories during the period $0 \leq t \leq 0.2$ sec ($\sim 10$ collisions) are shown for three densities in a 50-hr equilibrated run in the top frames of Figures 4.14–4.16. Instantaneous defect structures for the same region, separated in time by 0.1 sec during the time interval for the frame above, are shown in the center and bottom frames. These figures give a good impression of the similarilty of the dynamics of the system as the density is decreased from the crystal to the intermediate region and on to the high-density fluid.

*Crystal.* Close dislocation pairs and triples, tight quadrupoles or hexapoles of five- and sevenfold disclinations, were observed in the crystal just before melting as shown in Figures 4.11 and 4.14. The average concentration of unpaired dislocations in the seven equilibrated runs was $\sim 2 \times 10^{-4}$, very close to the limiting concentration imposed by the wedge density gradient. The defects in the crystal very close to melting occurred in well-separated clusters of size $2 - 4a$. There is a gradual increase in the concentration of five- and sevenfold disclinations as the density is lowered from the density of maximum order in the crystal. (Above this, the increase in defects is due to the inclusion of grain boundaries mentioned previously.) At the melting density $n_2$, the concentration of five- and sevenfold disclinations (an upper bound for the dislocation concentration) is $\sim 5\%$, and the concentration of vacancies is $\sim 1\%$. The individual particle motions (Figure 4.14) come in two varieties on this time scale: those that are vibrating around their equilibrium positions, and those that are experiencing highly correlated particle exchange in the plane, or "hopping." This has been associated with dislocation motion [49]. Highly correlated hopping loop motions have also been seen in molecular dynamics [55] and Monte Carlo trajectories [53] of crystals near melting.

*Intermediate Hexatic Region.* In the intermediate region between $n_1$ and $n_2$ (Figures 4.11, 4.12, 4.15), the defect structure and the individual particle trajectories are similar, but more complex and clustered than in the crystal. As the density is decreased into the intermediate region, there is a rapid increase of five- and sevenfold disclinations at $n_2$, followed by a smooth fast rise to a concentration of $\sim 8\%$ at $n_1$. The defect clusters have become much more numerous and have a wider distribution of separations. The average size of a defect cluster is still roughly $2 - 4a$ but some of the clusters have coalesced into longer strings. The strings are not closed into grain-boundary-like loops, at least on the scale of the images. On average for all the runs, roughly two orders-of-magnitude more free dislocations are observed upon entering the intermediate region, including any dislocations that may be attached to a cluster of other defects but can not be paired

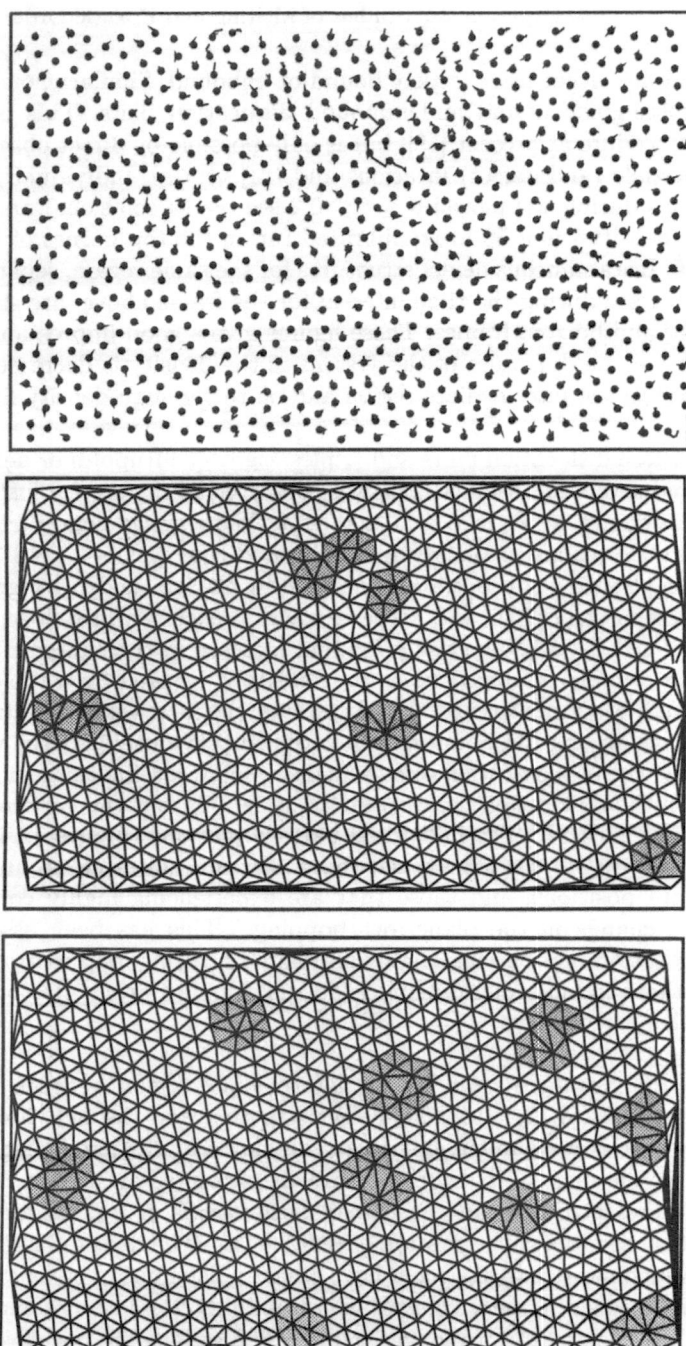

FIGURE 4.14. Top: particle trajectories over 10 collision times for a crystal at $n = 1.003n_2$. Center and bottom: Delaunay triangulations with disclination neighborhoods shaded for two times separated by five collisions, showing defect motions.

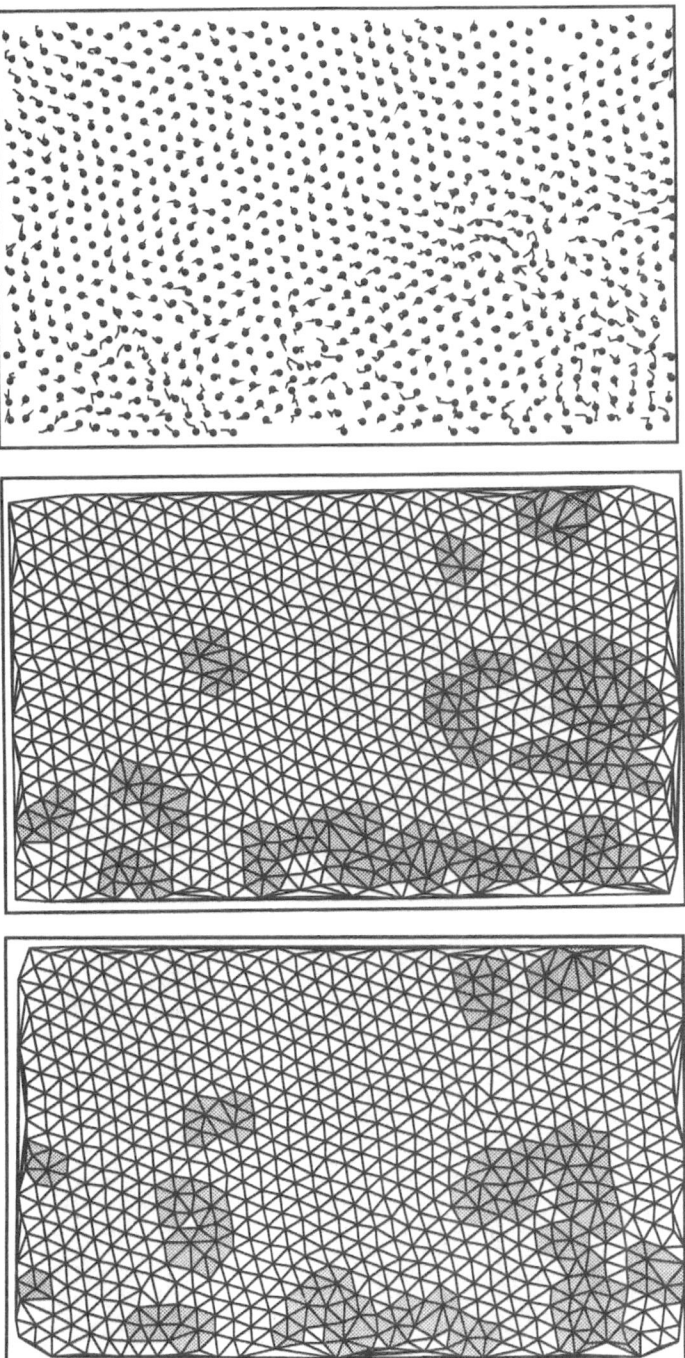

FIGURE 4.15. Top: particle trajectories over 10 collision times for a hexatic at $n = n_1 + 0.33(n_2 - n_1)$. Center and bottom: Delaunay triangulations with disclination neighborhoods shaded for two times separated by five collisions, showing defect motions.

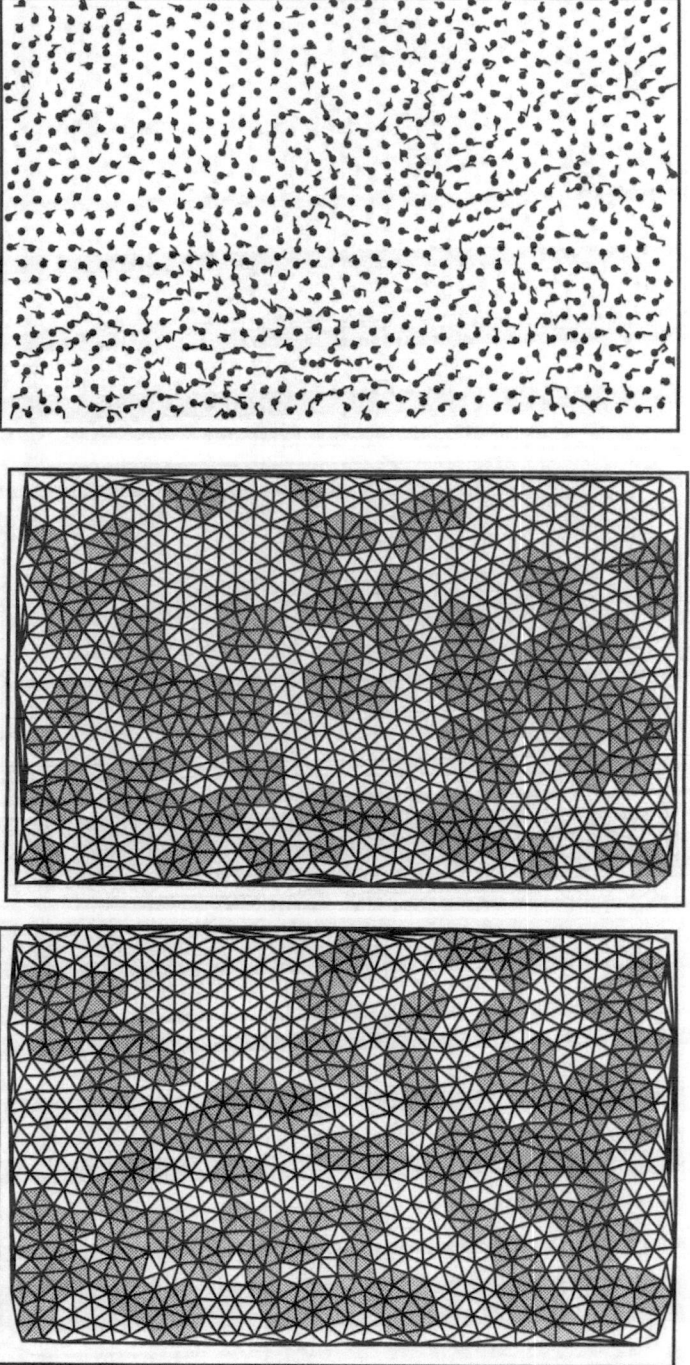

FIGURE 4.16. Top: particle trajectories over 10 collision times for a fluid at $n = 0.98n_1$. Center and bottom: Delaunay triangulations with disclination neighborhoods shaded for two times separated by five collisions, showing defect motions.

within the image to another of equal and opposite Burgers vector [49]. At lower densities closer to the fluid near $n_1$, the defect clusters are larger, more branched, and more closely spaced. In addition, a small density of individual disclinations separated by more than $2a$ from one of opposite charge (loosely bound disclination pairs) was detected. There is a good correspondence between the defects in snapshots of the same region separated by less than the correlation time $\tau$, which is roughly comparable to the dislocation pair life time. There is also a one-to-one correspondence between the "hopping" trajectories and defect cluster motion [49].

*Fluid.* In the high density fluid $n \leq n_1$ (Figures 4.12 and 4.16), particle motions and defects are similar to but even more complex than those in the hexatic. Correlated "hopping" was still observed around ordered regions that have size $\sim \xi$ and life time $\sim \tau$. By this density, the hopping paths and defect strings completely encircle the ordered regions, and thus do resemble grain boundaries in an instantaneous snapshot. These "grain boundaries," however, have very short life times $\sim \tau$, close to one collision time, and are comprised not only of clusters of dislocations, but also loosely bound or separated disclinations. When this happens, the orientation of the underlying topological lattice is completely disrupted in regions larger than $\xi$. Well-defined tightly bound dislocation pairs are still observed even in the high density fluid, where the translational and orientational correlation lengths are on the order of $4a$. However, there are many loosely bound dislocations and disclinations as well. At densities below about $0.9\,n_1$, the concentration of four- and eightfold disclinations grow to $\sim 1\%$, while that of the five- and sevenfold disclinations grows to nearly 15%. There is little correspondence of the defect structure between snapshots as the time interval between snapshots rapidly approaches $\tau$.

## Summary

In this first series of experiments Murray and Van Winkle found qualitative consistency with a two-step melting transition through a hexatic phase. We could say little about the order of the transitions or the scaling of the correlations because of the relatively large gradient of density compared to the extent of the intermediate hexatic region, although the exponent $\eta_6$ was consistent with the KTHNY-predicted value of $1/4$ at the fluid-hexatic transition. We performed a detailed study of the statistics and motions of topological lattice defects in their system, correlated with the microscopic particle trajectories over $\sim 10$ collision times. It is clear that both the observed orientational and translational ordering transitions are correlated with defect concentrations in the system.

In the experiment no clear two-phase separation was observed. The defect structure in the intermediate region is less grain-boundary-like than that of the dislocation vector simulation of Saito [10] with $E_c/K = 0.7/4\pi$, in which he found a distinctly first-order melting transition. However, there are also loosely bound disclination pairs and unbound disclinations in the real experimental system that are not included in Saito's model that will complicate the simple dislocation description of grain boundaries. It is difficult to make an argument that the complex defect structure observed in the experiment in the intermediate region resembles a network of grain boundaries or a fractal two-phase coexistence proposed by Udink and Frenkel [56] to explain the orientational ordering transition and lack of perfect scaling they found in a $10^4$ Lennard–Jones particle simulation occurring within the two-phase region of a much smaller simulation. A better picture for the defect structure in the Murray–Van Winkle experiment is the following: In the crystal, the ordered regions dominate, sprinkled with a low density of small isolated and well-separated paired dislocations. In the intermediate region, the dislocation pair density increases by about an order of magnitude, and the unbound dislocation density increases by about two orders of magnitude. The defect clusters have a core radius of $\sim 1 - 2a$. They first grow into long strings and then become branched as the density is lowered. The fluid transition, at which the translational and orientational decay become comparable in space and time, is reached at a density at which the branched defect clusters percolate across the images for a distance of $\sim 30a$.

Not only are the experimental correlation functions in the experiment qualitatively consistent with KTHNY theory, but also the observed defect structure, with the caveat that the experiment observes microscopic topological defects of the system, and not the renormalized defects on infinite distance scales that are driving the transitions in the theory. The total dislocation concentrations just at melting are rather low, $< 5\%$, which is consistent with the KTHNY arguments that require a low-density gas of defects. A very rough estimate for the dislocation core energy at melting can be obtained from the concentrations of microscopic dislocation pairs, with the bold assumption of Boltzman statistics and only pairwise interactions. From the Murray–Van Winkle experiment one finds $E_c/K \sim 0.13 \pm 0.7$, or $(3 < E_c < 5)$, which is borderline, but a bit too large compared to the limit $E_c/K < 0.06$ $(E_c < 2.84)$ set by Chui [9] for strongly first-order grain boundary melting.

The possible problem of disorder caused by an external random potential on the system (e.g., dirt on the glass walls or aggregate patches or dust in the suspension) in this experiment is minimal, as correlation lengths as high as $90a$ were observed in the crystal. Also, the system was large, $\sim 10^4$ particles, and was allowed to equilibrate for $\sim 10^7$ collisions before data were taken. The system needs to be studied with a smaller gradient for better density resolution, more averaging to improve the correlation functions, and larger images followed over longer time periods ($\sim 10^2$ collisions) to get a better idea of the patchiness and defect topology in the intermediate region before a serious quantitative comparison with the KTHNY predictions can be made. However, all of the experimental observations are qualitatively consistent with these predictions.

## 4.2.2    Floating Monolayers on a Liquid Surface

An obviously attractive system for the study of two-dimensional melting is
that of a layer of micrometer-size spheres constrained to be on the free water
surface. Pieranski [59] first demonstrated that two-dimensional triangular-
ordered arrays of spheres could be obtained on a water surface in 1980.
He used charge-stabilized polystyrene latex spheres of diameter 0.245 $\mu$m
suspended in water. A drop of suspension was introduced into a glass dish
so that it spread to a layer $\sim$ 3 $\mu$m thick. He then observed the water/air
interface with a microscope objective from below, through the 3 $\mu$m thick
suspension. The spheres at the interface exhibited very strong optical contrast so they could be imaged easily due to the very different ratios of
index of refraction, 1.6 between those of polystyrene and air, compared to
1.2 between those of polystyrene and water. He noticed that the spheres
at the interface were trapped in potential wells much larger than $k_B T$, as
they were never observed penetrating the bulk.

The energy scale of the surface well depth for a sphere of diameter $d$
is roughly $E_{well} \sim \pi d^2 \gamma_{w/a}$, where $\gamma_{w/a}$ is the water/air surface tension
72 erg/cm$^3$. This energy is roughly $10^{-7}$ erg, or at least $10^5 k_B T$. Therefore, the surface layer of spheres is constrained to the water surface and
is a good two-dimensional system, if we assume there are no large amplitude ripples on the surface. Pieranski also suggested that, since the spheres
were half-immersed in water, only half of the surface sulfate groups would
dissociate, giving each of the spheres a net vertical dipole moment of size
roughly $Z\lambda_D$. This would cause long-range, repulsive dipole-dipole interactions between the spheres. A repulsive screened monopole interaction will
also remain [60], corrected for the presence of a dielectric interface. In addition, there will be a long-range capillary interaction of the dimples in
the water surface surrounding the spheres [61] which can be either attractive or repulsive, depending on contact angle, as well as Van der Waals
forces and hydrodynamic forces. The sphere-sphere interactions are even
less characterized in this geometry than in three-dimensional solution or
between glass plates, but will have a longer-range component.

Pieranski proposed this system as a good analog of two-dimensional
atomic systems to study melting, but it was not until 1988 that Armstrong,
Mockler, and O'Sullivan [61] were successful in performing a careful and
detailed study of the melting transition of such a monolayer as it is isother-
mally expanded and compressed in an ultraclean Langmuir trough. The
basic problem with interfacial fluid-air layers is that of surface impurities
and sphere aggregate clusters, which are extremely difficult to overcome.
The experiment of Armstrong et al. was still greatly adversely affected by
this problem, despite their care. They have some evidence for separate orientational and translational ordering transitions in one of the systems they

looked at (2.88-$\mu$m diameter spheres), but the two transitions are separated by a factor of nearly 3 in density and they were never able to obtain a translational correlation length larger than $\sim 10 - 20a$ in the most ordered phase. The implication of this is that their system at high densities is a "quenched" hexatic, due to defect-induced free dislocations [30], rather than a crystal. The rate of expansion or compression of the system may also have been far too fast for the system to equilibrate, as I will discuss later. In a second system in which they studied smaller (1.01-$\mu$m diameter) spheres, they found no evidence for two separate orientational and translational ordering transitions and experienced worse surface contamination problems.

## Experimental Geometry and Techniques of Analysis

Armstrong et al. spread a monolayer of polystyrene sulfate latex spheres on the surface of a preswept $19 \times 3.8 \times 1$ cm$^3$ anodized Al Langmuir trough mounted on a visible microscope by first suspending the spheres in methanol and then depositing them by allowing several drops of the sphere-methanol suspension to flow down a glass rod penetrating the water surface. The methanol subsequently evaporated or dissolved, leaving a surface layer of spheres.

Two size spheres were used in experiments: two compression-expansion cycles for 2.88-$\mu$m diameter spheres and one compression cycle each for two separate preparations of 1.01-$\mu$m diameter spheres. The $d = 2.88$-$\mu$m spheres were deposited at a relatively high density onto half the trough and allowed to equilibrate for a day, after which the density was homogeneous across the surface. The aggregate cluster concentration was estimated to be about 0.003 at the beginning of the expansion-compression cycles. The 1.01-$\mu$m spheres were analyzed under compression only due to the build-up of rafts of aggregate (50–100 particles) at higher sphere densities. The 1.01-$\mu$m samples had initial aggregate densities before compression of $\sim$ 0.01. They nevertheless were able to analyze one of the compression cycles and for the other use the gravity-imposed density gradient at the trough meniscus to scan a series of images along the gradient. For comparison, the concentration of extraneous defects observed in the Murray–Van Winkle wedge just as melting was $\sim$ 0.0003. After several days the accumulating aggregate density in the Langmuir trough caused large clustered defects and vacant regions on the surface even of the 2.88-$\mu$m spheres, presumably caused by the build-up of surface contaminants, rendering the sample life a few days in these experiments.

A major advantage of looking at such a system in a Langmuir trough is the opportunity of applying external pressure and watching the evolution of the system, and even individual defects as the system expands or contracts. The 2.88-$\mu$m sample was compressed or expanded in steps corresponding to 1% density change, then allowed to equilibrate for about 60 min between steps. Anderson et al. saw no difference if the system was allowed to rest for several hours to overnight. The whole cycle of compression-expansion was videotaped through a microscope, and selected images of 600–6000 particles in the center of the trough were subsequently digitized with a frame grabber with resolution $512 \times 480$ pixels$^2$. The analysis of the translational and orientational order and defect densities in the images was similar to that of Murray and Van Winkle. Armstrong et al. found no hysteresis for the 2.88-$\mu$m sphere sample upon compression and expansion, and a similar density for the melting transition for the 1.01-$\mu$m sample in the bulk of the trough compared to the meniscus density gradient. They use this and also the fact that they observe defect rearrangements on times scales faster than the allowed equilibration times in the experiments to argue that the systems are not far from equilibrium. The statistical equilibration time $\tau_{eq}$ for the system should be the longest period required for a relevant defect to traverse a relevant length scale in the system. At the melting transition, this would be the time required for a dislocation to climb across a correlation length, or

$$\tau_{eq} \sim c_v \, \xi^2 / D_g, \tag{4.13}$$

where $c_v$ is the concentration of vacancies and interstitials of the system, and $D_g$ is the diffusion coefficient of dislocation glide, proportional to the inverse of the sphere diameter. For Armstrong et al.'s 2.88-$\mu$m sphere sample, if we assume $c_v \sim 1\%$ similar to that of Murray and Van Winkle, the equilibrium time would be roughly a factor of $2.88/0.3 \times (100 \, \mu\text{m}/20 \, \mu\text{m})^2 \sim 240$ larger than the $\sim 10$ hr of Murray and Van Winkle, or $2 \times 10^3$ hr, which is far longer than the 1–10 hr allowed experimentally. For the 1.01-$\mu$m samples, the corresponding factor would be $\sim 8$ times longer than that of Murray and Van Winkle, or 80 hr. If dislocation glide, rather than climb, were the relevant motion, these times would be reduced by two orders of magnitude; however in two runs of Murray and Van Winkle equilibrated for only 3 hr we found inconsistent transition densities (Figure 4.8) compared to the runs equilibrated for 20 hr or longer. Therefore, I am inclined to believe Eq. (4.13) is substantially correct.

## Summary

Anderson et al. were the first to attempt a detailed study of the melting transition of a two-dimensional layer of colloidal spheres trapped by surface tension on the surface of water in a Langmuir trough, in which they could be easily compressed or expanded by mechanical means. This, in principle, is a very nice system for studying the approach to equilibrium and the phase diagram of a rigidly two-dimensional layer of particles basically interacting with long-range dipole-dipole forces. Unfortunately, their experiments were plagued by extraneous surface contamination of the trough which caused the rapid build-up of aggregate particles and impurity-induced free dislocations in their system. In addition, their expansion and contraction cycles are probably too fast (by one to three orders of magnitude) for the system to be in adiabatic conditions or equilibrium. This problem is aggravated by the relatively rapid build-up of surface contaminants compared to the statistical equilibration time for the large spheres used in the experiment.

The aggregate and dust particle concentration is crucially important to the freezing of the two-dimensional layer, as each extraneous particle can serve as a nucleus for free dislocations. If this happens at low concentration of impurities, these dislocations can induce extrinsic hexatic [30] rather than crystalline order. This appears to be the case for the highest-compression 2.88-$\mu$m spheres. At high enough impurity concentrations, the impurity-induced dislocations will coalesce into a close network of grain boundaries and completely disorder the system. This appears to be the case for the 1.01-$\mu$m spheres in this experiment. For the 2.88-$\mu$m spheres, Armstrong et al. observed a bond-orientational ordering transition at low density ($0.03\ d^{-2}$) and a large increase in the translational correlation length at a higher density ($0.1\ d^{-2}$). However, the correlation lengths never became longer than $10$–$20a$, so that the ordered phase is probably best described as an extrinsic hexatic as mentioned above. Perusal of the defect maps of the phase is consistent with this conclusion, as there is a relatively high density of free dislocations ($\sim 1\%$) at this density. The lower-density transition seems to be a classic example of a continuous fluid-hexatic transition, as they observed no hysteresis on expansion and compression. For the 1.01-$\mu$m spheres, they observed a simultaneous translational and orientational ordering at a density of $0.014\ d^{-2}$, and a maximum translational correlation length of $\sim 4a$. This solid is quite disordered due to a network of grain boundaries caused by surface contamination, as observed in the defect plots.

It seems to me to be very difficult to reduce the surface contamination in such an experiment and increase the experimental equilibration times by even a factor of 10, due to the contamination problem. Thus, this technique as it stands seems to have insurmountable experimental problems for the study of two-dimensional melting. One possibility to reduce ambient surface contamination would be to trap the spheres at the interface of two very clean fluids, such as water and oil. If this were done in a Langmuir trough, one could still retain the capacity for compressing and expanding the system. By using spheres smaller in diameter by roughly a factor of 10, one could in the future reduce the equilibration time by an order of magnitude and still have the spheres perfectly visible with an ordinary high-power microscope objective. The advantages gained by the possibility of measuring true thermodynamic compression-expansion cycles in such an experiment make it worthwhile for future work.

## 4.2.3   Expansion between Rigid Plates

Tang, Armstrong, Mockler, and O'Sullivan performed an experiment in 1989 [63,64] in which a highly charged polystyrene colloidal sphere mono-layer suspended in water between two charged glass plates was observed to melt as the plate separation was mechanically expanded. In this experiment, they were able to circumvent the severe contamination problems of their earlier floating monolayer experiment, and at the same time retain the free expansion driving force. The gradient in plate separation and thus presumably sphere density in this expansion experiment was about 20 times smaller than that in the Murray–Van Winkle wedge experiment. They conclude that the correlation functions computed from a series of images of the system as it expands are consistent with the KTHNY scenario for two-step melting through an intermediate hexatic phase between densities $N_a$ and $N_b \sim 0.92\,N_a$ with long-range orientational order and short-range translational order. However, they state that direct observations of the time development of the defect creation and evolution of the system are inconsistent with this picture. In particular, they claim not to observe dislocation pair unbinding at $N_a$, nor disclination unbinding at the lower density $N_b$. They can better describe the defects in the intermediate region where $N_b < n < N_a$ as more consistent with two-phase equilibrium in which regions of solid surround liquid, and they surmise that vacancy creation by the annihilation of dislocation pairs is important in nucleating regions of isotropic fluid.

### Experimental Geometry and Data Analysis

Tang et al. used the inverted pyramid geometry shown in Figure 4.17 to create a small two-dimensional reservoir surrounding their two-dimensional layer in the thinnest part of the sample cell between two optical flats. They mechanically regulated the bending of a window on their sample cell against a viton 'O' ring by clamping and unclamping the periphery of the window to create the slow expansion of the inner sample region of the cell as the window flexed. The inner region had a gap thickness $t$ which had a gradient smaller than $2 \times 10^{-5}$ radians as determined by interferometry.

They performed two expansion runs, both with $d = 1.01$ $\mu$m highly charged polystyrene latex colloidal spheres. In Figure 4.18 is shown the decrease of the total particle count in the viewing region for the two runs with time. The first run had an initial gap separation of $\sim 4d$ and gradually decreased the center density from $n = n_1 = 0.057$ (in units of inverse diameters squared) in a linear fashion to roughly half the original density in going from ordered solid to disordered liquid over a period of 7 hr. The second run had a gap separation of $\sim 2.5d$ and decreased its density from $n = 0.035$ to roughly three-quarters of the original in going from ordered solid to liquid over a period of 20 min. Before the expansions both monolayers crystallized over a period of a week and were essentially free of

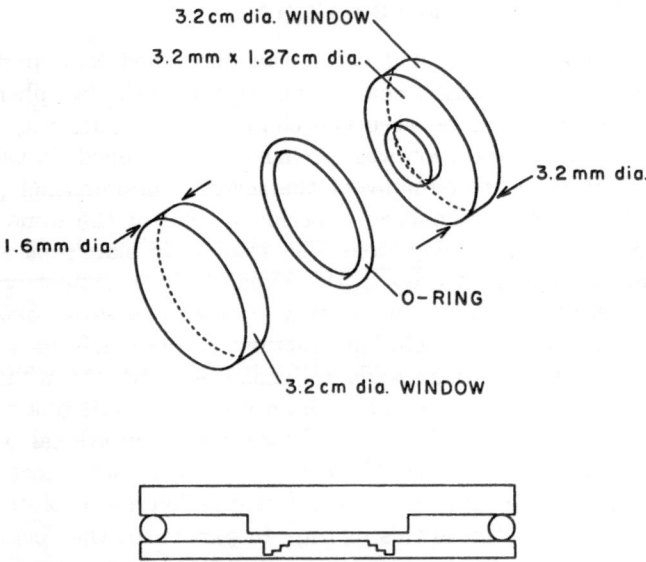

3.2 cm dia. WINDOW

3.2 mm x 1.27 cm dia.

3.2 mm dia.

1.6 mm dia.

O-RING

3.2 cm dia. WINDOW

FIGURE 4.17. Experimental cell of Tang et al. used in the expansion experiment.

vacancies, but had a small frozen in-dislocation density which limited the translational correlations in their most ordered state to less than $35a$. The results from both runs are consistent.

An area of $277 \times 222$ $\mu m^2$ in the center of the cell was videotaped during the entire expansion sequence, and selected frames were subsequently digitized and analyzed for correlation functions and topological defects. Tang et al. calculated the pair correlation functions $g(r)$ and $g_6'(r)$ from the locations of the sphere centers in a similar fashion to Murray and Van Winkle. They also performed a Voronoi polyhedron analysis to monitor defects. For fits to $g(r)$, Tang et al. used the first two terms in the Bessel function series

$$g(r) = \frac{1}{2\pi} \int_0^{2\pi} g(\mathbf{r}) \, d\phi = \sum_{i=0}^{i=\infty} A_i J_0(G_i r) r^{-\eta_i}, \qquad (4.14)$$

or, approximately,

$$g(r) \approx A_0 J_0(G_0 r) r^{-\eta_0} + A_1 J_0(G_1 r) r^{-\eta_i}. \qquad (4.15)$$

They also replaced the power laws by decaying exponentials in the expressions above for determining fits to exponential decay. They defined the translational correlation length $\xi$ and exponent $\eta$ as $\xi_0$ and $\eta_0$, respectively.

## Static Correlation Functions and Defect Statistics

In Figure 4.19 are plotted $g(r)$ and $g_6'(r)$ for the slower expansion run. It is apparent from the curves that there is long-range orientational order

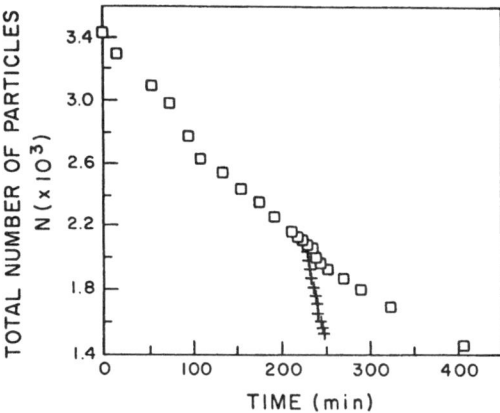

FIGURE 4.18. Total number of particles in the imaging field vs. time for the two expansion runs of Tang et al.

and short-range translational order between the curves marked $n_{11}$ and $n_{16}$ in the figure. In Figure 4.20 are shown the translational and orientational correlation lengths $\xi$ and $\xi_6$ and the power-law exponents $\eta$ and $\eta_6$ determined from fits to these correlation functions as discussed in the preceding section. Also shown is the total concentration of sixfold coordinates spheres during the run. Marked at the bottom of the figure are the densities $N_b = n_{16} = 0.033$ and $N_a = n_{11} = 0.036$, between which there is long-range orientational order and short-range translational order. At $N_b$ the concentration of sixfold coordinated spheres is 0.8, in agreement with that found by Murray and Van Winkle for the highest-density fluid (Figure 4.13). At $N_a$, the concentration of sixfold spheres is 0.96, also in agreement with that found by Murray and Van Winkle for the crystal right at melting. The density width of the intermediate region is approximately 8%, also in reasonable agreement with the Murray and Van Winkle wedge experiment. The orientational exponent $\eta_6$ is very close to $1/4$ at $N_a$ as well, in excellent agreement with KTHNY and the earlier experiment. However, the translational correlation length in the solid before melting hangs up at $\sim 10a$ ($\eta \sim 1$) for a considerable range of density, $0.036 < n < 0.046$. Tang et al. state that this is due to the original frozen-in defect concentration in the crystal, but from their figures at $n_1$, $\xi = 35a$, and $\eta \approx 0.3$. My interpretation of this hang-up in the defects and correlation length in the low-density crystal is that the density expansion is far too fast for the defects of the system to equilibrate, even in the slow run in which it took 3.5 hr to decrease the density to the intermediate region $N_a$. Following the discussion in Section 2.2, the equilibration time $\tau_{eq}$ for a dislocation to climb a correlation length in the system is proportional to the vacancy concentration. As the concentration of vacancies was stated to be originally extremely small in the experiment, equilibration would be considerably hampered in the crys-

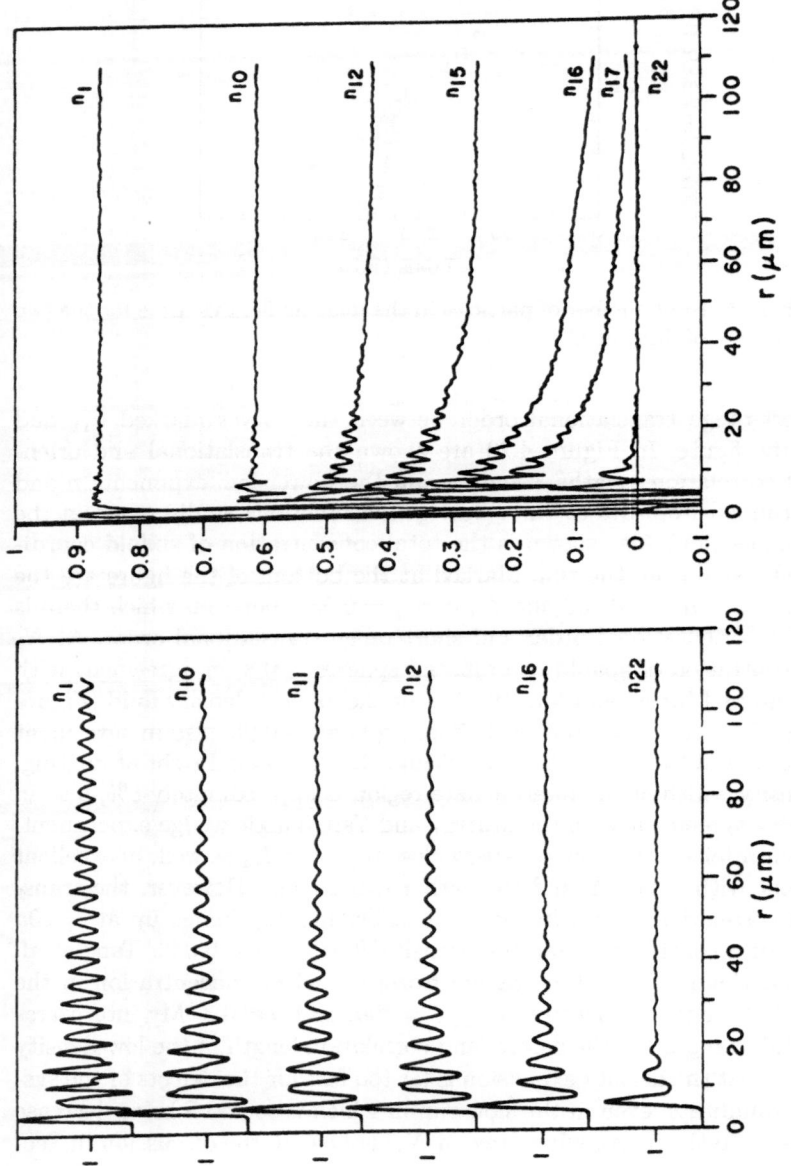

FIGURE 4.19. Left: $g(r)$ and right: $g_6'(r)$ for the slow expansion run of Tang et al. Each curve is marked with the density of the system.

tal. I will estimate an equilibration time for the system in the intermediate region $N_b < n < N_a$ in the following section.

### Microscopic Particle and Defects and Their Motions

In Figures 4.21 and 4.22 are shown defect maps and individual particle trajectories for the slower expansion run. The defect maps in Figure 4.21 are shown as Voronoi polyhedra around the sphere centers if the coordination of the sphere is not six, and can be directly comparable to those of Murray and Van Winkle in Figures 4.11 and 4.12, as five- and sevenfold coordinated spheres are clearly marked. In the crystal from $n_1 < n < n_{11}$, there exist closely bound dislocation pairs and occasionally a few unbound dislocations which limit the translational correlation length as mentioned previously. In the defect map at $n = n_{10}$, e.g., there are two or three clusters of defects with net Burgers vector, roughly 0.1% concentration. By $n = n_{11}$, at the beginning of the intermediate region, there are visible at least eight or nine such clusters, so that the concentration has increased to 0.4%. By $n = n_{16}$, at the border between the intermediate region and the isotropic fluid, the number of defect clusters with net Burgers vector has increased roughly fivefold, to a concentration of roughly 2%, while at the same time, there appear a few unbound disclinations. Although these statistics seem perfectly consistent with the KTHNY picture of dislocation pair unbinding followed by disclination pair unbinding, Tang et al. state that they never saw individual dislocation pairs of disclination pairs unbind as they were watching the expansion. Interestingly enough, then, the topological defects on the statistical average appear to be quite consistent with KTHNY, despite the lack of microscopic agreement with the theory. Perhaps this is due to the fact that the defects in the theory are actually renormalized objects at an infinite length scale and not necessarily the microscopic ones in a particular experiment. Then again, I have reason to believe that the dislocations in the system are not in equilibrium during even the slow expansion run, as I will detail in the following section. When we compare the defect maps in Figure 4.21 and Figures 4.11 and 4.12, it appears that those in the higher-density portion of the intermediate region in the expansion experiment, e.g., in Figure 4.21 at $n = n_{12}$, are considerably more clumped together than those in the wedge experiment, e.g., $n = 0.173$ in Figure 4.11. This could be a real qualitative difference, a statistical fluctuation, or it may be a result of the possibly nonadiabatic expansion of the system.

The particle trajectories displayed in Figure 4.22 were determined by exponentially weighting 17 sec of videotape and averaging. From these trajectories, we can get a rough estimate for the dislocation glide diffusion coefficient $D_g \sim$(distance moved)$^2$/17 sec in the intermediate region from the direct association of the "hopping" trajectories with dislocation glide [49]. At $n = n_{11} = N_a$, at the crystal-intermediate region boundary $D_g \approx 1$ $\mu m^2$/sec, and at $n = n_{16} = N_b$, at the intermediate region-fluid boundary,

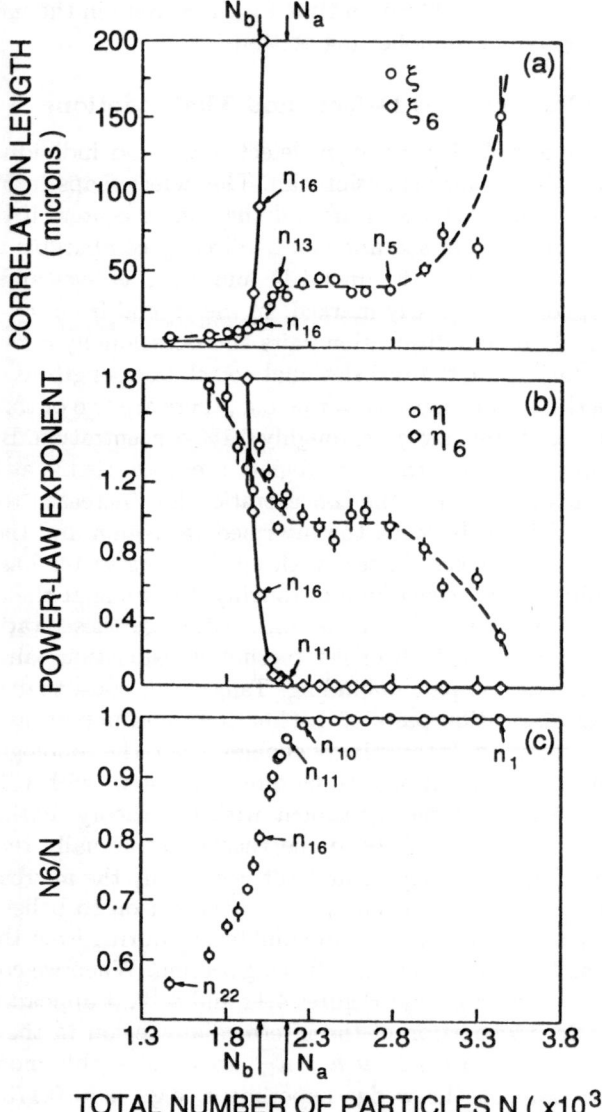

FIGURE 4.20. Top panel: translational (solid line) and bond-orientational (dashed line) correlation lengths for the slow run of Tang et al. Center panel: exponents from algebraic fits to the correlation functions. Bottom panel: ratio of sixfold coordinated spheres. Various densities are marked on each of the curves.

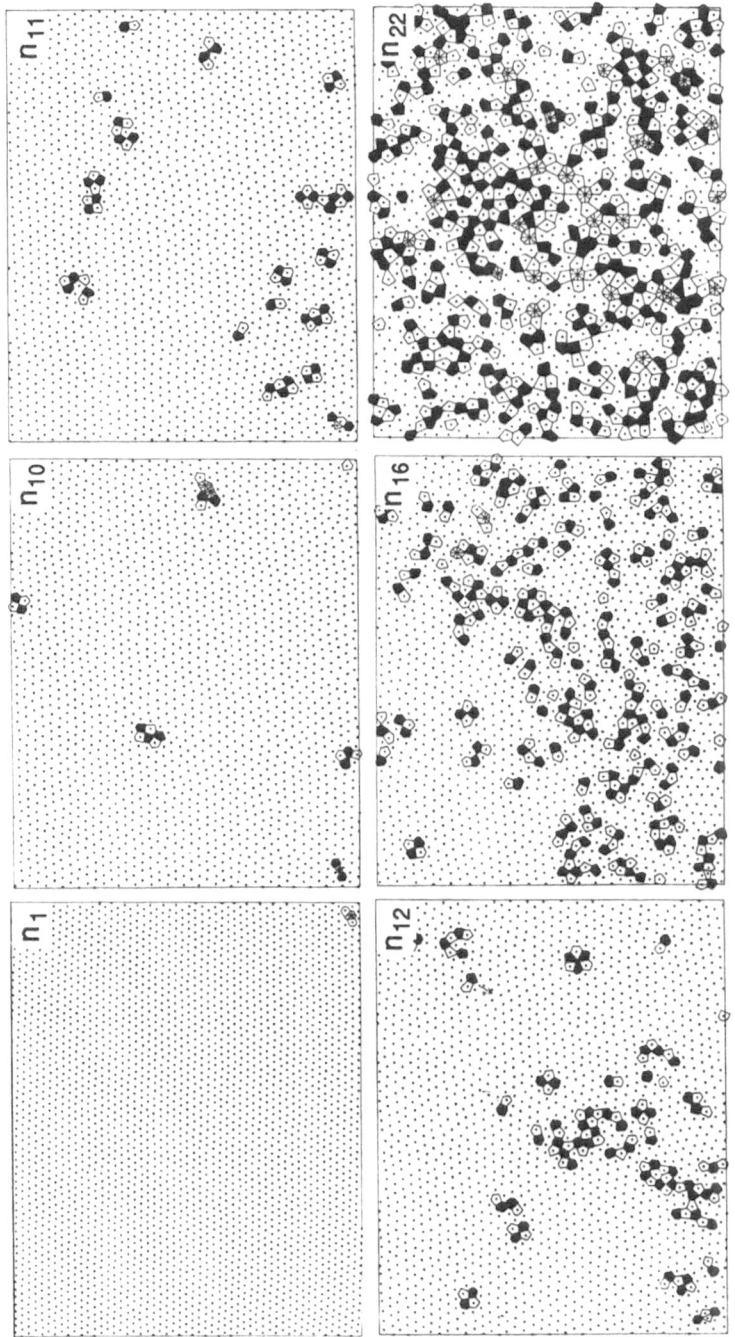

FIGURE 4.21. Defect plots obtained from Voronoi analysis of snapshots of the system of Tang et al. at various densities marked on the images. Sixfold coordinated spheres are depicted as dots, sevenfold as white, and fivefold as black Voronoi polyhedra.

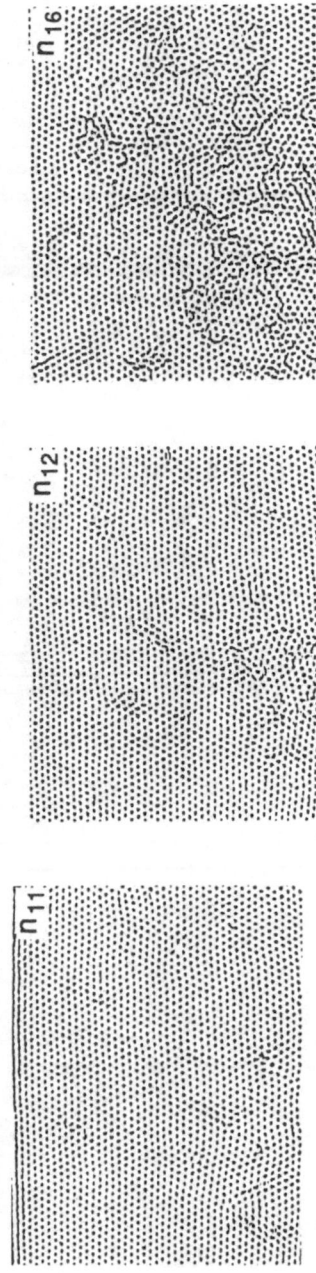

FIGURE 4.22. Trajectories obtained by exponentially weighting 17 min of videotape and averaging at various densities as marked in the slow expansion run of Tang et al.

$D_g \approx 8.5 \ \mu\mathrm{m}^2/\mathrm{sec}$. These numbers can be directly compared to those obtained from dislocation glide in the wedge with $0.305$-$\mu\mathrm{m}$ diameter spheres. (Figures 4.15 and 4.16 show trajectories in the hexatic intermediate region near the crystal and fluid, respectively.) Near the crystal, Murray and Van Winkle estimate $D_g \approx 2.8 \ \mu\mathrm{m}^2/\mathrm{sec}$ and near the fluid, $D_g \approx 25 \ \mu\mathrm{m}^2/\mathrm{sec}$. These values are roughly a factor of 3 larger for the smaller spheres, consistent with $D_g$ being proportional to the inverse of the sphere diameter, as is the Stokes drag of the water on each sphere. From Eq. (4.13) and Figure 4.20, we can now estimate the time needed for a dislocation to climb a correlation length in the expansion experiment in the intermediate region, where presumably there exist a finite number of vacancies and interstitials. Unfortunately, Tang et al. do not state the vacancy or interstitial concentration $c_v$ in their expansion experiment for $N_b < n < N_a$. However, if we take as a reasonable estimate $c_v \sim 0.01$, then $\tau_{eq} \approx 70$ hr at $n = N_a$ and $\tau_{eq} \approx 1.5$ hr at $n = N_b$. Thus, the dislocations in the system near the first transition at $n = N_a$ and certainly in the expanded crystal for $n > N_a$ are not in equilibrium, and it is not apriori clear what this effect will have on the defect map of the system. Near the intermediate region-fluid transition at $n = N_b$, however, and in the fluid, for $n < N_b$ the system is closer to equilibrium. This will certainly be the case if $c_v$ is larger there.

**Summary**

Tang et al. have performed a nice experiment in which they expanded a crystal between glass plates by slowly increasing the plate separation, maintaining a negligible density gradient in the system, and also minimizing external contamination effects. They find in a region of about 8% extent in density long-range orientational order and short-range translational order, with orientational and translational correlation functions in good agreement with KTHNY predictions, and statistical properties of the defects that are consistent with the KTHNY mechanism. It appears that the expansion rate of the system is about an order of magnitude or two too fast to be adiabatic. This is the most likely explanation for the observed hang-up in the correlation length and defects of the system in the low-density crystal during the expansion. It could also explain the inconsistency of the observed microscopic defects in the intermediate region with the KTHNY picture of dislocation pair and disclination pair unbinding. In subsequent experiments one could use smaller spheres to speed up dislocation glide and slower expansion rates by at least a factor of 10, in order to make quantitative comparisons with the equilibrium KTHNY theory. Of course, this geometry and experiment allow for direct observation of driven nonequilibrium effects, which is quite an interesting subject in itself.

### 4.2.4   Dipole Holes in Ferrofluid

Monodisperse polystyrene spheres dispersed in ferrofluid in a magnetic
field produce a system of monodisperse magnetic voids interacting with
dipole-dipole forces which can be controlled in strength and direction by
the external magnetic field. The ferrofluid itself is a colloidal suspension
of ferromagnetic particles dispersed in a solvent, usually oblong magnetite
particles with length $\sim$ 0.01–0.1 $\mu$m dispersed in kerosene [65]. A.T. Skjeltorp and co-workers have made use of an ingenious double colloidal system
of 1–100 $\mu$m-diameter polystyrene spheres dispersed in either kerosene-
or water-based ferrofluid to model many-body condensed matter phenomena for a number of years [27]. In particular, they have performed two
separate two-dimensional melting experiments. In the first experiment in
1987 [66], Skjeltorp used a system of 1.9-$\mu$m diameter polystyrene spheres
dispersed in a 4-$\mu$m thick layer of ferrofluid, with applied dc magnetic
fields of $\mathbf{H}$ = 10–100 G perpendicular to the layer. In this experiment the
polystyrene spheres were sufficiently small that their Brownian motion established the real temperature of the system. Reducing the magnetic field
strength, and thus the net perpendicular dipole moments of the magnetic
voids created by the spheres, reduces the repulsive sphere-sphere interaction
strength in relation to the system temperature sufficiently to melt the system. In the second experiment in 1990 [67], G. Helgesen and A.T. Skjeltorp
used larger polystyrene spheres of size $\approx$ 10 $\mu$m in a thin layer of ferrofluid
roughly twice the diameter thick and applied rotating in-plane magnetic
fields in addition to a small dc magnetic field perpendicular to the plane.
This produces a system of rotating dipole pairs of particles which form a
triangular lattice below a certain critical frequency of field rotation. This
lattice "melts" when the rotating field frequency becomes large enough to
break up the pairs to form lattice defects. In both experiments they found
evidence for separate bond-orientational and translational ordering transitions from the calculated correlation functions, but concluded from the
actual map of lattice defects that the disorder in the lattice grows in from
grain boundaries which are always present in the initial crystal (perhaps
outside the viewing region), and are furthermore far too clumped together
into grain-boundary-like structures to be consistent with KTHNY predictions. I will discuss the second in-plane rotating field experiment first as
the data are more extensive and briefly return to the earlier perpendicular
field experiment later.

### Experimental Geometry and Particle Interactions

The experimental geometry of both experiments is shown in Figure 4.23.
The sample of polystyrene spheres dispersed in ferrofluid is confined between two glass plates in the $x, y$ plane within a region of thickness roughly
twice the sphere diameter, surrounded by three Helmholtz coils used to
produce magnetic fields along the $x$, $y$, and $z$ axes. The interaction en-

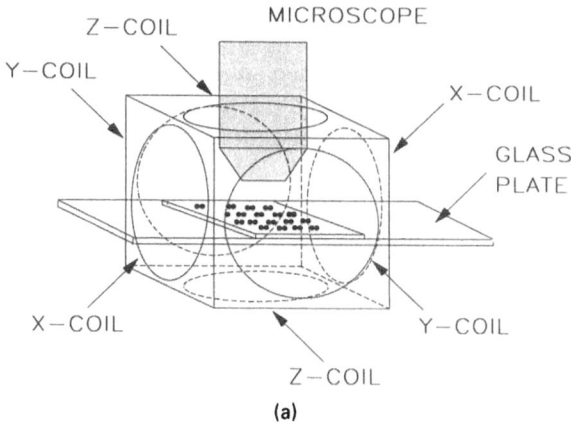

(a)

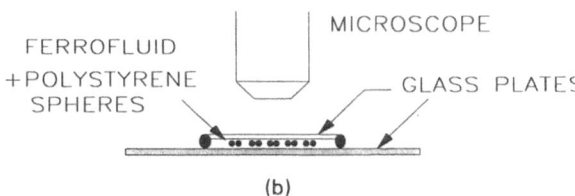

(b)

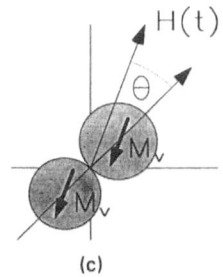

(c)

FIGURE 4.23. (a) and (b) The experimental geometry used by Skjeltorp. (c) A depiction of a sphere pair with in-plane dipole moments.

ergy between two magnetic holes is equivalent to the interaction between two magnetic spheres with magnetic moment $\mathbf{M} = -V\chi_{\text{eff}}\mathbf{H}$, where $V$ is the volume of ferrofluid displaced by the sphere, $\chi_{\text{eff}}$ the effective volume susceptibility of the ferrofluid, and $\mathbf{H}$ the applied magnetic field [66]. The interaction energy of two spheres whose centers are separated by a distance $r > d$ is given to first order by

$$E_{d-d} = M^2(1 - 3\cos^2\theta)/r^3, \qquad (4.16)$$

where $\theta$ is the angle from the line between the sphere centers to the direction of the magnetic field $\mathbf{H}$. In the presence of a field $H_{x,y}$ in the $x - y$ plane of the layer, the spheres line up in chains oriented along the field. When $H_{x,y}$ is slowly rotating with frequency $f_H$, the chains attempt to follow the field rotation. However, due to the viscous torque on the chains, they break up into short segments of two to four spheres, which can follow the rotation of the field with a phase lag $\phi$. The equation of motion for a single pair of spheres can be written [67]

$$\frac{1}{2\pi}\frac{d\phi}{dt} = f_H - f_c\sin(2\phi), \qquad (4.17)$$

where $f_c$ is a critical frequency given by the magnetic and viscous parameters of the system. For $f_H < f_c$, the motion of the pair will follow the field rotation. For $f_H > f_c$, the phase lag angle $\phi$ may exceed $\pi/4$ and the magnetic torque $\sim f_c\sin(2\phi)$ on the pair decreases and changes sign, so that for short periods of time the rotation direction of the pair is opposite in direction to that of the field rotation. The time intervals between periods of counter-rotation become shorter as $f_H$ increases [67]. By applying a weak dc field $H_z$ along the $z$ axis normal to the sample plane, Helgesen and Skjeltorp managed to create a triangular lattice of rotating pairs and could adjust $H_{x,y}$, $H_z$, and $f_H$ so that all pairs were able to follow the rotating field with a uniform phase lag.

Two spheres with dipole moments oriented in the plane will have a net attraction, while if the dipole moments are oriented perpendicular to the plane, they will repel one another. The combination of $H_{x,y}$ and $H_z$ creates a minimum in the pair potential [67] and thus determines a lattice constant for the ordered system. The rotation instability for a regular lattice of rotating pairs will take place at a critical frequency $f_c' < f_c$. Skjeltorp and co-workers found, however, that for $f_H > f_c'$, the counter-rotation of all the pairs does not take place simultaneously. Instead, at random sites the rotation of one pair is opposite to the rotation of its neighbors for short time intervals. This introduces random fluctuations and defects in the lattice due to the break-up of the counter-rotating pair. The rotating pairs of spheres serve as the constituant "molecules" of the two-dimensional "crystal" formed when $f_H < f_c$. The intermolecular interactions are quite complicated [66] and include not only the long-range magnetic forces but also image dipoles due to the ferrofluid-glass interfaces of the confining plates.

## Defects and Static Correlation Functions

Helgesen and Skjeltorp used $H_{x,y}/H_z = 0.7$ in the melting experiment and increased $f_H$ in steps, equilibrating for 5–30 min at each frequency before videotaping the structure of the system through a visible microscope objective. They also used a He–Ne laser to record the diffraction from the same region of the sample. The image area contained $\sim$ 1500 pairs, with the actual sample size 20–50 times larger. After equilibration at $f_H < f_c' \approx 0.43$ Hz, only a few isolated dislocations exist in the lattice, and occasionally, grain boundaries. In Figures 4.24 and 4.25 are shown defect maps of their system determined from a Voronoi polyhedron analysis at various driving frequencies. In Figure 4.26 are shown the static pair correlation function $g(r)$ and bond-orientational correlation function $g_6(r)$ for the images in Figure 4.24. In Figure 4.24, the system initially started out at $f_H = 0.40$ Hz with 0.5% free dislocation density, while in Figure 4.25, two grain boundaries are visible in the center of the imaging region. This system comprised their initial two-dimensional "crystal" state; however, due to the frozen-in dislocations or grain boundaries the translational correlation length should be about $50a$. When the system starts out with a small dislocation density, it is in actually a frozen-in nonequilibrium hexatic state [30] similar to, but with an order-of-magnitude fewer free dislocations than, the extrinsically induced hexatics discussed in Sections 4.2.2 and 4.2.3. When a sufficiently high dislocation density is frozen into the system, the dislocations align into a series of small-angle grain boundaries. For $f_H > f_c'$, pairs randomly break up to produce defects in the lattice, which clump together at $f_H = 0.47$ Hz, as in the center panel of Figure 4.24, and nucleate at pre-existing grain boundaries as shown in Figure 4.25. At even larger field rotation frequencies, e.g., if $f_H = 0.50$ Hz, the defects percolate across the entire image as shown in the bottom panels of the figures and the average diffraction pattern becomes essentially that of an isotropic fluid.

Helgesen and Skjeltorp calculate the static pair correlation function and bond-orientational correlation functions for their snapshots of "molecule" configurations in this system, shown for the snapshots in Figure 4.24 in Figure 4.26. I assume that the snapshots were taken with a long exposure time or were averaged over the rotation period of the pairs. The correlation functions are only shown out to $r = 10a$ in the figure, although they presumably were computed from the locations of all 1500 particles in the imaging area. The correlation functions at $f_H = 0.4$ Hz are consistent with $\xi \approx 50a$ and $\eta_6 \approx 0.02$ expected [30] for the dislocation density of 0.005 observed in Figure 4.24. For $f_H = 0.47$ Hz, Helgesen and Skjeltorp state that the translational correlation length $\xi \approx 22a$; however, my fits to their curve find $\xi \approx 10a$. For this frequency, they find $\eta_6 \approx 0.3$, which is a

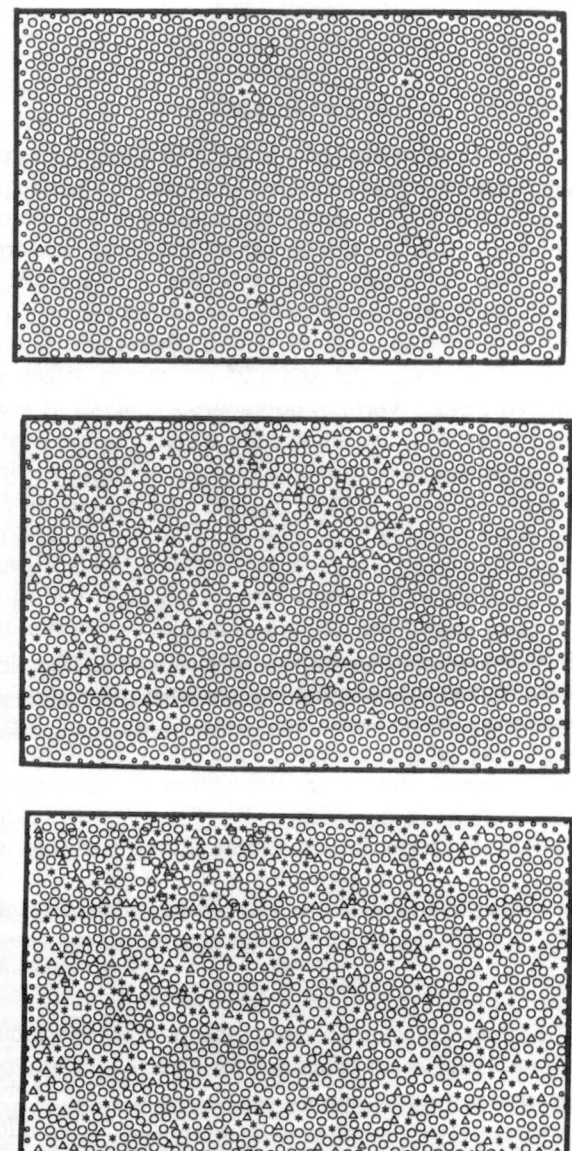

FIGURE 4.24. Defect plots of snapshots of one of the Helgesen and Skjeltorp runs. Top: $f_H = 0.43$ Hz; center, $f_H = 0.47$ Hz; bottom, $f_H = 0.60$ Hz. Sixfold coordinated molecules are depicted as circles, sevenfold as asterisks, and fivefold as triangles.

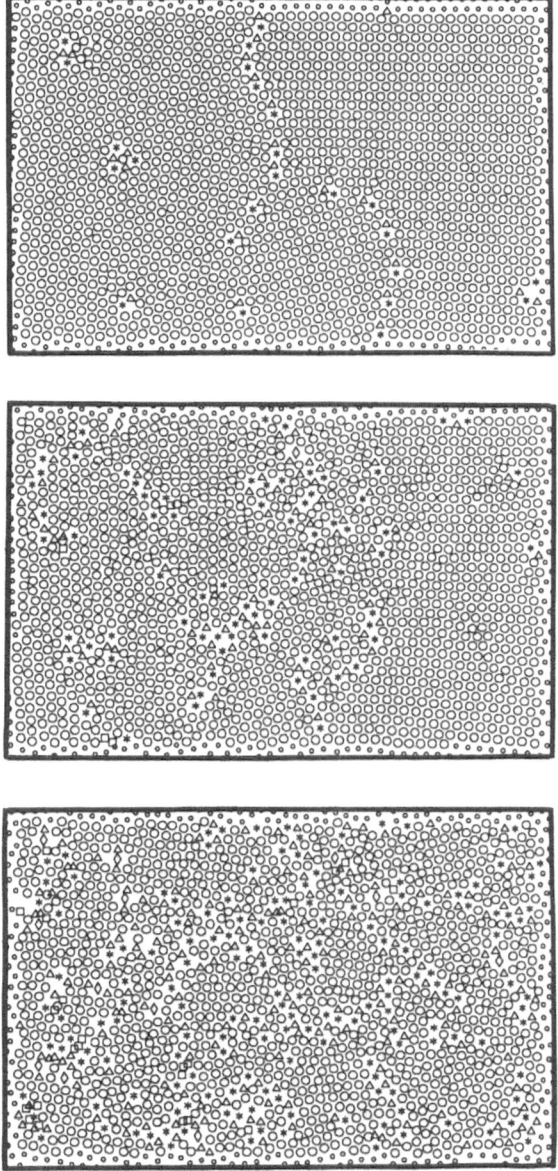

FIGURE 4.25. Same as Figure 4.24, but the top panel initially started out with grain boundaries.

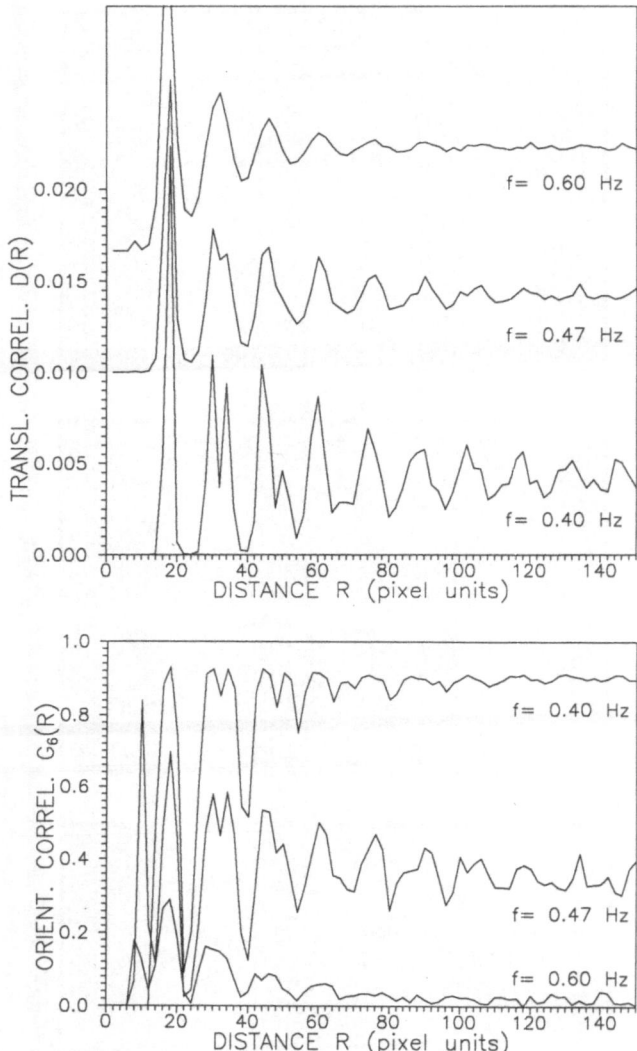

FIGURE 4.26. Correlation functions from center locations of the images depicted in Figure 4.24 from Helgesen and Skjeltorp. Top: $g(r)$ and bottom: $g_6(r)$.

bit large for but still consistent with a KTHNY hexatic very close to the fluid transition. However, the defect configuration of Figure 4.24 is far too clumped to resemble those found in the wedge experiment hexatic region, except possibly very close in density to the fluid.

### Discussion and Summary

From the published papers of Skjeltorp et al. we are not given any idea of the time scale for dislocation motion in the system; however, they state that they could observe both the creation and motion of defects. Due to the large sphere sizes used in this system and the increased viscous and magnetic drag compared to the case of the charged colloidal spheres in water, one would expect considerably smaller dislocation glide diffusion coefficients, and correspondingly much larger intrinsic equilibration times than the 5–30 min allowed for equilibration in the melting experiment, as discussed at length in Section 4.2.3. Thus, I have serious doubts that in either of Skjeltorp's beautiful experiments he is measuring the equilibrium defect configurations. For this system of $\sim 10$-$\mu$m diameter spheres in rotating magnetic fields, the temperature of the system may not be very well defined and is most likely no longer determined by Brownian motion, but could be measured by observing individual pair motions in the fluid phase.

Helgesen and Skjeltorp denote the defect clumps in their images in this frequency region as grain boundaries. It would be constructive to know the life time of one of their grains in this region and most important the dislocation glide and climb diffusion coefficients. A collection of small-angle randomly oriented grains will have orientational correlations decaying as fast as its translational correlations, on the scale of the grain size. The "grains" in the magnetic experiments definitely have overall longer-range orientational correlations than translational ones, with implications that the individual grains are oriented with respect to each other. For comparison to the Murray–Van Winkle wedge experiment, it would be constructive to probe carefully the frequency region 0.43 Hz $< F_H <$ 0.47 Hz in which a lower total defect and free dislocation density hexatic might be expected to form. If Helgesen and Skjeltorp have been unable to reach such a low defect region, then perhaps they have created a system in which the dislocation core energy $E_c$ and the dislocation interaction energy $K$ are radically different from that of the Yukawa interacting particles. As I mentioned in Section 4.1, the ratio $E_c/K$ is an indicator of the ease of dislocation creation compared to pair break-up. When this ratio is small, it is expected that there will be a sharp first-order transition and grain boundary proliferation exactly at melting, before the dislocation pairs unbind. If this were the case in Helgesen and Skjeltorp's experiment, one would then interpret the 0.47 Hz configuration in Figures 4.24 and 4.25 as a two-phase coexistence region which would not be inconsistent with their data, including the correlation functions. If this were a true two-phase region, then $g_6(r)$

would have two separate decays, a relatively slow one for $r < \xi$ showing the ordered regions, and a considerably faster one for $r > \xi$, the true decay of orientational correlations of the system. I will show this in Section 4.2.5. As $g_6(r)$ is only shown for $r < \xi$ in Figure 4.26, it is not clear whether they do observe a break between two regimes of behavior.

Helgesen and Skjeltorp give the same interpretation to the results for the earlier Brownian-motion-dominated experiment [27] in which only a perpendicular field $H_z$ is applied to the system and gradually decreased in magnitude, reducing the single sphere dipole-dipole interaction of Eq. (4.16) with respect to $k_BT$ so that thermal fluctuations melted the lattice. Snapshots of the defect configurations at various field strengths in this experiment are shown in Figure 4.27, which shows that they are qualitatively similar to those in the rotating pair experiment shown in Figure 4.24. For the Brownian-motion-dominated experiment, I have the same comments and questions as stated above.

Dipole holes in ferromagnetic fluid created in a double colloidal system of monodisperse polystyrene spheres suspended in ferrofluid is a fascinating system to use for modeling condensed matter many-body phenomena and especially nonequilibrium phenomena. For example, Helgesen and Skjeltorp note that in the rotating pair experiment the rotation of the pairs of large spheres produces a regular dual-flow pattern in the ferrofluid background, which becomes chaotic when some of the pairs begin to counterrotate and break up. This could be a controllable method of studying the route to chaos in a system of driven fluid rolls.

## 4.2.5    Wedge Geometry Revisited: Comparison of Melting in Two and Three Dimensions

In order to answer some of the outstanding questions remaining from the first series of experiments studying two-dimensional melting in a wedge geometry summarized in Section 4.2.1, Murray, Sprenger, and Wenk performed another series of gradient experiments [68] to study melting of both two-dimensional and three-dimensional colloidal crystals. In this new set of experiments, polystyrene sulfate colloidal spheres of diameter 0.305 $\mu$m and titratable surface charge $2 \times 10^5$ electronic charges were used. Thus, the sphere diameter remained the same as in the earlier wedge experiment, but the sphere bare charge was roughly five times larger. For these spheres, the 3D FCC–BCC phase transition is observed at a bulk FCC nearest-neighbor separation $a = (4.4 \pm 0.09)d$ with no added salt, which was close to the density of the three-dimensional reservoir in contact with the two-dimensional crystal. The two-dimensional melting was observed at $a = (4.3 \pm 0.05)d$, an in-plane density roughly a third of the earlier two-dimensional melting experiments discussed in Section 4.2.1, and at a layer thickness $t = 7.5d$, about twice as large as that of the earlier experiment. Mapping the ob-

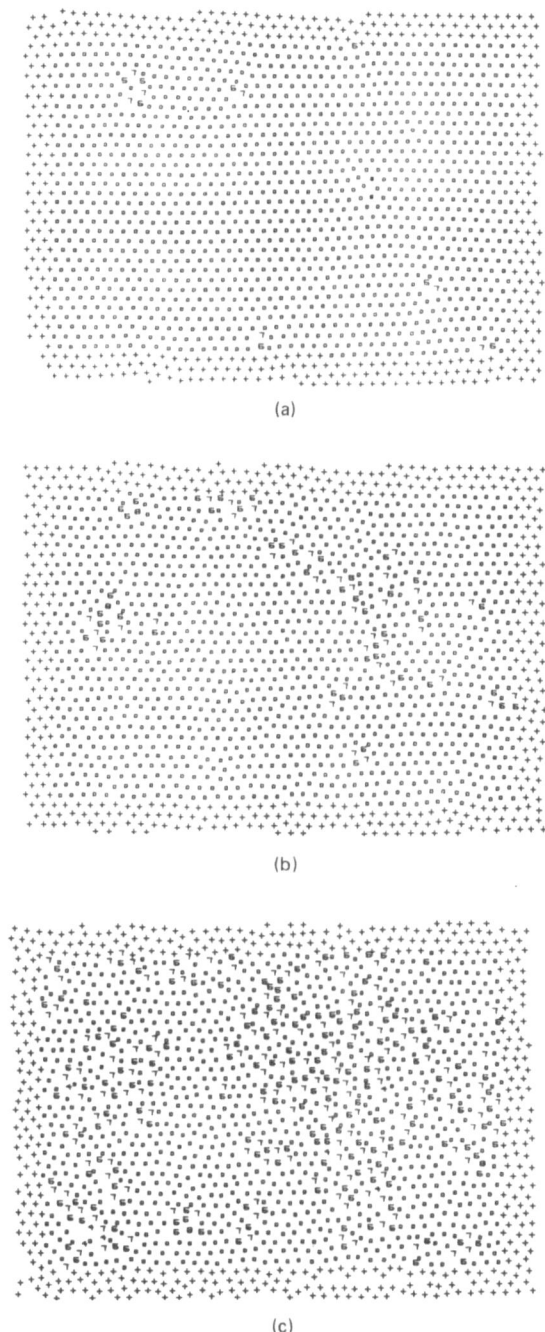

(a)

(b)

(c)

FIGURE 4.27. Defect plots of snapshots from the Brownian-motion-dominated dipole hole experiment of Skjeltorp. Top: crystal at $H_z = 100$ G; center: intermediate region at $H_z = 20$ G; bottom: fluid at $H_z = 13$ G. Here, sixfold coordinated spheres are depicted as squares, seven- and fivefold are labeled by 7 and 5, edges are depicted by +.

served FCC–BCC transition onto the Yukawa phase diagram [36] yields an effective charge of $Z^* \approx 1000$ electrons and effective screening parameter $\Lambda \approx 2.93$, in comparison to the values $Z^* \approx 700$ electrons and $\Lambda \approx 3.5$ for the earlier experiment.

In this series of experimental runs, we made a detailed comparison between a three-dimensional ($\geq$ 200-layer) FCC crystal in equilibrium with its melt along a density gradient parallel to a glass surface and a two-dimensional triangular crystal in equilibrium with its melt along a very similar density gradient along a wedge. Care was taken in the ten two-dimensional runs on separate wedges and seven three-dimensional runs along different gradients that at least one month was allowed for chemical equilibration of the screening ions along the sphere density gradient, and at least 24 hr were allowed for mechanical equilibration of the spheres after a small perturbation of the wedge angle. In the statistical sense both two-dimensional and three-dimensional density gradients were "equilibrated" in that a dislocation could climb across an imaging region at melting before we took snapshots of the system. In order to ensure as similar an interaction potential between the spheres in the two-dimensional and three-dimensional runs as possible, the experiments were performed in the same sample cell within days of each other. This was accomplished by having a 200-$\mu$m thick step in the center of the top cell window in the viewing region, outside of which the three-dimensional runs were made. At the same time, the bottom window of the cell was bowed to create a wedge angle of $5 \times 10^{-4}$ radians in the thin region of the cell for the two-dimensional runs.

For both two-dimensional and three-dimensional runs, the imaging area used was $\Delta x \times \Delta y \times \Delta z = 59 \times 46 \times 0.04$ $\mu$m$^3$ or $45 \times 35 \times 1$ spheres. The extent of the sample with an in-plane density within 1% of that in the imaging region was at least $35 \times 450$ spheres. For the three-dimensional runs, we imaged the first layer of colloid adjacent to the bottom glass surface. The three-dimensional colloidal fluid is strongly layered with at least three sharp layers next to the glass surface [28]. On the time scale of the video images taken at each of about 10 positions along the density gradients in each run (five images separated in time by 0.05 sec, or a total of about eight collisions in this experiment), a vast majority of the spheres remain within the first layer adjacent to the glass. As in Section 4.2.1, a collision time is defined here [28] as the time needed for a sphere to diffuse $0.1a$ with the Stokes diffusion coefficient, making a small correction for the extra drag on a single sphere due to the presence of the nearby solid glass wall(s). In the three-dimensional runs, we scanned the imaging volume along the first layer of the colloid adjacent to the glass along the gradient. For the two-dimensional runs, the procedure was similar to that used in the earlier wedge experiment. Figure 4.28 shows the in-plane density for representative three-dimensioal and two-dimensional runs, determined by counting spheres within the imaging region. The three-dimensional density gradient has two sharp breaks at $n_a$ and $n_b$, while the two-dimensional gradient is

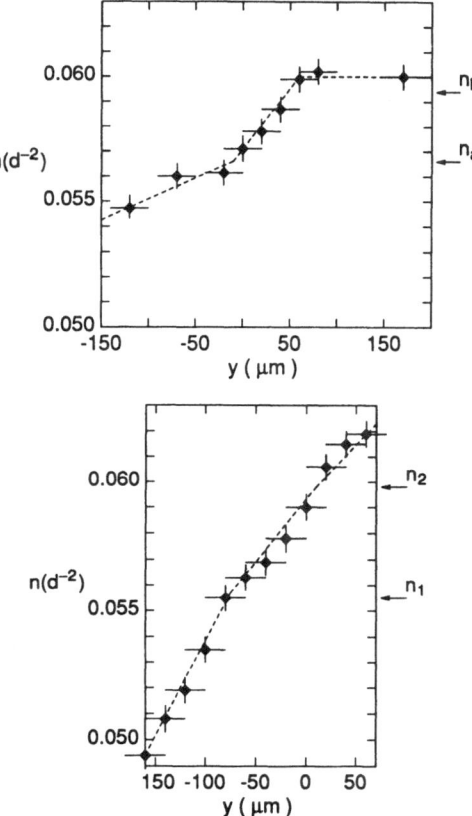

FIGURE 4.28. Density vs. distance moved along imposed density gradient in the Murray et al. second experiment. Top: three-dimensional run; $n_a$ and $n_b$ mark the extent of the intermediate two-phase region. Bottom: two-dimensional run; $n_1$ and $n_2$ mark the extent of the intermediate hexatic region.

much more smooth, with possible slope changes at $n_1$ and $n_2$. Densities above $n_b$ in the three-dimensional runs and $n_2$ in the two-dimensional run are crystalline, and below $n_a$ and $n_1$ are fluid, as will be shown from correlation functions later. The intermediate region $n_a < n < n_b$ in the three-dimensional run spans 5% in in-plane density, and in the two-dimensional run $n_1 < n < n_2$ spans 8% in density. Note that melting takes place at the same in-plane density, $n \approx n_c = 0.06 \pm 0.0005$, in both the second dimension and the third dimension.

### Defect Statistics

Maps of the defects, Delaunay triangulations of center locations in snapshots of the colloid at various average densities in the crystal and the intermediate region, are shown in Figure 4.29 for both the three-dimensional and two-dimensional run. In the top panels, the three-dimensional crystal at $n_c$ is less defected than the two-dimensional crystal. In the second two panels from the top in the figure, obvious from first glance, is an abrupt interface between a highly defected region and an ordered crystalline region in the three-dimensional run at $n = 0.98\ n_c$, while in the two-dimensional run this panel at a density of $0.974\ n_c$ exhibits slightly more defects and larger clumps of defects than the two-dimensional crystal, but no obvious interface. The interface in the three-dimensional run is not perpendicular to the translation of the imaging region along the macroscopic average density gradient. The interface is present in the entire set of images taken between $n_a = 0.95\ n_c$ and $n_b \approx n_c$, as shown in Figure 4.29.

In Figure 4.30 are shown trajectory plots during the time interval of eight collisions at the densities $n = 0.98\ n_c$ for the third dimension and $n = 0.974\ n_c$ for the second dimension corresponding to the second panels from the top in Figure 4.29. For the three-dimensional case, spheres to the left of the interface running from top left to bottom center of the image exhibit liquidlike motion, while only a few small regions to the right of the interface, associated with dislocation pairs and triples in the crystal, show more than $0.1a$ distance moved in this time. This is a classic example of an interface between highly mobile fluid and rigid crystal, as would be expected from a first-order phase transition. By contrast, the two-dimensional trajectories in the region $n_1 < n < n_2$ show motion very similar to those associated with defects in the three-dimensional crystal, but in slightly larger regions of the image. The defect clumps of 10–20 disclinations in the lower left and right in Figure 4.29 for the two-dimensional run at $n = 0.974\ n_c$ move around and change shape entirely in the course of several seconds and are always surrounded by regions of order, while the interface in the three-dimensional run fluctuates in position but remains, separating fluidlike and crystalline regions.

For the two-dimensional runs, the lower-density snapshots in the intermediate region $n_1 < n < n_2$ have larger and more branched defect clumps as the density approaches that of the fluid. Dislocation pairs appear and annihilate on the time scale of several collisions. With equal probability they attach themselves to the existing clumps of defects or appear in the center of an ordered region between the existing clumps. The defect clumps, $4–10a$ in extent, also move and change shape constantly on a similar time scale. The defects do not appear anything like grain boundaries surrounding the ordered regions; rather it seems a better picture to describe the clumped defect regions as surrounded by regions of ordered spheres that have considerable orientational order. This is completely consistent with the defect statistics found in the earlier wedge run of Murray and Van Winkle described in Section 4.2.1, despite the much lower density and the larger extent of the intermediate phase for this later run. The life times of the ordered regions can be obtained quantitatively by calculating cross-time

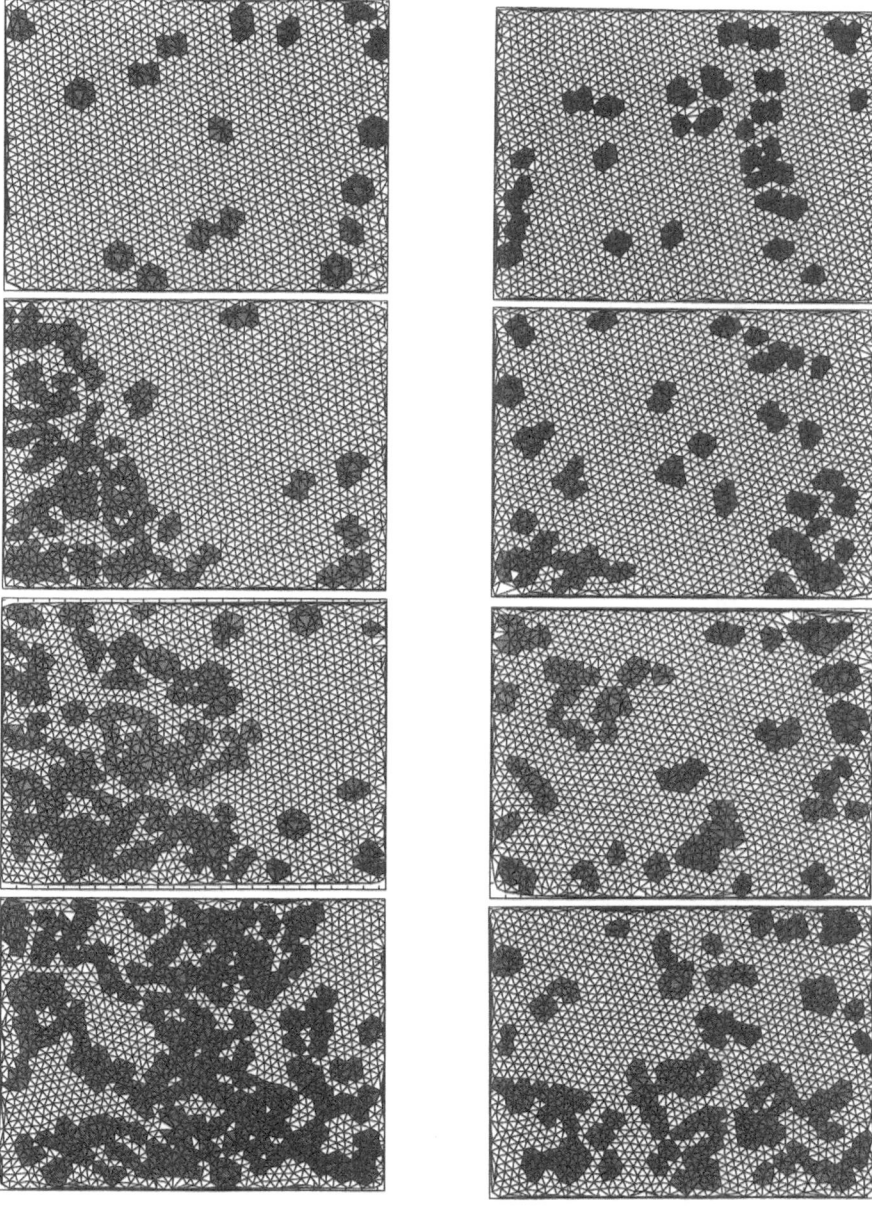

FIGURE 4.29. Defect maps from snapshots along the density gradient in the Murray et al. second experiment. Sphere centers are represented as vertices. Defects or non-sixfold coordinated neighborhoods are shaded. Left: three-dimensional run. Density from top to bottom: crystal, $n = 0.0599 \equiv n_c$; and intermediate region, $n = 0.98\,n_c$, $0.965\,n_c$, and $0.953\,n_c$. Right: two-dimensional run. Density from top to bottom: crystal, $n = 0.0606 \equiv n_c$, and intermediate region, $n = 0.974\,n_c$, $0.94\,n_c$, and $0.93\,n_c$.

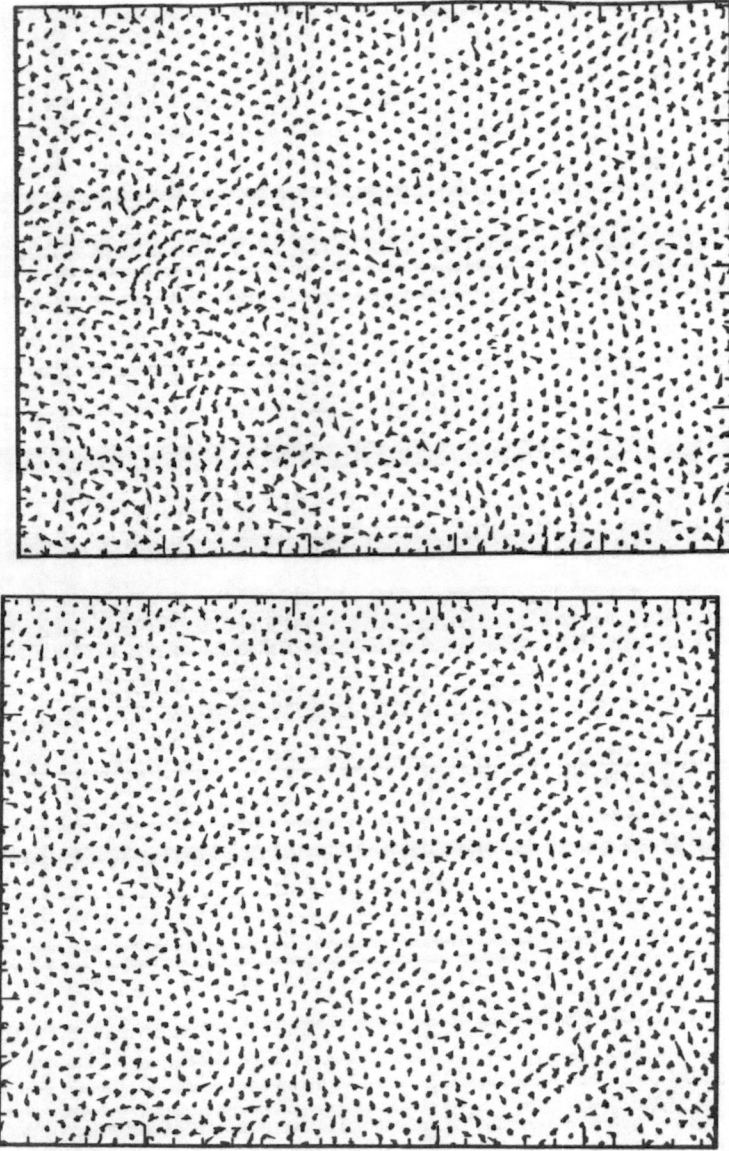

FIGURE 4.30. Trajectories for the period $t = 0.2$ sec, or about eight collisions, for the three-dimensional (top) and two-dimensional (bottom) intermediate regions in the Murray et al. second experiment. Densities of the colloid are $n = 0.98\,n_c$ for three dimensions and $n = 0.974\,n_c$ for two dimensions.

correlations of successive images [28] and matching the gross rearrangement time of the defects. On the other hand, the defects in three-dimensional images at $n \approx n_a$ do nearly completely surround the ordered regions, which are for the most part no longer oriented with each other. They are best described as low-angle grain boundaries, but the life time of the microscopic structure is also only a few collisions [51].

## Static Correlation Functions

In Figure 4.31 are shown the static pair and bond-orientational correlation functions calculated for the snapshots of both the three-dimensional and two-dimensional runs in Figure 4.29. As in Section 4.2.1, the pair correlation functions were fit to an exponentially decaying, broadened, pair correlation function for a perfect hexagonal lattice, in order to pull out an effective translational correlation length. The translational correlation length obtained in this way for each snapshot is marked by an arrow on each curve in Figure 4.31. From these correlation functions, we can pick out the images corresponding to the lowest curves $a$ and $b$ for both runs as fluid, as the orientational correlations fall off with the characteristic translational correlation length. Note that for the three-dimensional fluid, the correlation length never exceeds $2a$ and $g_6(r)$ decays exponentially, while for the high-density two-dimensional fluid with density close to $n_1$, the correlation length is nearly $3a$ and $g_6(r)$ decays exponentially at first but develops a large tail at high $r$.

The highest curves, $f$, exhibit crystalline order: $g_6(r)$ is constant to within experimental accuracy, and the translational correlation lengths are $\xi \sim 90a$ for the three-dimensional crystal and $\xi \sim 40a$ for the two-dimensional crystal. In order to test the HN predictions of the two-dimensional crystalline algebraic exponent $\eta_G$ in Eq. (4.9), we calculated directly the correlation function of the translational order parameter $\rho_G = e^{i\mathbf{G}\cdot\mathbf{r}}$, $g_G(r)$. This is shown for the first three reciprocal lattice vectors of the system for $n = n_c$ in Figure 4.32. We find that $g_G(r)$ is consistent with very slow power-law decay, where the power $\eta_G$ scales as $G^2$ and $\eta_0 = 0.327 \pm 0.025$. This is in excellent agreement with the predictions [2,5] of two-dimensional crystal stability which have $1/4 \leq \eta_{G_0} \leq 1/3$, and it is good evidence that at $n_c$ we have a genuine two-dimensional crystal which becomes unstable at Poisson's ratio $\sigma = 1$.

In the intermediate region for both three-dimensional and two-dimensional runs, $g(r)$ exhibits shorter range translational order than the solid and longer-range than the fluid. In addition, $g_6(r)$ has intermediate behavior as well. For the three-dimensional run in which there is two-phase separation apparent in the defect maps (curves $c$–$e$ in the left-hand panel), the translational correlation length $\xi$ is the size of the crystalline portion of the image. The three-dimensional $g_6(r)$ curves cannot be fit with a single decaying curve. There is a clear break in $g_6(r)$ at $r = \xi$, as can be better seen

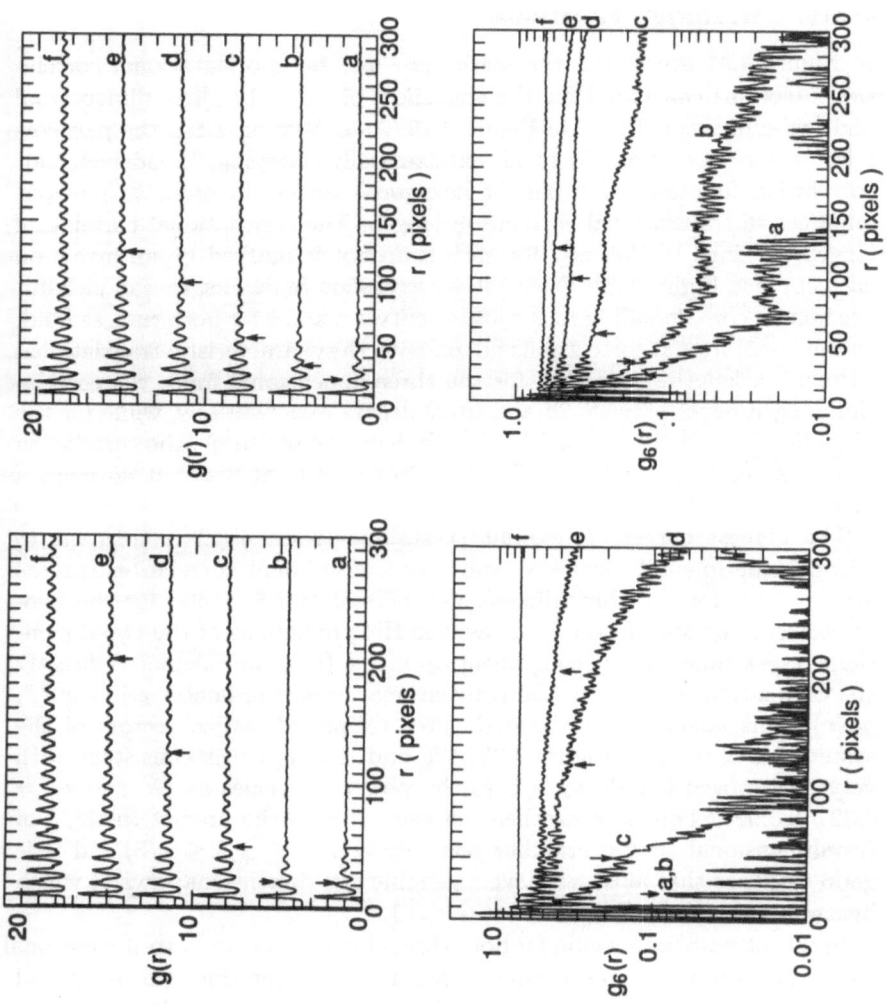

in Figure 4.33, in which fits to exponential decay for $r < \xi$ and $r > \xi$ are also shown. For $r > \xi$, the bond-orientational correlation length $\xi_6 \approx \xi$, as shown in Figure 4.34. This is not the case, however, for the two-dimensional curves in the intermediate region (curves $c$–$e$ on the right-hand side of Figures 4.31 and 4.33) which exhibit oscillations due to fluctuations, but do not have a clean break at $r = \xi$. The two-dimensional $g_6(r)$ curves in this region can be fit equally well by a single decaying exponential or a power law. Much more averaging is needed for the curves in order to distinguish the forms of the decay. Fits to exponential decay for $r > \xi$ yield the values for $\xi_6$ shown in Figure 4.34 for the two-dimensional case, and fits to power laws are shown in Figure 4.33. For density $n > n_1$ just beyond the fluid, $\eta_6 \approx 0.25$, in excellent agreement with KTHNY predictions. The actual count of free dislocations in the images increases gradually between $n_2$ and $n_1$ and never exceeds 3%.

**Summary**

From the evidence given above, I conclude that the two-dimensional intermediate region does not exhibit two-phase separation or grain-boundary-like behavior as does the three-dimensional intermediate region. The two-dimensional intermediate region is consistent with the KTHNY hexatic both in its correlation functions and defect statistics. Furthermore, although the density extent of the hexatic region is larger than that of the earlier experiment, and the densities of transition much less, both the correlation functions and defect statistics are consistent with the Murray–Van Winkle experiment results. Thus, we have a clear demonstration of an equilibrium hexatic phase in two separate systems of spherical particles on a smooth substrate. If the two-dimensional melting transition were first-order similar to that in the third dimension, then I would expect a distinct two-phase separation and fluid-crystal interface to form in the density gradient, as was demonstrated in the three-dimensional runs. The fact that an obvious interface did not form is evidence, but not absolute proof, that the two transitions observed in the two-dimensional system are continuous.

---

FIGURE 4.31. Correlation functions from the Murray et al. second experiment. Top panels: $g(r)$ and bottom panels: $g_6'(r)$. The arrows on each curve mark the translational correlation length $\xi$ for that density. Left: three-dimensional run. The densities for the curves are as follows: (a) fluid at $n = 0.0560$, (b) fluid at $n = 0.0561$, (c) intermediate region at $n = 0.0571$, (d) intermediate region at $n = 0.0578$, (e) intermediate region at $n = 0.0587$, (f) crystal at $n = 0.0599$. Right: two-dimensional run. The densities for the curves are as follows: (a) fluid at $n = 0.519$, (b) fluid at $n = 0.535$, (c) intermediate region at $n = 0.0563$, (d) intermediate region at $n = 0.0569$, (e) intermediate region at $n = 0.0590$, (f) crystal at $n = 0.0606$.

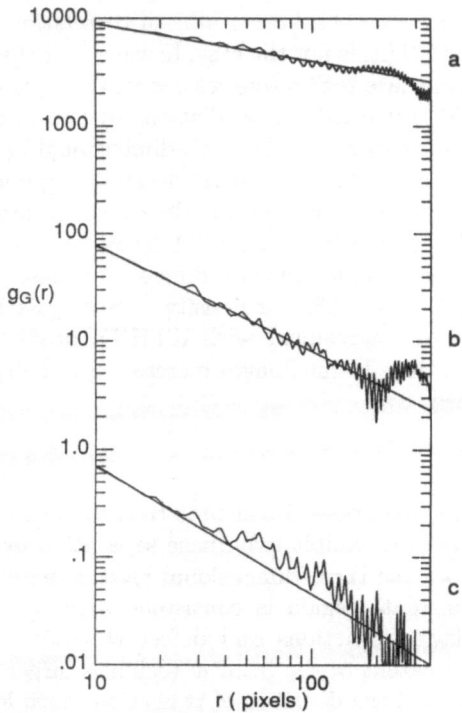

FIGURE 4.32. Translational correlation function $g_G(r)$ for the two-dimensional crystal at $n = n_c \equiv 0.0606$ in the Murray et al. second experiment. Top (a), $G = G_0$, the first reciprocal lattice vector; center (b), $G = \sqrt{3}\,G_0$; and bottom (c), $G = 2G_0$. The curves are displaced upward for clarity. Also shown are best-fit algebraic decay $r^{-\eta_0}$ for each $G$. From top to bottom, the exponents are 0.3, 1.0, and 1.2, respectively. Here, $1\,\mu\text{m} = 8.14$ pixels.

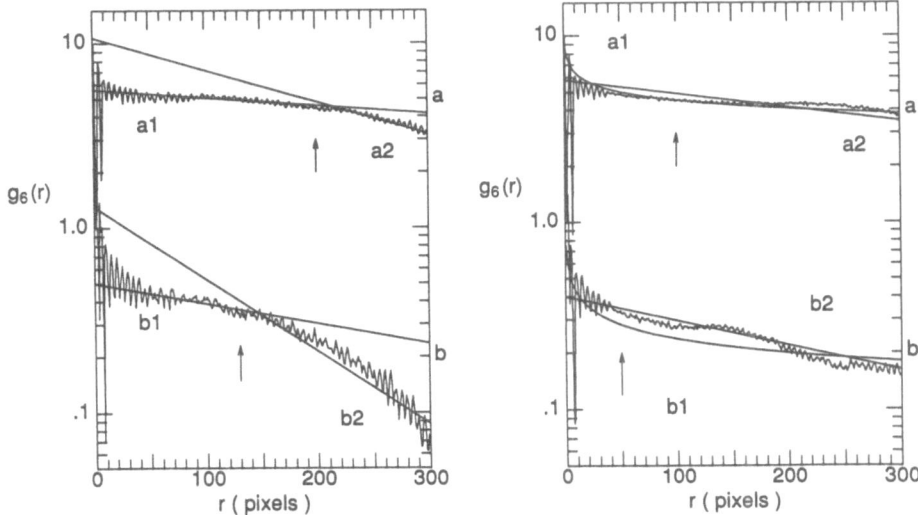

FIGURE 4.33. Orientational correlation functions for the Murray et al. second run in the intermediate region, shifted upward for clarity. The arrow next to each curve marks the translational correlation length for that density. Left: three-dimensional curves, with best fits to exponential decay for $r < \xi$ and for $r > \xi$. (a) $n = 0.0587$; curve $a1$ has decay length $\xi_6 = 4\xi$, and $a2$ has $\xi_6 = \xi$. (b) $n = 0.0578$; curve $b1$ has $\xi_6 = 3.2\xi$, and $b2$ has $\xi_6 = 0.89\xi$. Right: two-dimensional curves, with best fits to both exponential and algebraic decay for all $r$. (a) $n = 0.0569$; $a1$ has $\eta_6 = 0.15$, and $a2$ has $\xi_6 = 5.9\xi$. (b) $n = 0.0563$; $b1$ has $\eta_6 = 0.25$, and $b2$ has $\xi_6 = 6.7\xi$.

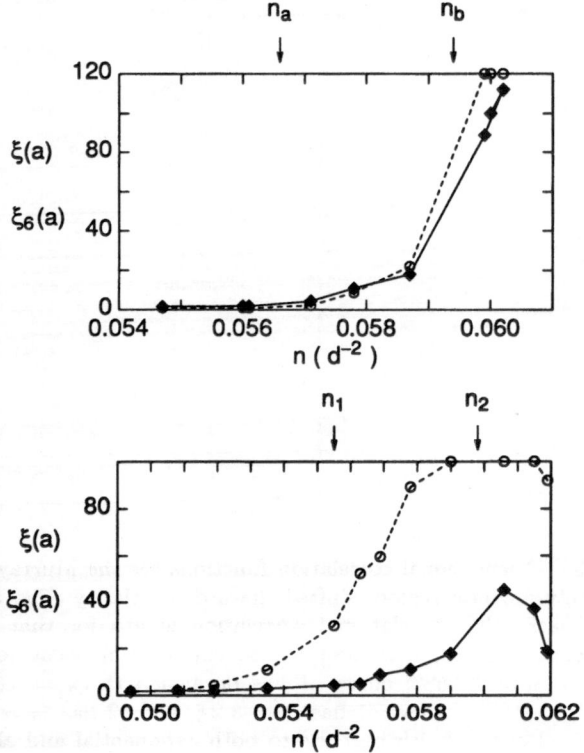

FIGURE 4.34. Translational (solid lines) and bond-orientational (dashed lines) correlation lengths for the Murray et al. second experiment. Top: three-dimensional run. Bottom: two-dimensional run.

# 4.3   Conclusions and Suggestions for Future Work

## 4.3.1   Equilibration and System Size

I believe that a long equilibration time ($\sim 10^7$ collisions) and possibly a large system ($\sim 10^4$ particles) are essential to obtain a reproducible equilibrium melting transition in a two-dimensional system. Insufficient equilibration may be one of the reasons that hysteresis has been observed in many simulations; they typically have two to three orders-of-magnitude less equilibration [17]. The use of periodic boundary conditions in small systems may also effectively drive the transition more first-order [69]. In recent years, simulations of over 10,000 particles interacting with hard and soft disks [53] and Lennard–Jones [70] potentials find system size dependences in orientational order very close to melting. In addition, Zollweg et al. [53] found a topological defect density growing logarithmically with system size, for both hard disks and particles interacting with repulsive $r^{-12}$ potentials, in agreement with Toxvaerd's [69] findings in his simulations of smaller Lennard–Jones systems. Zollweg et al. [53] find that equilibrium times appear to be growing with system size and possibly diverging very close to melting.

The colloidal crystal experiments discussed in this chapter all satisfy the system size requirement, basically that the size exceeds the area $\sim \xi^2$. However, a number of the experiments did not allow sufficient time to "equilibrate" the dislocations of the system. It is in these experiments in which a clumping of defects is observed, much like the results of the computer simulations. For the experiments in which sufficient time is allowed for defect equilibration, or time is allowed for a dislocation to climb a correlation length of the system right at melting, both the correlation functions and statistical defect maps of the system are consistent with KTHNY predictions.

## 4.3.2   Possibility for Studying Driven Nonequilibrium Phase Transitions

Not all real systems are in equilibrium. Important examples of nonequilibrium systems are glasses and splat-cooled amorphous metals. An advantage of using what I like to call "analog" molecular dynamics simulations such as colloidal crystals to model real atomic systems is the wide range of conditions it is possible to study with them. For example, one can study in detail the configuration of all the particles in the same system both in and driven far out of equilibrium with the colloids, while this is very difficult for a typical molecular dynamics experiment with current computation speeds available. This advantage has not yet been adequately taken advantage of in experiments, but I predict that the modeling of driven nonequilibrium systems will be of great interest in the future.

### 4.3.3    Relevant Energy Scales

In the colloidal crystal experiments studying melting of two-dimensional systems discussed in this chapter, insufficient attention has been placed on the experimental measurement of the important energy scales of the problem: $Kk_BT$, the interaction energy of two dislocations, and $E_c k_B T$, the energy required to create a dislocation core. It would be useful to find systems in which the ratio of these two energies is vastly different and can be controlled. In this way, one could obtain from experiments the critical value that may separate KTHNY behavior from first-order melting, as has been done for the Laplacian roughening model [11]. One possible interesting experiment along this line of thought is to compare the melting transition of two layers of colloidal crystal between parallel plates, which occurs with two very different symmetries: triangular and square [43], which should be nearly identical in all respects except for the dislocation core energy.

### 4.3.4    Other Predictions of KTHNY

So far, none of the colloidal crystal experiments have looked at other predictions of the KTHNY theory, e.g., the scaling of the shear modulus or specific heat or exponents of the system with temperature near the two transitions. No experiment has yet studied the collective dynamics, such as sound waves and possible hexatic modes [71], that should exist if the transition were consistent with KTHNY. This is difficult for the colloidal systems as the particles are macroscopic and the modes of the system are mostly overdamped. Some clever experiments should be devised to look into these questions.

### 4.3.5    Substrates and Their Effects

The effect of underlying substrates on the melting transition has also not yet been adequately pursued. Colloidal crystal experiments are almost ideal for such a study. Indeed, this was the primary reason that I became interested in studying two-dimensional melting. I predict that a number of beautiful experiments will be done in the near future. Colloidal crystal imaging experiments will also be very useful for the study of such interesting nonequilibrium situations as two-dimensional domain growth on a substrate and for modeling three-dimensional crystal growth from the melt.

## 4.3.6   Other Analog Molecular Dynamics Experiments

Other systems that should be taken advantage of for "analog" molecular dynamics direct-imaging experiments are "soft" spheres, or sterically stabilized monodisperse polymer particles mentioned previously in Section 4.1 and monodisperse systems of colloidal particles of different shapes, such as ellipsoids, to model liquid crystal systems. Some systems have been utilized already to study bond-orientational order in two dimensions, but are beyond the scope of this chapter. These are Bitter imaging patterns of flux lattices in high $T_c$ superconductors [72] and magnetic bubble domains in garnets [73]. I predict that many more systems will be used in the future to elucidate a large range of many-body phenomena.

# References

[1]  J.M. Kosterlitz and D.J. Thouless, *J. Phys. C* **6**, 1181 (1973).

[2]  D.R. Nelson, in *Phase Transitions and Critical Phenomena*, Vol. 7, edited by C. Domb and J.L. Leibowitz, Academic Press, London, 1983, p. 1.

[3]  N.D. Mermin and H. Wagner, *Phys. Rev. Lett.* **17**, 1133 (1966).

[4]  N.D. Mermin, *Rev. Mod. Phys.* **51**, 591 (1979).

[5]  B.I. Halperin and D.R. Nelson, *Phys. Rev. Lett.* **41**, 121 (1978); D.R. Nelson and B.I. Halperin, *Phys. Rev. B* **19**, 2547 (1979).

[6]  F.P. Preparata and M.I. Shamos, *Computational Geometry, an Introduction*, Springer-Verlag, New York, 1985.

[7]  A.P. Young, *Phys. Rev. B* **19**, 1855 (1979).

[8]  G. Aeppli and R. Bruinsma, *Phys. Rev. Lett.* **22**, 2133 (1984).

[9]  S.T. Chui, *Phys. Rev. B* **28**, 167 (1983).

[10]  Y. Saito, *Phys. Rev. B* **26**, 6239 (1982).

[11]  K.J. Strandburg, S.A. Solla, and G.V. Chester, *Phys. Rev. B* **28**, 2717 (1983).

[12]  H. Kleinert, *Phys. Lett.* **95A**, 381 (1983).

[13]  B. Joos and M.S. Duesbery, *Phys. Rev. B* **33**, 8632 (1986).

[14]  D.S. Fisher, B.I. Halperin, and R. Morf, *Phys. Rev. B* **20**, 4692 (1979).

[15] A.J.C. Ladd and W.G. Hoover, *Phys. Rev. B* **26**, 5469 (1982).

[16] T.V. Ramakrishnan, *Phys. Rev. Lett.* **41**, 541 (1982); T.V. Ramakrishnan and M. Yussouff, *Phys. Rev. B* **19**, 2775 (1979).

[17] K.T. Strandburg, *Rev. Mod. Phys.* **60**, 161 (1988).

[18] R.S. Pindak, D.E. Moncton, S.C. Davey, and J.W. Goodby, *Phys. Rev. Lett.* **46**, 1135 (1981).

[19] M. Cheng, J.T. Ho, S.W. Hui, and R.S. Pindak, *Phys. Rev. Lett.* **61**, 550 (1988).

[20] D.R. Nelson and J. Toner, *Phys. Rev. B* **24**, 363 (1981).

[21] W.B. Russel, D.A. Saville, and W.R. Schowalter, *Colloidal Dispersions*, Cambridge University Press, Cambridge, 1989.

[22] A review of the use of colloidal spheres to study condensed matter physics is given in P. Pieranski, *Contemp. Phys.* **24**, 25 (1983).

[23] A. Homola and R.O. James, *J. Coll. and Int. Sci.* **59**, 123 (1977).

[24] For example, manufactured by the Dow Chemical Corporation, Midland, Mich.

[25] L. Antl, J.W. Goodwin, R.D. Hill, R.H. Ottewill, S.M. Owens, S. Papworth, and J. Waters, *J. Coll. Surf.* **17**, 67 (1986).

[26] P.N. Pusey and W. van Megan, in *Physics of Complex and Supramolecular Fluids, Exxon Monographs*, edited by S.A. Safran and N.A. Clark, Wiley, New York, 1987, p. 673; P.N. Pusey and W. van Megan, *Phys. Rev. Lett.* **59**, 2083 (1987).

[27] A.T. Skjeltorp, *Phys. Rev. Lett.* **51**, 2306 (1983).

[28] C.A. Murray, W.O. Sprenger, and R.A. Wenk, *J. Phys. Cond. Matt.* **2**, SA385 (1990).

[29] C.A. Murray and D.H. Van Winkle, *Phys. Rev. Lett.* **58**, 1200 (1987).

[30] D.R. Nelson, M. Rubinstein, and F. Spaepen, *Phil. Mag. A* **46**, 105 (1982).

[31] E.M. Chudnovsky, *Phys. Rev. B* **40**, 11355 (1989).

[32] J.Z. Larese, L. Passell, A.D. Heidemann, D. Richter, and J.P. Wicksted, *Phys. Rev. Lett.* **61**, 432 (1988).

[33] W. Hess and R. Klein, *Advances in Phys.* **32**, 173 (1983).

[34] E.B. Sirota, H.D. Ou-Yang, S.K. Sinha, P.M. Chaiken, J.D. Axe, and Y. Fujii, *Phys. Rev. Lett.* **62**, 1287 (1989).

[35] Y. Monovoukas and A.P. Gast, *J. Coll. and Int. Sci.* **128**, 533 (1989).

[36] M.O. Robbin, K. Kremer, and G. Grest, *J. Chem. Phys.* **88**, 3286 (1988).

[37] E.J.W. Verwey and T.G. Overbeek, *Theory of the Stability of Lyophobic Colloids*, Elsevier, New York, 1948.

[38] S. Alexander, P.M. Chaikin, P. Grant, G.J. Morales, P. Pincus, and D. Hone, *J. Chem. Phys.* **80**, 5776 (1984).

[39] M. Baus and J.P. Hansen, *Phys. Rep.* **59**, 1 (1980).

[40] A good review is found in J.P. Hansen and J.P. McDonald, *Theory of Simple Liquids*, Academic Press, London, 1976.

[41] C.A. Murray, D.H. Van Winkle, and R.A. Wenk, *Phase Trans.* **21**, 93 (1990).

[42] P. Pieranski, L. Strzlecki, and B. Pansu, *Phys. Rev. Lett.* **50**, 900 (1983); B. Pansu, P. Pieranski, and L. Strzelecki, *J. Phys.* **44**, 531 (1983).

[43] D.H. Van Winkle and C.A. Murray, *Phys. Rev. A* **34**, 562 (1986); D.H. Van Winkle and C.A. Murray, *J. Chem. Phys.* **89**, 3885 (1988).

[44] N.A. Clark, B.J. Ackerson, and A.J. Hurd, *Phys. Rev. Lett.* **50**, 1459 (1983).

[45] D. deFontaine, K.A. Jackson, and C.E. Miller, *Am. J. Phys.* **37**, 789 (1969).

[46] R.K. Iler, *The Chemistry of Silica*, Wiley, New York, 1979.

[47] D. Hone and E. Chang, *Bull. Am. Phys. Soc.* **33**, 668 (1988).

[48] R. Evans, *Microscopic Theories of Simple Fluids and Their Interfaces*, in *Les Houches Session XLVIII, 1988, Liquids at Interfaces*, edited by J. Charolin, J.F. Joanny, and J. Zinn-Justin, Elsevier, Amsterdam, 1989, Chap. 1 and references therein; R. Evans and M.B. Marconi, *J. Chem. Phys.* **86**, 7138 (1987).

[49] C.A. Murray and R.A. Wenk, *Phys. Rev. Lett.* **62**, 1643 (1989).

[50] J.P. Hansen and J.P. McDonald, *Theory of Simple Liquids*, Academic Press, London, 1976, Chap. 7.

[51] C.A. Murray, W.O. Sprenger, and R.A. Wenk, (unpublished).

[52] A.F. Bakker, C. Bruin, and H.J. Hilhorst, *Phys. Rev. Lett.* **52**, 449 (1984).

[53] J.A. Zollweg, G.V. Chester, and P.W. Leung, *Phys. Rev. B* **39**, 9518 (1989).

[54] R. Collins, in *Phase Transitions and Critical Phenomena*, Vol. 2, edited by C. Domb and M.S. Green, Academic Press, London, 1972, p. 271.

[55] P. Choquard and J. Cleroin, *Phys. Rev. Lett.* **50**, 2086 (1983); B.J. Alder, D.M. Ceperley, and E.L. Pollock, *Int. J. Quantum Chem. Sym.* **16**, 49 (1982).

[56] C. Udink and D. Frenkel, *Phys. Rev. B* **35**, 6933 (1987).

[57] J.D. Weeks, *Phys. Rev. B* **24**, 1530 (1981).

[58] F.R.N. Nabarro, *Theory of Crystal Dislocations*, Oxford University Press, London, 1967.

[59] P. Pieranski, *Phys. Rev. Lett.* **45**, 569 (1980).

[60] J.C. Earnshaw, *J. Phys. D.* **19**, 1863 (1986); A.J. Hurd, *J. Phys. A* **18**, L1055 (1985).

[61] D.Y.C. Chan, J.D. Henry, Jr., and L.R. White, *J. Coll. and Int. Sci.* **79**, 410 (1981).

[62] A.J. Armstrong, R.C. Mockler, and W.J. O'Sullivan, *J. Phys. Cond. Matt.* **1**, 1707 (1989).

[63] Y. Tang, A.J. Armstrong, R.C. Mockler, and W.J. O'Sullivan, *Phys. Rev. Lett.* **62**, 2401 (1989).

[64] Y. Tang, A.J. Armstrong, R.C. Mockler, and W.J. O'Sullivan, *Phase Trans.* **21**, 75 (1990).

[65] S.W. Charles and R.E. Rosensweig, *J. Magn. Mat.* **39**, 190 (1983).

[66] A.T. Skjeltorp, *J. Magn. Mat.* **65**, 195 (1987).

[67] G. Helgesen and A.T. Skjeltorp, *Physica A* **170**, 488 (1991).

[68] C.A. Murray, W.O. Sprenger, and R.A. Wenk, *Phys. Rev. B* **42**, 688 (1990).

[69] S. Toxvaerd, *Phys. Rev. A* **24**, 2735 (1981).

[70] C. Udink and J. van der Elskin, *Phys. Rev. B* **35**, 279 (1987).

[71] A. Zippelius, B.I. Halperin, and D.R. Nelson, *Phys. Rev. B* **22**, 2514 (1980).

[72] C.A. Murray, P.L. Gammel, D.J. Bishop, D.B. Mitzi, and A. Kapitulnik, *Phys. Rev. Lett.* **64**, 2312 (1990); D.G. Grier, C.A. Murray, C.A. Bolle, P.L. Gammel, and D.J. Bishop, D.B. Mitzi, and A. Kapitulnik, *Phys. Rev. Lett.* **66**, 2270 (1991).

[73] R. Sheshadri and R. Westervelt, *Phys. Rev. Lett.* **66**, 2774 (1991).

# 5

# Faceting in Bond-Oriented Glasses and Quasicrystals

*Tin-Lun Ho*

## 5.1 Introduction

### 5.1.1 What Have Quasicrystals Brought Us?

If there is any important lesson to be learned from the development of quasicrystals [1], it would be that one should be suspicious of conventional wisdoms. The field of quasicrystals is full of paradox. Nearly all its major developments are marked by puzzles questioning current wisdom. There have been many occasions upon which we have thought we understood the nature of quasicrystals, but were soon confronted with contradicting phenomena. In fact, up to this day, the real physical mechanism for the formation of quasicrystals remains a mystery. On the other hand, through the study of quasicrystals, we have both widened and deepened our understanding of many fundamental concepts and processes in condensed matter physics.

The recent success in generalizing many symmetry-related concepts in conventional solid-state physics [2], and the recent efforts in understanding quasicrystal growth [3], are good examples of the interesting ideas and challenging questions that quasicrystals "force" us to consider. The former revises the conventional symmetry classification of solid matter to include symmetries that are supposed to be forbidden. The latter explores the mechanism of solid growth to understand why nature manages to achieve with great ease many seemingly impossible structures. Often, the fundamental issues of the subject go beyond its own boundaries. For example, the growth of quasicrystals must be closely related to the growth of crystals with large unit cells. (I submit that it is impossible to understand one without understanding the other.) Quasicrystal facets are phenomena of a similar type. We are forced to look at the problem with a broader perspective. As we shall see, the problem of quasicrystals faceting inevitably brings up a number of fundamental issues in quasicrystal and crystal growth.

## 5.1.2    The Problem of Quasicrystal Facets

The discovery of large (micron-size) quasicrystal facets came at the time when two popular models for quasicrystals were in heated competition—the perfect tiling model and the glass model [1]. (A detailed comparison of these two models can be found in the previous chapter.) In the perfect tiling model, the underlying structure has both perfect bond-orientational order and long-range (quasiperiodic) position order. These structures can also be viewed as a projection of a slab of a six-dimensional crystal into three-dimensional space [4]. This model has become very popular not only because it gives a good account for the observed diffraction patterns, but also because it is conceptually very attractive. It circumvents the classical theorem forbidding icosahedral symmetry in crystals by simply replacing periodicity by quasiperiodicity. The model, however, has a problem. It requires atomic units to match each other in a manner so specific that it seems impossible to achieve with *realistic* atomic interactions. There is so far no solution to this "problem of formation."

The experimental challenge to the perfect tiling model comes from the X-ray measurements [5], which show that many real quasicrystals are not so perfect after all. In fact, before 1989, all quasicrystals studied through X-ray measurements were found to be glassy. Their correlation lengths deduced from the width of the X-ray peaks are of the order of a few hundred Å's. This feature appears to be so common that it has been referred to as "universal disorder" of quasicrystals. At about the same time when these X-ray experiments were performed, Stephens and Goldman [6] pointed out that a random collection of icosahedral units can also produce diffraction patterns similar to the observed ones. Their model received immediate attention, for it seems more natural for the atomic units to form a random aggregate than to satisfy the very specific tiling matching rules. It was then a competition between two time-honored traditions in physics: the pursuit of the beautiful versus faith in the natural.

Large quasicrystal facets were first discovered in Al-Li-Cu [7], which is also known to have a correlation length of a few hundred Å's. For a while this discovery was regarded as evidence for the perfect tiling model [8] because facets are commonly viewed as signs of long-range positional order (LRPO), originating from the regularity of the lattice planes. If quasicrystals were disordered over a few hundred Å's, it would be highly unlikely for them to maintain coherence on the micron scale, and thus no facets would be expected. On the contrary, the appearance of facets would be a natural consequence of the description of quasicrystals as perfect tilings, for it is known that they have LRPO. However, the X-ray results do show a disordered structure. We are therefore facing a paradox. Either the X-ray measurements failed to reveal the positional order of QCs (which can be quickly dismissed by recalling the impeccable foundation of X-ray diffraction), or the conventional wisdom on the relation between facets and LRPO fails to convey the truth.

The state of affairs took an unexpected turn around March 1989 when several groups reported the discoveries of a number of essentially perfect quasicrystals [9]. The X-ray diffraction peaks of these quasicrystals are instrumentally sharp, implying a micron-sized lower bound on the correlation length. The diffraction patterns of these quasicrystals also have a body-center-cubic (BCC) Bravais lattice in $k$-space, unlike the simple cubic type previously found. All these quasicrystals are faceted. Although the existence of facets in these systems is hardly surprising because they possess LRPO, the existence of these perfect quasicrystals themselves is totally unexpected. At the time of seemingly no escape from "universal disorder," nature once again surprised us with a perfect beauty which was commonly thought to be impossible. Not only do these observations dismiss the generality of the "universal" disorder, they have also elevated the problem of quasicrystal growth into a serious paradox, fervently demanding a solution. (Recently there have been interesting attempts [3] to demonstrate at least in principle the possibility of growing perfect quasicrystals. However, the proposed growth rules are not yet realistic, nor do the resulting growth shapes reflect the symmetry of the system.)

The purpose of this chapter is to reexamine the origin of facet formation, to discuss the roles of positional order and bond-orientational order, and to show that under certain conditions positionally disordered systems can develop equilibrium facets. Let me say at the outset that the problem of facet formation in disordered systems is *not* completely solved. Our demonstration illustrates a theoretical possibility rather than representing the real structure of disordered quasicrystals. There are still interesting and important issues to be settled, and they are (I believe) ultimately connected to quasicrystal growth. What we present here is a sufficient condition for facet formation and a demonstration of the realization of this condition [10, 11]. Even though we cannot show that this condition is satisfied by real disordered quasicrystals, the form of this condition and the resulting surface energies are sufficiently simple and natural that one cannot resist believing that they carry a certain degree of truth.

(Recently, there have been many discussions on surface roughening of quasicrystals [12–14]. We shall not include them here because these studies focus on the quasiperiodicity rather than bond-orientational order or spatial disorder of the system.)

## 5.1.3    Organization of This Chapter

The organization of this chapter is as follows. Section 5.2 is a brief discussion of the experimental findings concerning quasicrystal facets. Section 5.3 reviews the conventional view of facet formation. We shall present Landua's explanation and begin to question the role of LRPO. The famous

Wulff construction is also briefly reviewed. In Section 5.4, we discuss facet formation in systems with perfect orientational order. We begin by reexamining the origins of facet formation. The reexamination immediately leads to a weak sufficient condition for facet formuation that has no apparent relation to LRPO. These studies imply that disordered systems with perfect bond-orientational order can have equilibrium facets, provided they satisfy this sufficient condition. Next, we study the equilibrium shapes implied by this condition, using a very useful but not yet popularized reformulation of the Wulff construction (the Herring algorithm). The equilibrium shape of the simplest bond-oriented system with icosahedral symmetry turns out to be a triacontahedron. In Section 5.5, we construct tiling models for perfect and disordered quasicrystals in order to study the effects of disorder on the surface energy. For these models, the sufficient condition discussed in Section 5.4 is always satisfied. The surface energy of the perfect quasicrystal remains unchanged even when a substantial amount of spatial (but not orientational) disorder is introduced into the system. These examples show that, at least in principle, facets can exist in disordered systems. In Section 5.6, we discuss and further explore the reasons for faceting in disordered systems. We shall end with discussions of the effects of "orientation preserving" topological defects, and the relation between facet formation and quasicrystal growth.

## 5.2   Quasicrystal Facets

In recent years, very large quasicrystal facets have been observed in Al-Li-Cu [7], Al-Fe-Cu [15], and Ga-Mg-Zn quasicrystals [16]. These quasicrystals are distinct from previous ones in that they are grown from the melt by slow cooling, are much larger in size, and appear to be stable. The cooling rates can be $10^6$ times slower than those in rapid quenching. The grain size of Al-Li-Cu ranges from 0.1 to 1.0 mm. That of Al-Fe-Cu can go up to 2.0 cm.

All these quasicrystals have beautiful gain shapes. The grain shape of Al-Li-Cu QC is a triacontahedron (see Figure 5.1). Those of Al-Fe-Cu and Ga-Mg-Zn are dodecahedrons (see Figure 5.2). The Al-Li-Cu quasicrystals appear to be disordered. Their X-ray peak widths imply a correlation length of about a few hundred Å's, which is two to three orders of magnitude smaller than the observed facet size. Only very recently has it been found that the Al-Fe-Cu quasicrystal is essential perfect, with at least micron-size correlation length.

Facets were also observed in decagonal quasicrystals when they were discovered [17]. Decagonal quasicrystals are periodic stackings of two-dimensional quasiperiodic structures [17,18]. Their diffraction patterns have tenfold symmetry along the stacking direction, but exhibit crystalline symmetry in the other two perpendicular directions. Until very recently, the only

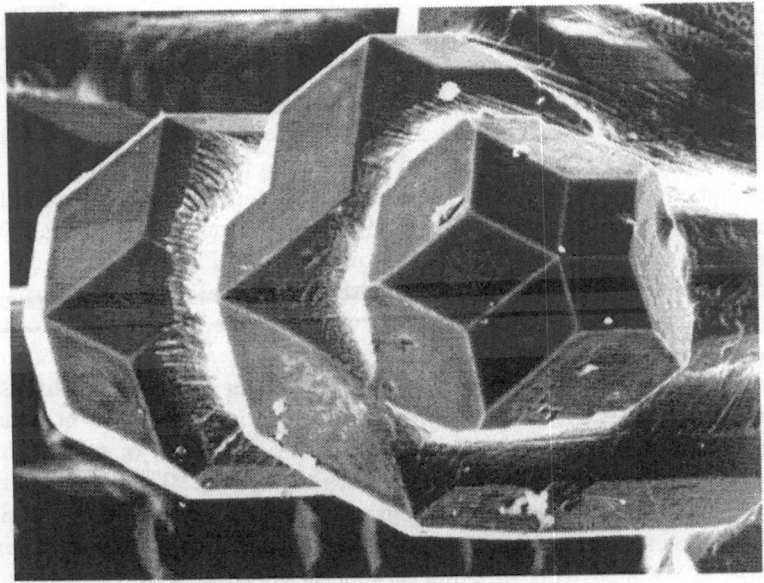

FIGURE 5.1. The grain shape of Al-Li-Cu icosahedral quasicrystal. The observed facets are the faces of a triacontahedron. (Courtesy of Frank W. Gayle, Reynolds Metals Co.)

decagonal quasicrystal for which a faceted grain shape had been reported was the Al-Mn decagonal quasicrystal. The grain shape is of the form of a decagonal prism. The symmetry axis is along the stacking direction which is also the fast growth direction. The Al-Mn decagonal quasicrystal is highly disordered, with correlation length comparable to or shorter than its icosahedral counterpart. Very recently, large stable decagonal quasicrystal has also been found in Al-Cu-Co [19]. The grain shape and fast growth direction are identical to those of the Al-Mn decagonal quasicrystal (see Figure 5.3). However, the correlation length is believed to be a great deal larger [19].

# 5.3    The Conventional View of Facet Formation

As early as the 16th century [20], crystal facets were thought to be manifestations of the lattice planes of the system. Since, in modern terms, lattice planes are a consequence of long-range position order (LRPO) which manifests itself as a set of $\delta$-function peaks in the diffraction pattern, it is generally thought that the existence of facets is a direct consequence of

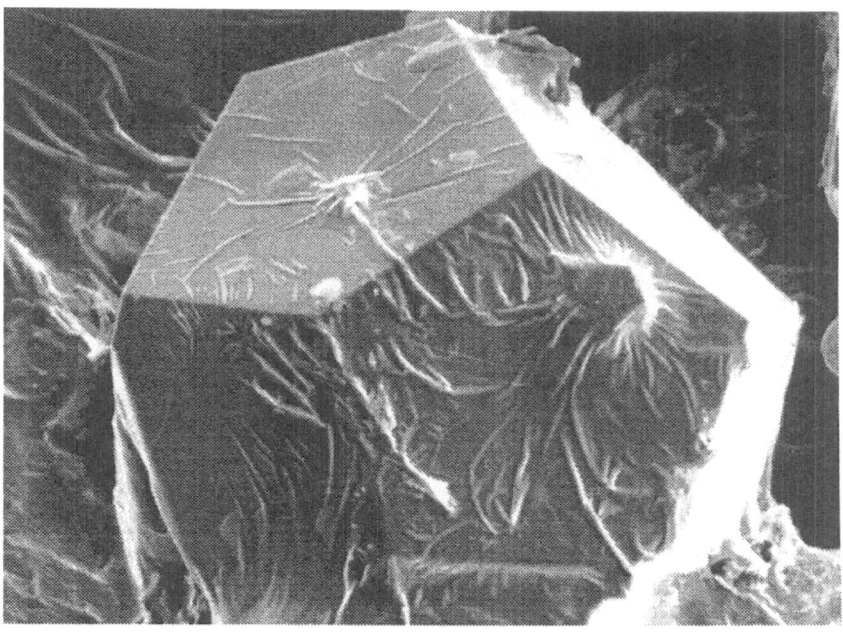

FIGURE 5.2. The dodecahedral grain shape observed in Al-Fe-Cu icosahedral quasicrystal. (Courtesy of Frank W. Gayle, Reynolds Metals Co.)

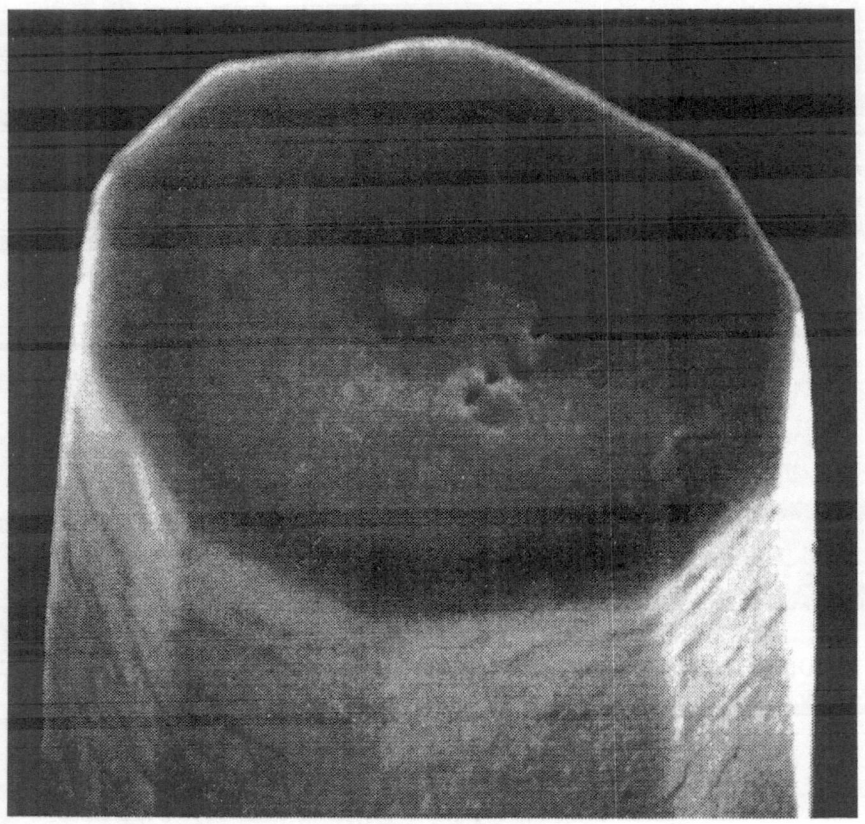

FIGURE 5.3. The grain shape of Al-Cu-Co decagonal quasicrystal: a decagonal prism. (Courtesy of A.R. Kortan, AT&T Laboratories.)

LRPO.

In this section, we shall review briefly some general knowledge of facet formation in crystals and begin to question the role of LRPO. Before proceeding, we would like to emphasize that despite the slow cooling rates, the observed quasicrystal grain shapes are growth shapes rather than equilibrium shapes, as indicated by their dendritic morphology. Although thermodynamic equilibrium is notoriously difficult to achieve in crystal growth [21], it is commonly believed that growth shapes are closely related to the $T = 0$ equilibrium shapes [22]. In particular, it is expected that the facets in the growth shape are the large facets of the $T = 0$ equilibrium shape. This can be explained pictorially as follows. Suppose the equilibrium shape of a crystal consists of a small facet and two bigger facets as shown in Figure 5.4. It is clear that smaller facets must have higher surface energies, as the total surface energy must be a minimum in equilibrium. One can imagine that the surface energy arises from a set of "dangling bonds" emerging from the surface. Surfaces with higher surface energy will have more dangling bonds per unit area and therefore have more ability to attract additional atoms. As a result, smaller facets will grow faster than larger ones. They will either shrink in size or grow out of existence, leaving the larger facets behind (see Figure 5.4).

In the following, we review the well-developed theory of the $T = 0$ equilibrium shape of crystals [21] in order to set up some necessary background. By comparison, the theory of growth shapes is much less well developed because of the complicated kinetic factors in growth processes. We focus on the $T = 0$ shape mainly for simplicity, bearing in mind the aforementioned relation between growth and equilibrium facets.

### 5.3.1  The Determination of the Equilibrium Shapes of Solids: The Wulff Construction

The equilibrium shape of a solid at temperature $T$ is determined by minimizing the surface free energy subject to the constraint of constant volume. The minimization procedure can be expressed as

$$\delta \int dS \left[ \gamma(\hat{\mathbf{n}}) - \lambda \hat{\mathbf{n}} \cdot \mathbf{r} \right] = 0, \tag{5.1}$$

where $\gamma(\hat{\mathbf{n}})$ is the surface free energy per unit area of a surface element $dS$ normal to $\hat{\mathbf{n}}$, $\lambda$ is a Lagrange multiplier, and $\mathbf{r}$ denotes the position of the surface element $dS$. The variation of the surface shape is performed by varying the element $\hat{\mathbf{n}} \, dS$. The equilibrium shape is given by

$$\lambda r(\hat{\mathbf{r}}) = \min_{\hat{\mathbf{n}}} \frac{\gamma(\hat{\mathbf{n}})}{\hat{\mathbf{n}} \cdot \hat{\mathbf{r}}}. \tag{5.2}$$

This minimization procedure was reformulated by Wulff into an elegant algorithm, known as the Wulff construction: (1) For each direction $\hat{\mathbf{n}}$, rep-

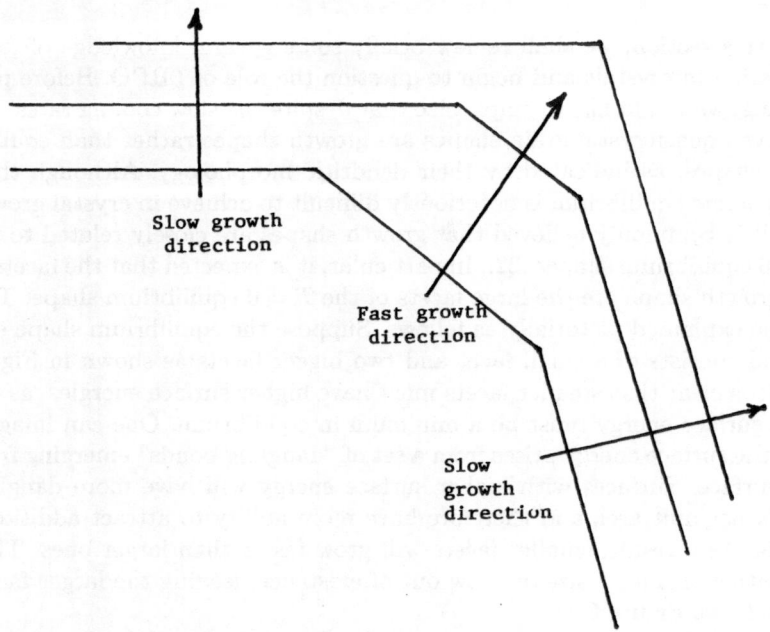

FIGURE 5.4. The shrinking of small equilibrium facets in growth processes: The smaller facet has higher energy. It tends to attract more atoms and therefore grows faster than the larger facets. In growth processes, it will shrink in size or grow out of existence.

resent the energy $\gamma(\hat{n})$ by a radial vector of length $\gamma(\hat{n})$ along $\hat{n}$. Repeating the process for all $\hat{n}$, one obtains a representation of $\gamma(\hat{n})$ as a surface in angular space, known as the "polar plot." (See Figure 5.5 for a two-dimensional version of this construction.) (2) At each point on the polar plot, construct a plane (called the stationary plane) normal to the radius vector $\hat{n}$ (represented by straight lines in Figure 5.5). The inner envelope of all these planes is the equilibrium shape of the solid.

The equivalence between the Wulff construction and the minimizing procedure is simply that along each direction $\hat{r}$, there is a family of stationary planes $\lambda \mathbf{r} \cdot \hat{n} = \gamma(\hat{n})$ (labeled by $\hat{n}$). Choosing the inner envelope in the Wulff construction amounts to selecting the lowest energy solution within this family of stationary planes for each direction $\hat{r}$ (see Figure 5.5).

Although the Wulff construction gives an intuitive representation of the minimizing procedure, it is not yet practically useful. In a later section, we shall present a reformulation of the Wulff construction (Herring algorithm) which is highly efficient for a large class of surface energies.

Before proceeding, there is a conceptual point which needs clarification. Since we shall be discussing equilibrium facets in disordered systems, it appears that we have implicitly assumed that the *true* equilibrium state of the system is disordered. This, however, is not necessarily the case. Even if the disordered state were not in thermodynamic equilibrium, surface equilibrium can still be achieved if the relaxation time of the bulk is much longer than that of the surface. In that case, the minimizing procedure that we just discussed is still applicable.

## 5.3.2   The Origin of Facets: Cusps in the Surface Energy

It is clear from the Wulff construction that smooth surface free energies $\gamma(\hat{n})$ can only lead to smooth equilibrium shapes. Faceted equilibrium shapes can only result from sharp angular variations (i.e., nonanalyticity) of the surface free energy.

It was pointed out by Landau that at $T = 0$ the surface energies of crystals have (linear) cusps [23]. It is then a simple application of the Wulff construction to show that the equilibrium shape is indeed faceted [21,23]. Landau's derivation relies heavily on the characteristics of the crystalline structure, leaving the strong impression that LRPO is essential for facet formation. In the following, we shall first repeat Landau's derivation. We then present another equally elementary derivation which gives the same answer but makes no use of LRPO or the existence of lattice planes.

### Landau's Derivation

Consider a two-dimensional crystal with a Bravais lattice as shown in Figure 5.6(a). Suppose the atoms only interact with nearest neighbors with

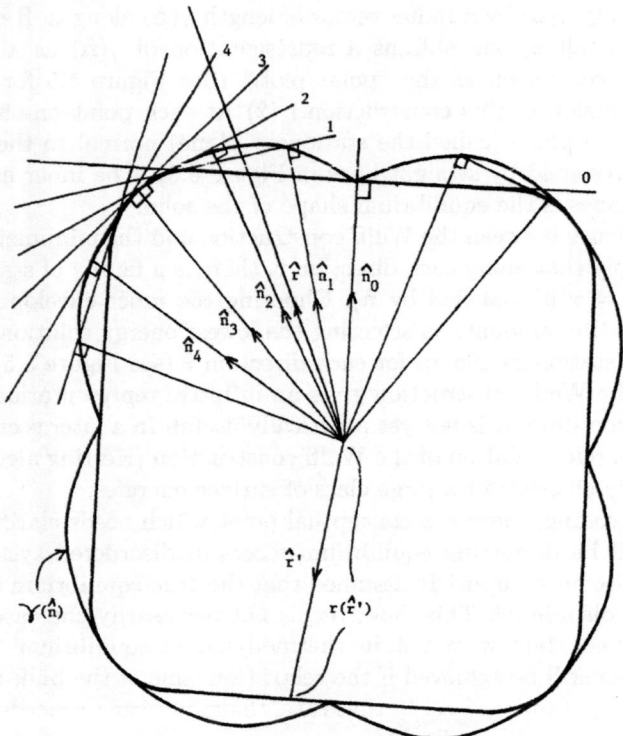

FIGURE 5.5. The Wulff construction for a two-dimensional solid: The outer solid
line represents the surface energy $\gamma(\hat{n})$. The stationary planes normal to $\hat{n}_i$, $i = 0$
to 4 are labeled $0, 1, \ldots, 4$. To find the equilibrium shape $r(\hat{r})$, we note that for
a given direction $\hat{r}$, by changing the radius $r(\hat{r})$, we sweep through an (infinite)
family of stationary planes. Here, we have only displayed the stationary planes
$0, 1, \ldots, 4$. Choosing the lowest-energy stationary plane (i.e., plane 0 for the direction $\hat{r}$) amounts to choosing the inner envelope produced by the stationary
planes, which is equilibrium shape and is represented by the inner solid curve in
the figure.

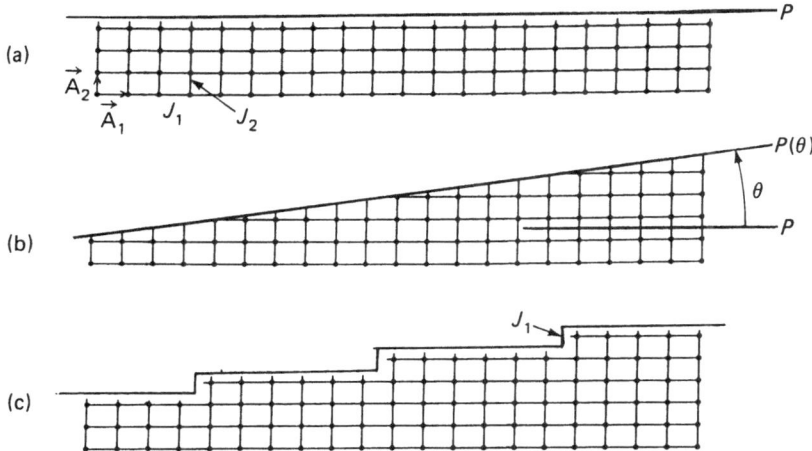

FIGURE 5.6. The surface energy of a lattice plane as a function of its orientation. (a) A stable surface $P$ of a crystal with a rectangular unit cell: The surface energy is the number of dangling bonds, represented as short solid lines attached to the surface atoms. (b) A tilted surface $P(\theta)$ and the corresponding cut bonds. (c) The surface $P(\theta)$ is made up of a sequence of original surfaces $P$ separated by steps with energy $J_1$.

interaction represented by bonds. The bond strengths for atoms separated by $\mathbf{A}_1$ and $\mathbf{A}_2$ are denoted as $J_1$ and $J_2$, respectively. It is easy to see that the plane $P$ parallel to $\mathbf{A}_1$ is a low-energy surface. It only cuts through the $J_2$ bonds and has surface energy $J_2/A_1$. A plane $P(\theta)$ making an angle $\theta$ with $P_1$ is composed of a sequence of equilibrium planes $P_1$ separated by "steps" cutting an additional $J_1$ bond [see Figure 5.6(b) and (c)]. It is geometrically obvious that for small $\theta$'s, the energy increase (i.e., the number of $J_1$ bonds cut) is linearly proportional to the total step height, and that the planes $P(\theta)$ and $P(-\theta)$ are degenerate. Hence, we have

$$\gamma(\theta) = \alpha J_1 |\theta| + \beta, \qquad (5.3)$$

where $\alpha$ and $\beta$ are constants.

### Alternate Derivation

Let $\sigma(\hat{\mathbf{n}})$ be a macroscopically flat surface with an average normal $\hat{\mathbf{n}}$ as shown in Figure 5.7. (One can simply choose $\sigma$ as a plane if one does not like it wobbly.) To calculate the surface energy (i.e., the number of cut bonds per unit area), note that all the $J_1$ cut bonds must originate from atoms inside the strip $\Omega[\sigma(\mathbf{A}_1, \hat{\mathbf{n}})]$, defined as the volume swept through by $\sigma(\hat{\mathbf{n}})$ as it is translated to $-\mathbf{A}_1$ (see Figure 5.7). If we denote the density of pairs of atoms connected by $\mathbf{A}_i$ as $\nu_i$, $i = 1, 2$, the total number of $J_1$ bonds cut is therefore $J_1 \nu_1 [|\hat{\mathbf{n}} \cdot \mathbf{A}_1 | \, (area \perp \hat{\mathbf{n}})]$, where the quantity inside

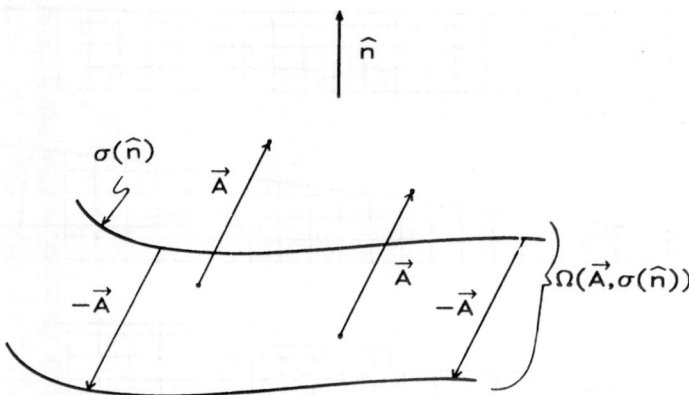

FIGURE 5.7. A simple way of counting the number of dangling bonds of an identical type (say $J$) emerging from a surface $\sigma(\hat{n})$: Simply slide $\sigma(\hat{n})$ along $-\mathbf{A}$, where $\mathbf{A}$ is the vector connecting a pair of atoms with bond $J$ and count the number of such pairs inside the swept volume $\Omega[\mathbf{A}, \sigma(\hat{n})]$.

the square bracket [...] is the strip volume $\Omega[\sigma(\hat{n}), \mathbf{A}_1]$. Hence, we have

$$\gamma(\hat{n}) = \sum_i J_i \nu_i |\hat{n} \cdot \mathbf{A}_i|. \qquad (5.4)$$

It is trivial to see that Eq. (5.4) agrees with Eq. (5.3) when $\hat{n}$ is almost perpendicular to $\mathbf{A}_1$. We now note that an expression of the form $|\hat{n} \cdot \mathbf{A}|$ represents a cusp.

The significance of this alternative derivation is that neither LRPO nor the existence of lattice planes have been invoked. There are, however, two key assumptions in arriving at Eq. (5.4): (1) The system has perfect bond-orientational order (defined more precisely in the next section), (2) the densities of neighboring pairs (connected by $\mathbf{A}_i$) in the strip $\Omega$ are given by their bulk value $\nu_i$. Although (2) seems reasonable and is certainly true for the example we considered, it need not be true in general. The reason is that the strip $\Omega$ is of atomic thickness. Condition (2) therefore requires essentially zero fluctuation in the pair density over atomic distances, a condition which may not always be met. In the next section, we shall give a more precise formulation of this condition.

## 5.4    Faceting in Perfect Bond-Oriented Systems

From now on, we shall only consider "perfect bond-oriented systems" (PBOSs), where neighboring atoms are attached to each other only along

a specified (finite) set of directions. It is clear that crystals, perfect quasicrystals, and icosahedral glasses (discussed in the previous chapter) are
PBOSs. It is also clear that PBOSs may or may not possess LRPO. The
precise definition of a PBOS is: Within the range of interaction $A^*$, the
pair correlation function of PBOSs

$$\nu(\mathbf{r}; \mathbf{A}) \equiv \rho(\mathbf{r})\rho(\mathbf{r} + \mathbf{A}), \tag{5.5}$$

where $\rho(\mathbf{r})$ is the density, is of the form

$$\nu(\mathbf{r}; \mathbf{A}) = \sum_\tau \delta(\mathbf{A} - \mathbf{A}_\tau)h(\mathbf{r}, \mathbf{A}_\tau), \qquad A < A^*, \tag{5.6}$$

where $\{\mathbf{A}_\tau\}$ is a set of specified vectors. The value of $h(\mathbf{r}, \mathbf{A}_\tau)$ is either 0 or
1, representing the appearance of a neighbor at $\mathbf{A}_\tau$. Equation (5.6) means
that any two atoms in the system are connected by a sequence of vectors, all
in the set $\{\mathbf{A}_\tau\}$. Without further specification for the occurrence function
$h$, there is no guarantee that the system will have LRPO. The existence of
LRPO means that the density Fourier transform is a set of $\delta$-functions,

$$\rho(\mathbf{k}) = \sum_\mathbf{G} \rho_\mathbf{G}\delta(\mathbf{k} - \mathbf{G}). \tag{5.7}$$

The diffraction peak at $\mathbf{G}$ reflects the fact that the atoms are arranged in
planes perpendicular to $\mathbf{G}$.

## 5.4.1   Sufficient Conditions for Facet Formation

We now explore the conditions under which PBOSs can have facets. We
shall show that perfect bond-orientational order as we have just defined it
is "almost" sufficient to produce facets. The additional condition required
for sufficiency turns out to be rather weak. The energy associated with a
surface $\sigma(\hat{\mathbf{n}})$ normal to $\hat{\mathbf{n}}$ dividing the system into two semiinfinite pieces
is

$$E_s[\sigma(\hat{\mathbf{n}})] = \tfrac{1}{2} \int d\mathbf{r}_> \, d\mathbf{r}_< \rho(\mathbf{r}_>)\rho(\mathbf{r}_<)J(\mathbf{r}_> - \mathbf{r}_<), \tag{5.8}$$

where $\mathbf{r}_>$ and $\mathbf{r}_<$ denote different sides of $\sigma(\hat{\mathbf{n}})$, and $J(\mathbf{r})$ is the interaction
between two atoms separated by $\mathbf{r}$. The factor $\tfrac{1}{2}$ indicates that the energy
is shared by both pieces of semiinfinite solid. Equation (5.8) can be written
as

$$E_s[\sigma(\hat{\mathbf{n}})] = \tfrac{1}{2} \int d\mathbf{A} \, J(\mathbf{A}) \int_{\Omega(\mathbf{A},\sigma)} d\mathbf{r} \, \nu(\mathbf{r}, \mathbf{A}), \tag{5.9}$$

where $\Omega[\mathbf{A}, \sigma(\hat{\mathbf{n}})]$ is the volume swept through by $\sigma$ as it is translated
by $-\mathbf{A}$ (see Figure 5.7). For PBOSs, Eq. (5.6) implies that this integral
becomes a discrete sum,

$$E_s[\sigma(\hat{\mathbf{n}})] = \tfrac{1}{2} \sum_\tau J(\mathbf{A}_\tau)\langle h\rangle_{\Omega_\tau}|\mathbf{A}_\tau \cdot \hat{\mathbf{n}}|S[\sigma(\hat{\mathbf{n}})], \tag{5.10}$$

where $\langle h \rangle_{\Omega_\tau}$ is the average of the pair correlation $h(\mathbf{r}, \mathbf{A}_\tau)$ over the strip $\Omega[\mathbf{A}_\tau, \sigma(\hat{\mathbf{n}})]$,

$$\langle h \rangle_{\Omega_\tau} \equiv \int_{\Omega_\tau} dV \, h(\mathbf{r}, \mathbf{A}_\tau) \Big/ \int_{\Omega_\tau} dV, \tag{5.11}$$

and $S[\sigma(\hat{\mathbf{n}})]$ is the projected area of $\sigma$ along $\hat{\mathbf{n}}$. The surface energy entering the Wulff construction is therefore

$$\gamma(\hat{\mathbf{n}}) = \min_{\sigma(\hat{\mathbf{n}})} E_s[\sigma(\hat{\mathbf{n}})]/S[\sigma(\hat{\mathbf{n}})] = \sum_\tau |\mathbf{A}_\tau \cdot \hat{\mathbf{n}}| J(\mathbf{A}_\tau) h^*(\hat{\mathbf{n}}, \mathbf{A}_\tau), \tag{5.12}$$

where the minimization is performed over the set of macroscopically flat surfaces normal to $\hat{\mathbf{n}}$, $h^*$ is the strip average of $h$ [Eq. (5.11)] for the optimum surface $\sigma^*(\hat{\mathbf{n}})$, and the factor $\frac{1}{2}$ has been absorbed into $J$ for convenience.

It is clear from Eq. (5.12) that a sufficient condition for facet formation is that the average pair correlation $h^*$ is smooth in $\hat{\mathbf{n}}$. In that case, $\gamma(\hat{\mathbf{n}})$ will have a set of cusps at $\{\mathbf{A}_\tau\}$. Of course, sharp angular variations in $h^*$ can generate new cusps.

The sufficient condition of smooth $h^*$ takes on a particularly simple form if the average $\langle h \rangle$ is independent of the surface $\sigma(\hat{\mathbf{n}}, \mathbf{A}_\tau)$ (i.e., independent of the shape and location of $\sigma$). In this case, it is easily seen that the strip average is identical to the bulk average,

$$\langle h \rangle_{\Omega_\tau} = h^*(\hat{\mathbf{n}}, \mathbf{A}_\tau) \equiv h_B(\mathbf{A}_\tau), \tag{5.13}$$

a condition which we shall refer to as the "simple sufficient condition." The surface energies of systems satisfying the simple sufficient condition can be written as

$$\gamma(\hat{\mathbf{n}}) = \sum_\tau J(\hat{\mathbf{A}}_\tau)|\hat{\mathbf{n}} \cdot \hat{\mathbf{A}}_\tau|, \tag{5.14}$$

$$J(\hat{\mathbf{A}}_\tau) \equiv \sum_{|\mathbf{A}_\tau|} J(\mathbf{A}_\tau) h_B(\mathbf{A}_\tau)|\mathbf{A}_\tau|. \tag{5.15}$$

If the system has a point group symmetry, the vectors $\mathbf{A}_\tau$ can further be grouped into symmetric stars. The energy Eq. (5.14) becomes

$$\gamma(\hat{\mathbf{n}}) = \sum_\alpha g_\alpha \sum_\tau |\hat{\mathbf{A}}_\tau^\alpha \cdot \hat{\mathbf{n}}|, \tag{5.16}$$

where $g_\alpha$ is the energy of the symmetric star $\{\mathbf{A}_\tau^\alpha\}$. In case there is only one symmetric star, Eq. (5.16) reduces to the particularly simple form

$$\gamma(\hat{\mathbf{n}}) = g \sum_\tau |\hat{\mathbf{A}}_\tau \cdot \hat{\mathbf{n}}|. \tag{5.17}$$

The main message is that PBOSs satisfying the simple sufficient condition are guaranteed to have cusps and therefore facets. What is not clear is whether this condition is, in fact, satisfied by disordered systems. In Section 5.5, we construct disordered quasicrystal models to demonstrate that this is, in fact, possible, thereby demonstrating the existence of equilibrium facets in spatially disordered systems. On the other hand, the validity of Eq. (5.14) definitely goes beyond such specific constructions. The derivation of Eq. (5.14) also suggests that this form of the surface energy is insensitive to LRPO.

If one were only interested in the $T = 0$ shape of PBOSs with icosahedral symmetry rather than concerned with the compatibility between facets and disorder, one could proceed from here by applying the Wulff construction to Eq. (5.16) for various icosahedral stars. The rest of this section is devoted to these studies.

## 5.4.2   The Herring Algorithm

The surface energies we study in this chapter [Eqs. (5.14), (5.16), (5.17)] are all of the form

$$\gamma(\hat{\mathbf{n}}) = \sum_\beta |\hat{\mathbf{n}} \cdot \mathbf{B}_\beta|, \qquad (5.18)$$

where $\{\mathbf{B}_\beta\}$ is a finite set of vectors not necessarily of the same length. We shall from now on refer to the $\mathbf{B}_\beta$ as "bonding vectors." For energies of the form Eq. (5.18), one can make use of an observation of Herring [24] to recast the Wulff construction in a more efficient form, which I shall call the "Herring algorithm." For reasons which are quite unclear, this method never makes its way into textbooks or the literature (except very recently [11,25]), though I suspect it is known to many. The algorithm is: (1) Represent all possible directions $\hat{\mathbf{n}}$ by the surface of a unit sphere $S_{\hat{\mathbf{n}}}$, i.e., each point on $S_{\hat{\mathbf{n}}}$ corresponds to a direction $\hat{\mathbf{n}}$ [see Figure 5.8(a)]. (2) For each direction $\mathbf{B}_\beta$, construct a plane $\mathbf{r} \cdot \mathbf{B}_\beta = 0$ passing through the origin normal to $\mathbf{B}_\beta$. The intersection between this plane and $S_{\hat{\mathbf{n}}}$ is a great circle on $S_{\hat{\mathbf{n}}}$. The great circles corresponding to all $\mathbf{B}_\beta$ divide the unit sphere into angular regions [see Figure 5.8(b)]. Within each angular region, the surface energy $\gamma(\hat{\mathbf{n}})$ takes the form

$$\gamma(\hat{\mathbf{n}}) = \hat{\mathbf{n}} \cdot \mathbf{D}, \qquad (5.19)$$

$$\mathbf{D} \equiv \sum_\beta \epsilon_\beta \mathbf{B}_\beta, \qquad (5.20)$$

where $\epsilon_\beta$ is $+1$ or $-1$, depending on whether the angular region is on one side or the other of the plane $\mathbf{r} \cdot \mathbf{B}_\beta = 0$. The vectors $\mathbf{D}$ are referred to as "diameter" vectors. The equilibrium shape is simply a convex polygon with the diameter vectors as polyhedral vertices [see Figure 5.8(c)]. (See also Figure 5.9 for a two-dimensional analog.)

The equivalence between the Herring algorithm and the Wulff construction can be seen as follows. In each angular region, the surface energy is of

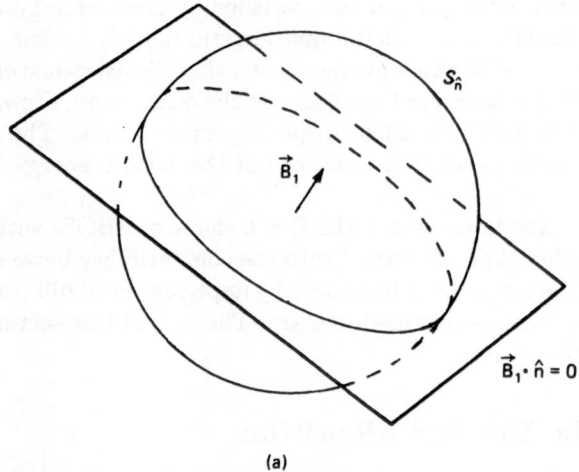

(a)

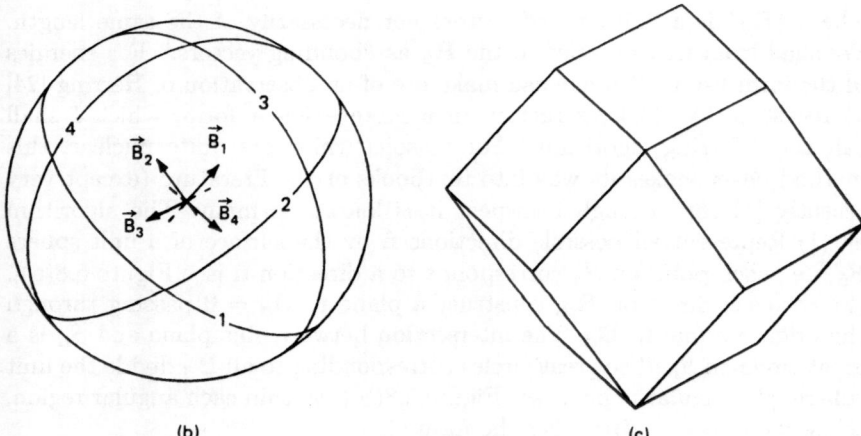

(b)                                        (c)

FIGURE 5.8. An illustration of the Herring algorithm: (a) The plane $\hat{n} \cdot \mathbf{B}_1 = 0$ cutting the unit sphere $S_{\hat{n}}$ at a great circle. (b) The great circles produced by all the normal planes $\hat{n} \cdot \mathbf{B}_i = 0$, $i = 1$ to 4, and the resulting angular regions. This is an example of a BCC crystal with one atom per unit cell. It has four bonding vectors. (c) The equilibrium shape obtained by connecting the diameter vectors (not shown) of neighboring angular regions.

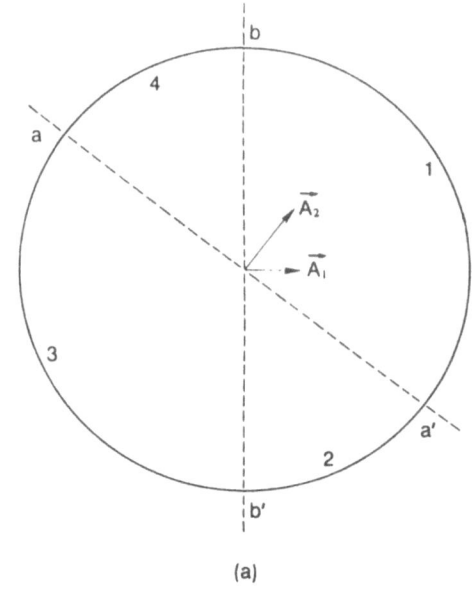

(a)

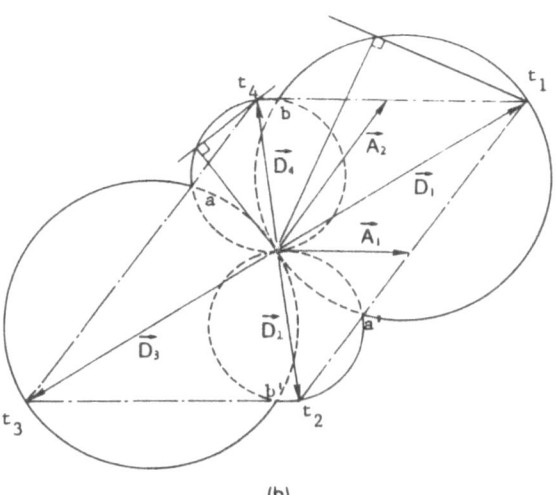

(b)

FIGURE 5.9. The construction of surface energy (or polar plot) and the application of the Herring algorithm for a two-dimensional crystal with two bonding vectors $\mathbf{A}_1$ and $\mathbf{A}_2$. (a) In two dimensions, the unit sphere $S_{\hat{\mathbf{n}}}$ reduces to a circle. The planes $\hat{\mathbf{n}} \cdot \mathbf{A}_1 = 0$ and $\hat{\mathbf{n}} \cdot \mathbf{A}_2 = 0$ divide the unit circle into four angular regions: 1, 2, 3, and 4. The great circles in Fig. 8.5(b) now become pairs of diametrically opposite points on the circle: $(a, a')$ and $(b, b')$. (b) The surface energy in angular region $i$ is a part of a circle that passes through the origin. For example, the surface energies in regions 1 and 2 are represented by the arcs $(a', t_1, b)$, and $(b', t_2, a')$, respectively, with corresponding diameter vectors $\mathbf{D}_1$ and $\mathbf{D}_2$. The equilibrium shape is obtained by joining neighboring diameter vectors, which is the parallelogram $t_1 t_2 t_3 t_4$.

the form of Eq. (5.19). As a function $\hat{n}$, this is just the surface of a sphere with $\mathbf{D}$ as a diameter vector joining antipodal points. The entire polar plot is therefore made up of different portions of spherical surfaces. (See Figure 5.9 for a two-dimensional analog.) The intersection of two spheres produces a groove. The intersection of three spheres produces a cusp. Within each angular region, all stationary planes in the Wulff construction are of the form (up to a scale factor) $\mathbf{r} \cdot \hat{n} = \gamma(\hat{n}) = \hat{n} \cdot \mathbf{D}$, implying that all stationary planes including those in the inner envelop (i.e., the equilibrium shape) must pass through $\mathbf{D}$. In other words, the intersection of three (or more) facets inside an angular region must appear in $\mathbf{D}$.

The Herring algorithm has the following immediate consequences:

(1) There is a one-to-one correspondence between the facets and the intersections of the great circles. [It is also easy to see from Figure 5.9(b) that the intersections of great circles are cusps of the surface energy.]

(2) Since the vertices of the $T = 0$ shape are the diameter vectors $\mathbf{D}$, the edges of the $T = 0$ shape are given by $\mathbf{D}' - \mathbf{D}$, where $\mathbf{D}'$ and $\mathbf{D}$ are diameter vectors of neighboring regions. If these two regions are separated by the plane $\hat{n} \cdot \mathbf{B}_\beta = 0$, then the vector $\mathbf{D}' - \mathbf{D}$ is simply $2\mathbf{B}_\beta$. This means that all the edges of the $T = 0$ shapes are given by the set $\{2\mathbf{B}_\beta\}$.

### 5.4.3    The Equilibrium Shape of Simple Bond-Oriented Systems with Icosahedral Symmetry

In this subsection we study the equilibrium shape implied by Eq. (5.17) with the bonding vectors $\{\mathbf{A}_\tau\}$ forming a single icosahedral star. The simplest icosahedral stars are the sets of vertex, edge, and face vectors of an icosahedron. These are vectors pointing from the center of the icosahedron to the vertices, along the edges of an icosahedron, and pointing from the center of the icosahedron to the center of the faces, respectively.

The equilibrium shapes in these cases are shown in Figures 5.10(a), (b), and (c). The equilibrium shape of the vertex set (constructed by the Herring algorithm) turns out to be a triacontahedron [Figure 5.10(a)], which is the simpliest icosahedral equilibrium shape in terms of least number of facets. The equilibrium shape of the edge set is a great rhombicosadodecahedron [(Figure 5.10(b)]. The fact that the facets of this polyheron along the fivefold axes are much larger than those along the two- and threefold axes means that the fivefold facets will grow at the expense of the two- and threefold ones in growth processes. If all the smaller facets disappear completely in the growth process, the growth shape will be a dodecahedron. [There is no established nomenclature for the polyhedron in Figure 5.10(c).]

For surface energies of the form of Eq. (5.16), the Herring algorithm has an interesting implication; namely, if the interactions are purely attractive (i.e., all $g_\beta > 0$), the equilibrium shape can be neither an icosahedron nor a dodecahedron. They can only be growth shapes. To see this, we first note that the icosahedron and dodecahedron are dual to each other, i.e., the vertices of one polyhedron correspond to the faces of the other and vice versa (see Figure 5.11). If the equilibrium shape were a dodecahedron or an icosahedron, then according to consequence (2) of the Herring algorithm, (see previous section), the faces (or facets) of these polyhedra must be in one-one correspondence with *all* the intersections of a set of great circles. However, such a set does not exist. If one tries to construct a set of great circles such that their intersections account for all the vertices of an icosahedron or a dodacahedron, one immediately discovers that there are always excess intersections not accounted for in the original set of icosahedron or dodacahedron vertices. It is also easy to see that increasing the number of stars merely increases further the number of great circles and hence the number of intersections, which certainly does not help to generate an icosahedral or dodecahedral equilibrium shape.

Recently, Ingersent and Steinhardt [25] pointed out that if symmetric stars with repulsive interactions ($g < 0$) are included in the surface energy, dodecahedral and icosahedral equilibrium shapes can result.

## 5.5    Equilibrium Shapes of Perfect and Random Quasicrystals

In the previous section, we have established a simple sufficient condition for facet formation. Although it seems reasonable that such a condition can be satisfied by glassy systems, one still has to come up with an explicit example. In this section, we use tiling models to study the surface energies of both perfect and disordered quasicrystals. We show that the sufficient condition is, in fact, satisfied in all these cases. These tiling models therefore demonstrate the possibility of equilibrium facets in disordered systems.

### 5.5.1    Tiling Model of Perfect and Disordered Quasicrystals

Years before the discovery of quasicrystals, deBruijn had come up with an ingenious way of generating aperiodic tilings [26] which turned out to be (in today's terms) a perfect quasicrystal. This method is now known as the dual method. It can be shown to be equivalent to projecting a slab of higher-dimensional crystal into a lower-dimensional space [27]. The method

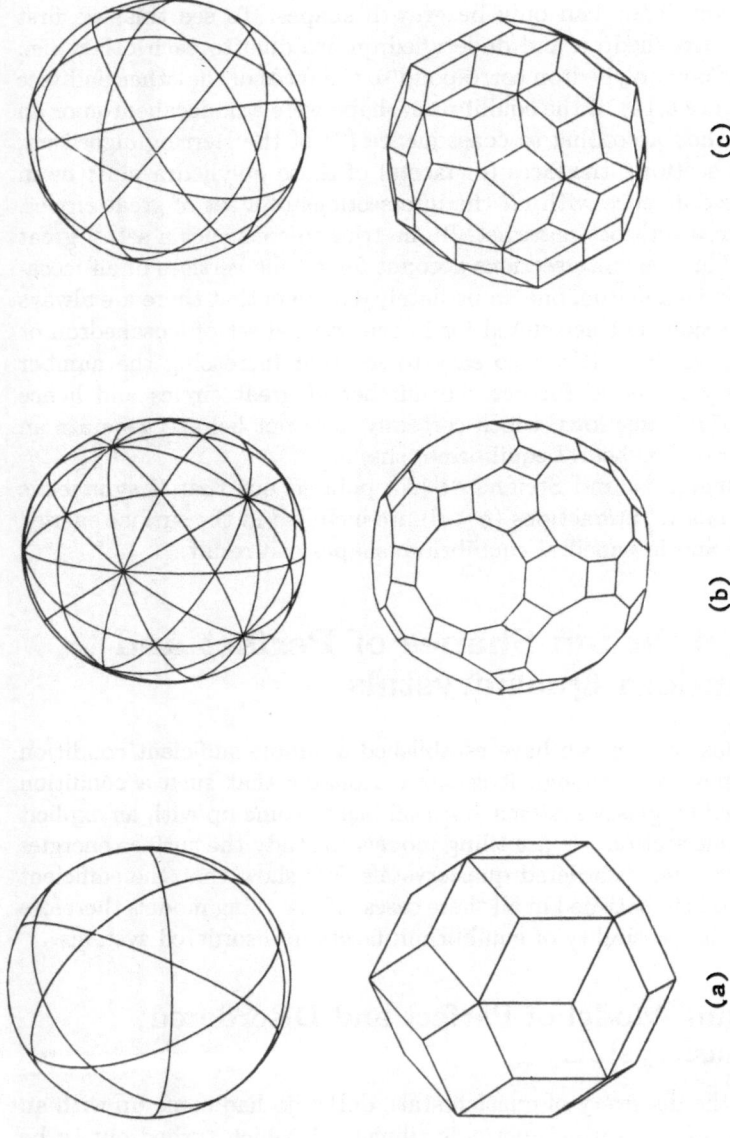

(a)

(b)

(c)

FIGURE 5.10. (a) through (c) The equilibrium shapes and the corresponding angular regions in the Wulff construction of PBOSs, with bonding vectors forming an icosahedral symmetric star that is along the five-, two-, and threefold axis of an icosahedron. Figure 5.9(a) is a triacontahedron, Fig. 5.9(b) is a great rhomicosadodecahedron.

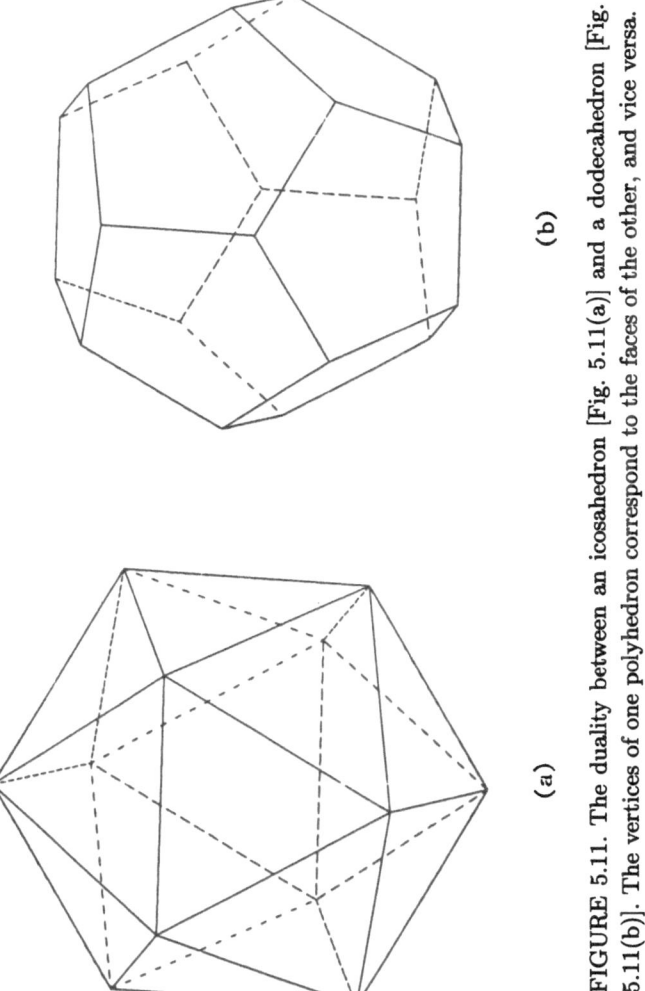

(b)

(a)

FIGURE 5.11. The duality between an icosahedron [Fig. 5.11(a)] and a dodecahedron [Fig. 5.11(b)]. The vertices of one polyhedron correspond to the faces of the other, and vice versa.

is easily generalized to arbitrary dimensions and arbitrary symmetry. For simplicity, we shall only consider the two-dimensional tilings with decagonal symmetry. The algorithm of the dual method is as follows:

(1) We start with a pentagonal star $(\mathbf{k}_1, \mathbf{k}_2, \ldots, \mathbf{k}_5)$, which is a set of vectors separated from each other successively by $2\pi/5$. We then construct in a two-dimensional "dual" space five sets of equally spaces lines, each normal to a star vector [see Figure 5.12(a)]. (The phase of each set specifying its position is arbitrary.) The superposition of all these lines, called a grid, partitions the dual space into polygonal regions.

(2) To construct a tiling, we pick an arbitrary polygonal cell in the grid and an arbitrary point in a second plane (the "physical" plane), which we shall say "corresponds" to the cell in the grid. Now we march around the grid. Everytime we cross a boundary to a new cell, we draw a line in the physical space of length $a$ along the direction of the outward normal from the old cell to the new one. The tip of this vector will be the point in the physical space *corresponding* to the new cell. It is easy to see that the structure we end up with, after every side shared by every pair of neighboring cells has been crossed at least once, does not depend on how we march through the grid [see Figures 5.12(a) and (b)]. The resulting tiling is referred to as a perfect quasicrystal tiling. It can be shown that this tiling has LRPO, i.e., its diffraction pattern is a set of $\delta$-functions [27].

The dual method has the following important features which are very useful in deriving the expression of the surface energy: (1) As we circle around an intersection of two grid lines (say, normal to $\mathbf{k}_i$ and $\mathbf{k}_j$), a rhombus (or tile) with edges parallel to $\mathbf{k}_i$ and $\mathbf{k}_j$ is generated in the physical space. Thus, we say every grid intersection *corresponds* to a tile. It is easy to see that there are only two types of tiles (with areas $|\mathbf{k}_1 \times \mathbf{k}_2|$ and $|\mathbf{k}_1 \times \mathbf{k}_3|$, respectively. (2) As we travel along a grid line in dual space, (say, normal to $\mathbf{k}_i$), we encounter a sequence of intersections between this line and the other four sets of lines. These intersections will induce in the physical space a sequence of tiles packed together along the edge parallel to $\mathbf{k}_i$. This sequence of tiles will be referred to as a track [see Figure 5.12(b)]. It is the regular arrangement of these tracks in the perfect quasicrystal that leads to LRPO.

To generate a disordered quasicrystal, one can randomly flip the tiles of a perfect tiling. A "flipping" of the tiles is shown in Figure 5.13, which corresponds to slightly distorting the grid lines in dual space. As pointed out by a number of authors [28], the LRPO of a quasicrystal is harder to destroy than one would have expected. A typical random tile flipping might not produce enough fluctuation to destroy the $\delta$-function peaks in the diffraction pattern. The explanation for such persistence of LRPO requires the use of the so-called "projection method" which is described in the chapter by A. Goldman. (See also [4,27,28].) Qualitatively speaking, a typical random tile flipping will result in a configuration in which the re-

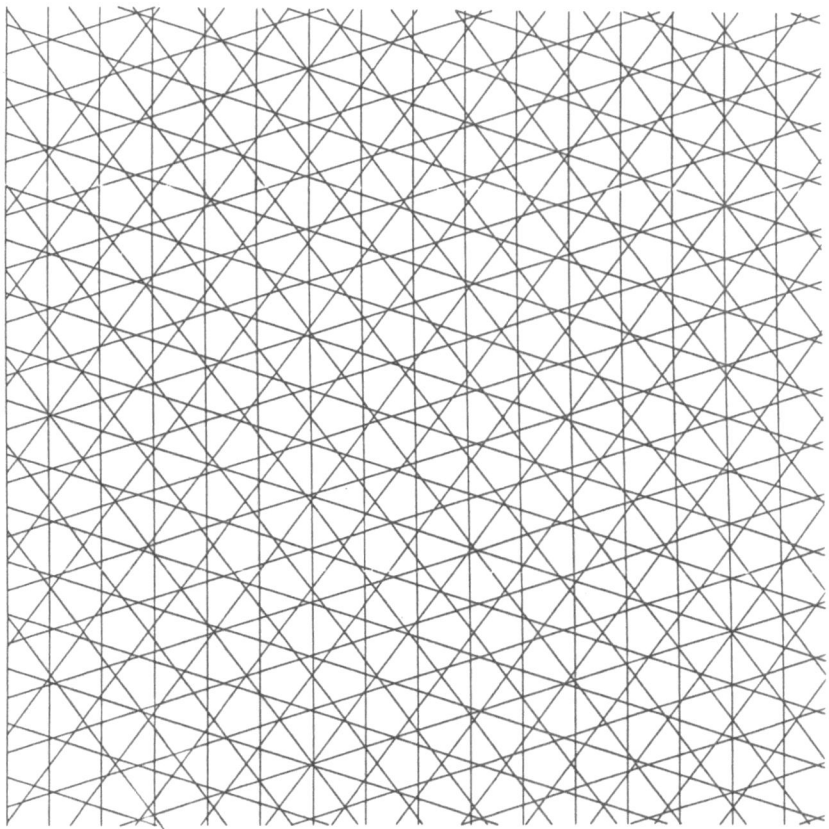

FIGURE 5.12. (a). The pentagrid in dual space in the dual construction.

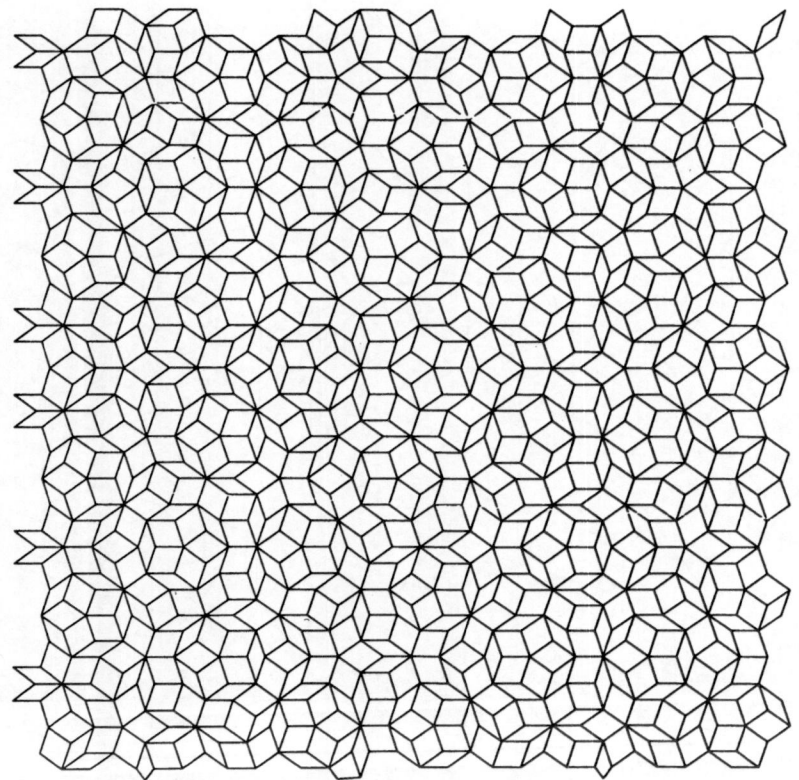

FIGURE 5.12. (b). The perfect quasicrystal corresponding to the pentagrid in Fig. 5.12(a).

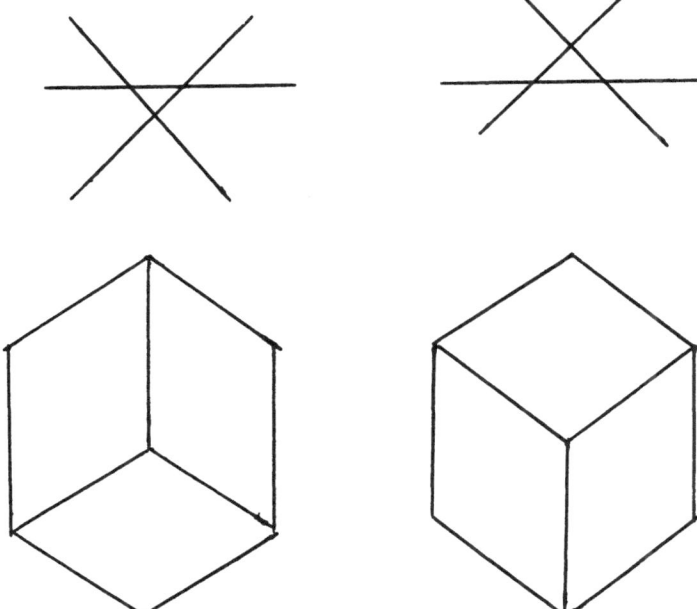

FIGURE 5.13. An illustration of the tile flipping process: It corresponds to a shift of the grid lines in dual space.

sulting distorted tracks are not too far away from their original positions, and as a result, there is still a strong presence of the original LRPO. One way to destroy LRPO is to produce a large excursion of the original tracks. In Refs. 10 and 11, the conventional dual method is modified such that the lines in each set are randomly rather than evenly spaced [see Figures 5.14(a), (b), and (c)]. In this way, the total distance in dual space between two lines which are very far apart (say, line 1 and line $N$, $N >> 1$) is simply the original distance plus a quantity which is determined by a random walk process, which can be very large. (It increases as $\sqrt{N}$.) It is shown in Ref. 11 that the diffraction peaks of these tilings all acquire widths, signifying the destruction of LRPO. These tilings are referred to as "quasiglasses." In the next subsection, we show that the surface energy of the system is insensitive to this randomizing process. As a result, the original facets remain, despite the destruction of LRPO.

## 5.5.2  The Surface Energy of Perfect and Disordered Quasicrystals

In order to study the equilibrium shape, we have to introduce energetics into the system. We can consider each tile to be a group of atoms and

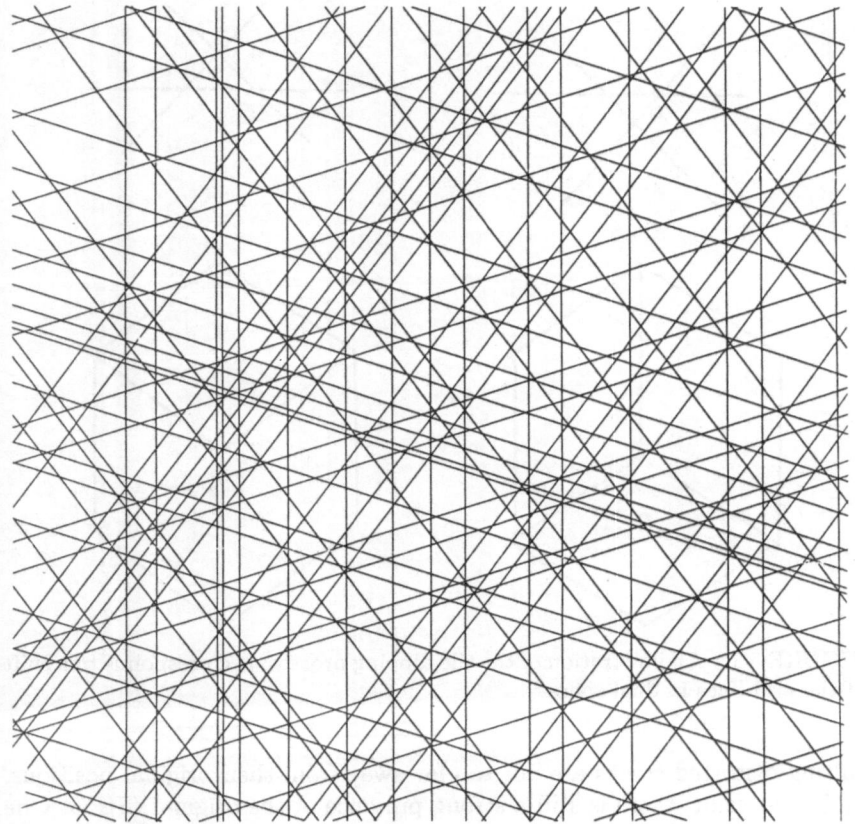

FIGURE 5.14. (a). A randomized pentagrid: The spacing between the grid lines in the pentagrid in Fig. 5.12(a) is randomized.

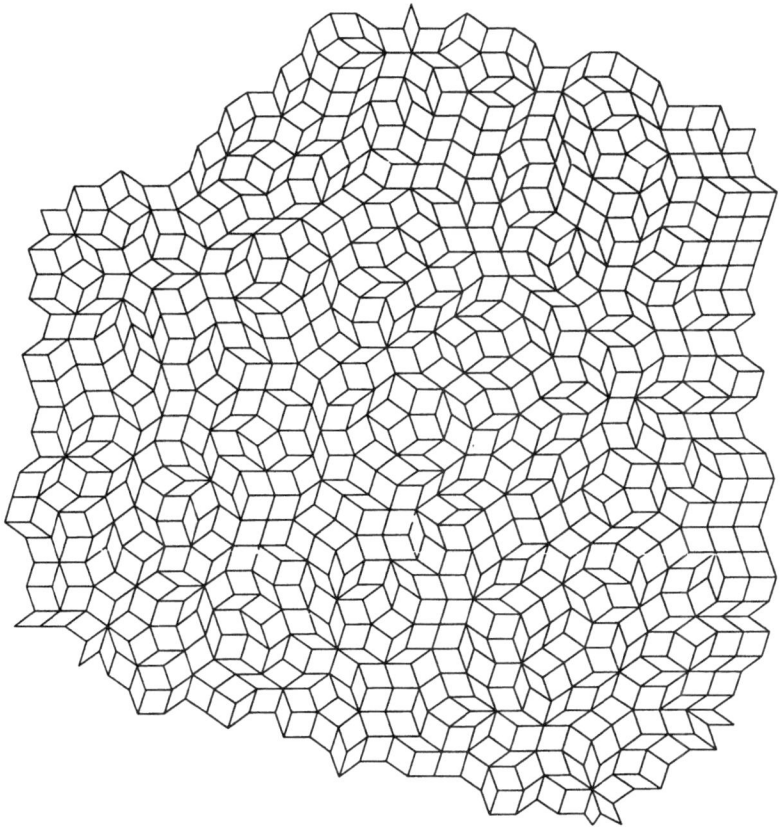

FIGURE 5.14. (b). The tiling (quasiglass) obtained from the randomized penta-grid in Fig. 5.14(a).

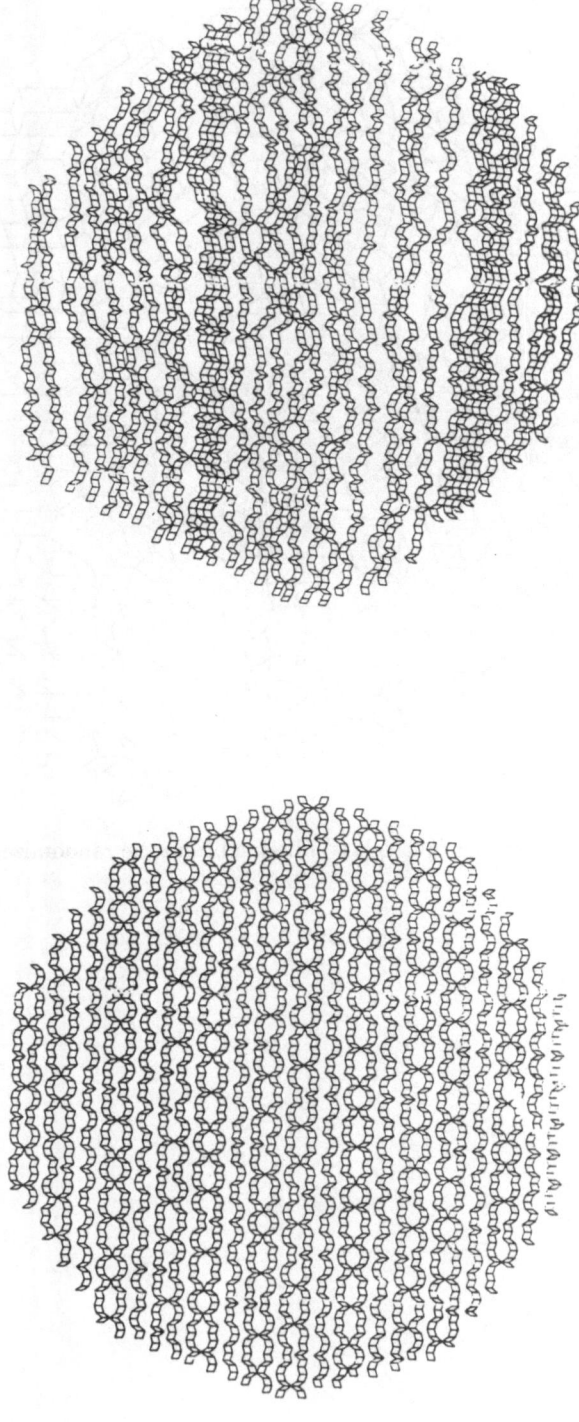

FIGURE 5.14. (c). Another illustration of the difference between the perfect quasicrystal and the quasiglass. The tracks corresponding to a set of parallel grid lines in dual space are plotted out. The tracks of perfect quasicrystals are straight lines on large scales. This is not the case of quasiglasses.

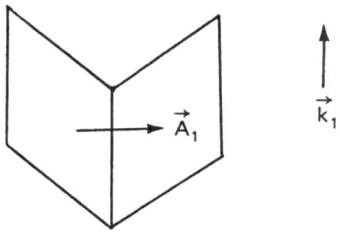

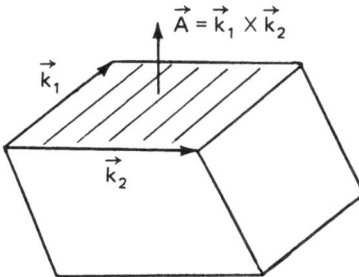

FIGURE 5.15. The relation between the bonding direction and the tiling vectors. The bonding vectors are normal to the surface of the tiles. In two dimensions we have $\mathbf{A}_1 = \hat{\mathbf{z}} \times \mathbf{k}_1$. In three dimensions we have $\mathbf{A} = \mathbf{k}_1 \times \mathbf{k}_2$.

assume that there is an attractive "bonding" energy $J$ when two tiles share a common edge. In this way, we can regard the bonding direction between two tiles as always perpendicular to the edges (see Figure 5.15). It is clear that we have five bonding directions, $\hat{\mathbf{A}}_i = \hat{\mathbf{z}} \times \hat{\mathbf{k}}_i$, $i = 1$ to 5, corresponding to the vectors $\hat{\mathbf{A}}_\tau$ in Section 5.4.1. The surface energy of a quasicrystal is simply the total number of exposed tile edges (dangling bonds) of the surface.

Let us first consider the case of a perfect tiling. Let $\mathbf{r}_a$ and $\mathbf{r}_b$ be the positions of the centers of two tiles widely separated. Let $\sigma$ be a surface (a sequence of tile edges) connecting (say, the upper left corners of) these two tiles. The "average" normal of this surface is defined to be $\mathbf{n} = \hat{\mathbf{z}} \times \hat{\mathbf{r}}$, where $\hat{\mathbf{r}}$ is the unit vector of $\mathbf{r}_b - \mathbf{r}_a \equiv \mathbf{r}$. Clearly, there are infinitely many surfaces connecting these two tiles. Let $\sigma^*$ be the one with the least number of tile edges, $(N^*)$. The surface energy in the direction $\hat{\mathbf{n}}$ is simply $\gamma(\hat{\mathbf{n}}) = JN^*/r$.

Next, we recall that a tile edge in real space joining two vertices 1 and 2 corresponds to in dual space the cell edge separating the polygonal cells $1'$ and $2'$, which are the images of 1 and 2. To travel in real space from $\mathbf{r}_a$ to $\mathbf{r}_b$ along a sequence of tile edges corresponds to hopping in dual space from one polygonal cell to a neighboring one through a sequence of cell edges. Let $\mathbf{s}_a$ and $\mathbf{s}_b$ be the centers of the cells corresponding to $\mathbf{r}_a$ and $\mathbf{r}_b$.

To determine the surface $\sigma$ in real space with the lowest surface energy, we only need to determine the path in dual space connecting $\mathbf{s}_a$ and $\mathbf{s}_b$ with the least number of cell crossing. The least number of cell edges one has to cross from $\mathbf{s}_a$ to $\mathbf{s}_b$ can be shown to be

$$N^* = |\mathbf{s}| \sum_{1=1}^{5} |\hat{\mathbf{n}} \cdot \hat{\mathbf{z}} \times \mathbf{k}_i|, \qquad (5.21)$$

where the quantity in the sum represents the frequency of grid intersections between the set of lines normal to $\mathbf{k}_i$ and the line joining $\mathbf{s}_a$ to $\mathbf{s}_b$, and $\mathbf{s} = \mathbf{s}_b - \mathbf{s}_a$. In Ref. 11, it is shown that to order $1/r$, $\mathbf{s}$ and $\mathbf{r}$ are identical. This means that the surface energy is

$$\gamma(\hat{\mathbf{n}}) = J \sum_{i=1}^{5} |\hat{\mathbf{n}} \cdot \hat{\mathbf{z}} \times \mathbf{k}_i|, \qquad (5.22)$$

which is of exactly the same form as Eq. (5.17), as we have pointed out that the bonding direction $\hat{\mathbf{A}}_i$ is the same as $\hat{\mathbf{z}} \times \hat{\mathbf{k}}_i$.

The generalization to the case of quasiglass turns out to be fairly straightforward. In Ref. 11, it is shown that because the spacing of the grid lines is randomized in a random walk fashion, the total number of intersections $N^*$ differs from that of the perfect case [Eq. (5.21)] by a quantity of the order of $\sqrt{s}$, which is of the order of $\sqrt{r}$. Therefore, the correction to the surface energy due to randomness is of the order of $r^{-1/2}$, which vanishes in the thermodynamic limit. This shows that the surface energy is independent of the randomizing process we introduced. The equilibrium shape of the perfect quasicrystal and the corresponding quasiglass is a decagon with facets normal to the star vectors $\mathbf{k}_i$, as implied by Eq. (5.22).

It is straightforward to generalize the dual method to three dimensions to construct quasicrystals or quasiglasses with icosahedral symmetry. The resulting surface energy is [10]

$$\gamma(\hat{\mathbf{n}}) = J \sum_{i,j} |\hat{\mathbf{n}} \cdot \mathbf{k}_i \times \mathbf{k}_j|, \qquad (5.23)$$

where $\mathbf{k}_i \times \mathbf{k}_j$ are the bonding vectors (see Figure 5.15). If $\{\mathbf{k}_i\}$ are the vertex vectors, then $\mathbf{k}_i \times \mathbf{k}_j$ are along the twofold axes. The equilibrium shape is identical to that of a PBOSs with edge vectors, i.e., Figure 5.10(b). The equilibrium shapes when $\{\mathbf{k}_i\}$ are the edge vectors and face vectors are shown in Figures 5.16(a) and (b), which will have triacontehedral and icosahedral growth shapes if the smaller facets are eliminated in growth processes.

(a)

(b)

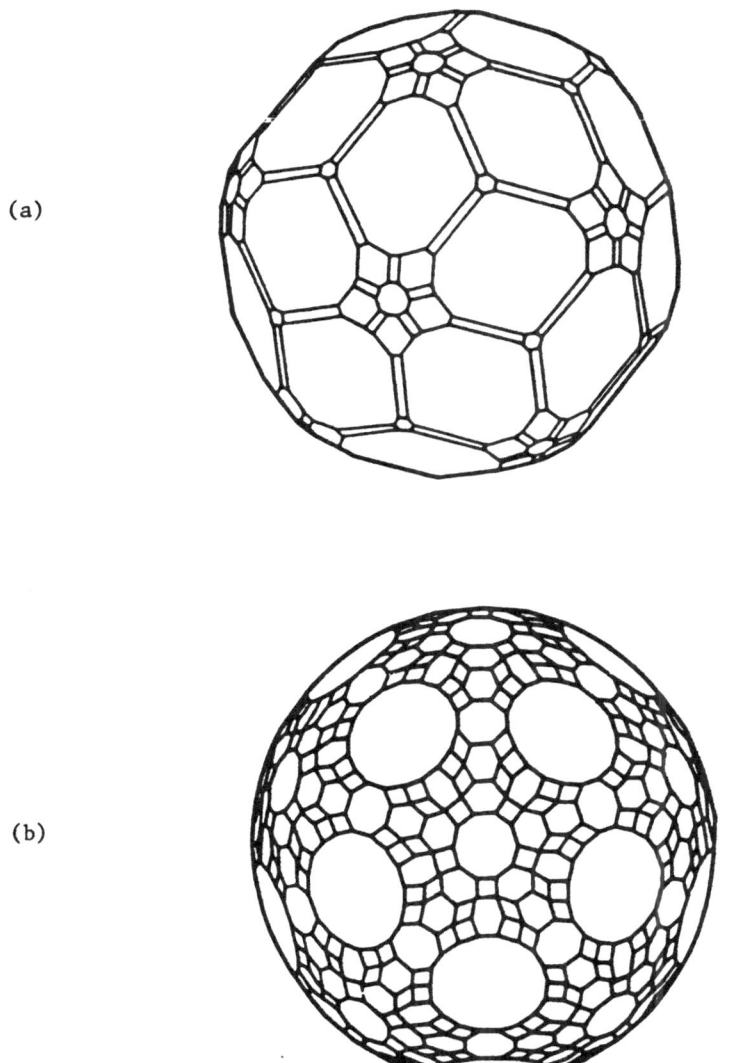

FIGURE 5.16. (a) and (b). The equilibrium shapes of quasicrystals and quasi-glasses with the star vectors $\{k_i\}$ along the two- and threefold axis of an icosahedron.

## 5.6    Final Remarks

The tiling models we have just considered show that contrary to conventional wisdom, facet formation can be compatible with spatial disorder. The reason is that for these models, facets arise from the planelike surfaces of the system (i.e., tracks) which have nonzero stiffness when tilted away from their equilibrium directions. This aspect is identical to that of a crystal. The tile flipping disorder in quasiglasses merely destroys the long-range quasiperiodic order over the scale of the correlation length. It does not destroy these surfaces themselves.

On the other hand, quasiglasses are unrealistic because they are still highly correlated structures, despite the loss of LRPO. The correlation comes from the fact that the underlying grid in dual space contains parallel grid lines, which automatically correlate pairs of tiles in one region with those very far away. A more realistic model would allow the grid lines to become very wobbly. However, are planelike surfaces really necessary for facet formation? While it might seem unlikely that a system without *well-defined* planelike structures in the bulk can have facets, it is interesting to note that the existence of such structures, like LRPO, has not been invoked in the derivation of the simple sufficient condition in Section 5.4. Moreover, real disordered quasicrystals probably have defects, which may cause tears in the original surfaces.

Here, we would like to mention an interesting difference between quasicrystal and crystal topological defects. Let us again consider systems in two dimensions. The topological defects of two-dimensional crystals are edge dislocations, which are usually characterized by a Burgers vector. It is geometrically obvious that an edge dislocation will destroy the *perfect* bond-orientational orders. The extent of destruction is given by the range of the strain field [see Figure 5.17(a)]. In the presence of an array of defects with identical Burgers vector, the crystal must undergo elastic distortion to relax the strain, and the equilibrium shape of the crystal will not be faceted [see Figure 5.17(b)].

However, for two-dimensional quasicrystals there are orientation preserving defects. Let us consider a grid in dual space that generates a perfect quasicrystal. Let $L_i$, $i = 1$ to 5, denote five infinite grid lines with distinct orientations meeting almost at a point $P$. By erasing half of each of these grid lines $L_i$ (from one end at infinity to point $P$), so that the remaining grid lines form a pentagonal star [see Figure 5.18(a)], and by assigning a tile to each intersection of two grid lines as done in the dual construction, we obtain a tiling corresponding to an orientation-preserving stable defect [Figure 5.18(b)]. The perfect bond orientation of this tiling can be seen from the fact that any path encircling $P$ in dual space will accumulate a vector of the form $\sum_{i=1}^{5} \mathbf{A}_i$, where $\mathbf{A}_i$ is a tiling vector. Since this sum is zero, there will be no (orientation-destroying) strains in the physical space. Bond-orientational order persists right down to the core of the defects. [In

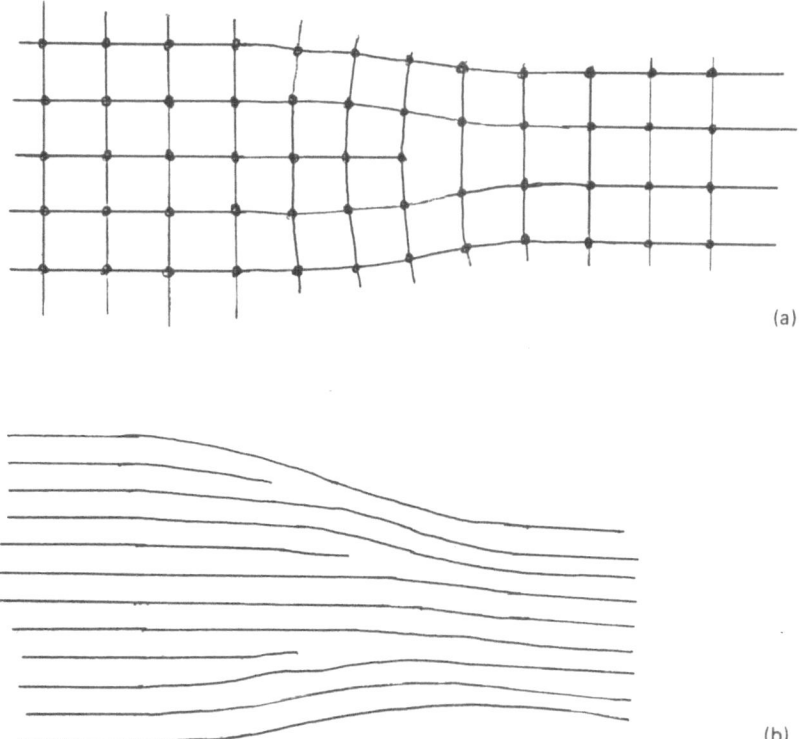

FIGURE 5.17. (a) An edge dislocation in a two-dimensional crystal. Bond-orientational order is destroyed in the neighborhood of the defect. (b) An array of defects of identical Burgers vectors will cause enormous deformations in crystalline planes.

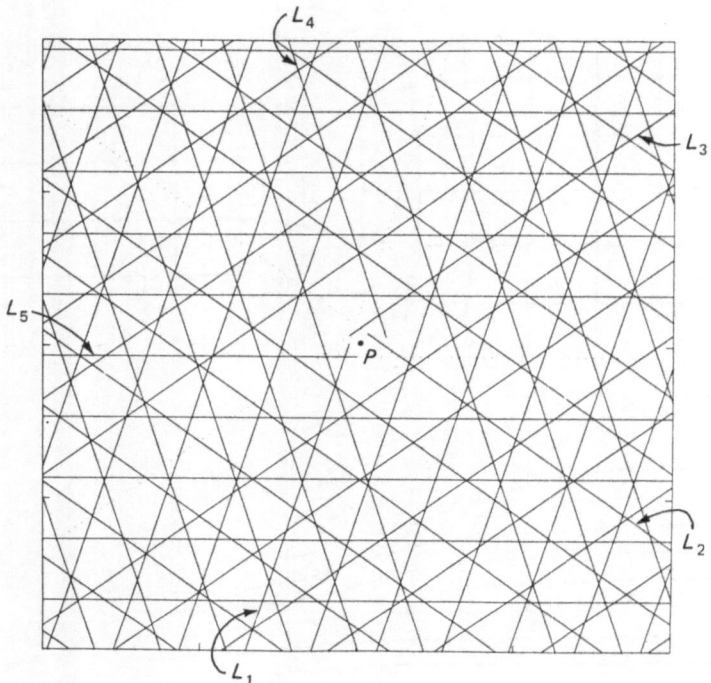

FIGURE 5.18. (a). The pentagrid in dual space that generates an orientation-preserving defect.

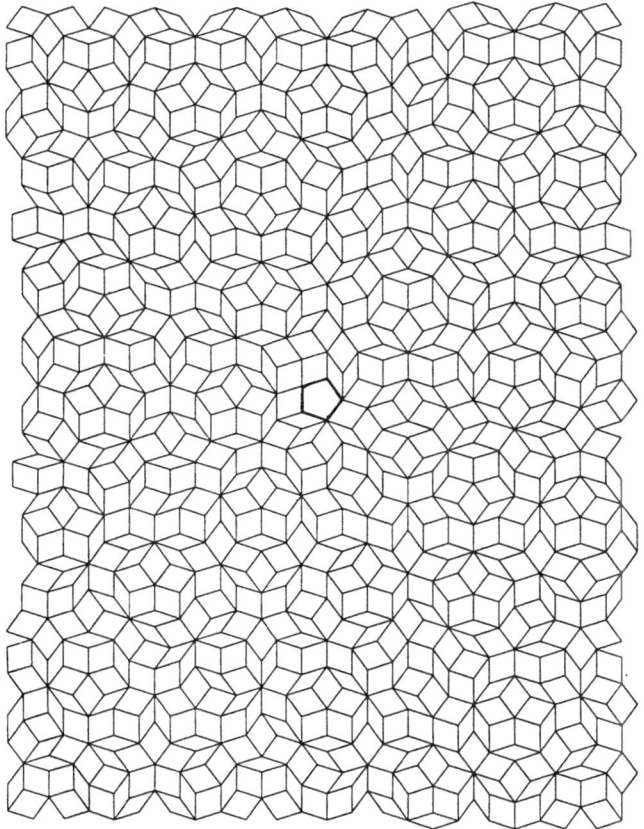

FIGURE 5.18. (b). The tiling corresponding to the pentagrid in Fig. 5.18(a).

the case of icosahedral quasicrystals, the tiling vectors $\mathbf{A}_i$ (say, taken to be the vertex vectors) are incommensurate, i.e., integer combinations of the form $\Delta\mathbf{A} \equiv \sum_{i=1}^{6} N_i \mathbf{A}_i$ do not vanish unless all integers $N_i = 0$. This means that there will be no orientation-preserving defects as in the two-dimensional case. On the other hand, unlike the Burgers vectors of ordinary crystals which have a minimum length, the Burgers vectors of quasicrystals $\Delta\mathbf{A}$ can be made arbitrarily small by appropriately choosing the integers $N_i$. As it turns out, the possibility of making such choices increases rapidly as one moves away from the defect. One therefore expects that the strain fields of these defects are of much shorter range compared to the crystal strain fields, and that bond-orientational order can be preversed to very high accuracy over large distances.] The effects of orientation-preserving defects on the equilibrium shape are not yet clear. What is clear is that these defects destroy tracks (or lattice planes). If the equilibrium shape continued to be faceted despite the introduction of a density of these defects, it would mean that even planelike structures are unnecessary for facet formation.

Finally, we must come back to the question of quasicrystal growth, since all the observed facets are growth facets. Here, we have a similar puzzle as in the equilibrium case. If we regard a faceted growth front as a highly ordered structure and an indication that atoms are added to the bulk in successive layers, how do these ordered layers manage to generate so much spatial disorder in the bulk? One possibility is that the growth front may not be as perfect as it appears. For example, if the growth is induced by a density of orientation-preserving (or almost orientation-preserving) defects with net zero Burgers vector, the surface may still be flat on a macroscopic scale, even if it is full of tears on the scale of dislocation separation. One must also note that the insensitivity of facet formation to bulk order is not an isolated problem. Whatever growth mechanism one invents to deal with this issue will also have to generate a perfect quasicrystal in the appropriate limit (i.e., absence of defects). The facet formation problem is therefore intimately connected to that of quasicrystal growth, and to the growth of crystals with large unit cells. The solutions of these two fundamental problems still elude us at the moment.

*Acknowledgment.* I would like to thank Katherine Strandburg for comments on the first draft of this review. This work is supported by NSF Grant No. DMR-87-17574.

# References

[1]  For a review of the development of quasicrystals before 1987, see *The Physics of Quasicrystal*, edited by P.J. Steinhardt and S. Ostlund, World Science, Singapore, 1987.

[2]  See N.D. Mermin, D. Rokhsar, and D.C. Wright, *Phys. Rev. Lett.* **58**, 2099–2010 (1987); D. Rokhsar, D.C. Wright, and N.D. Mermin, *Acta Crys. A* **44**, 197–211 and references therein (1988).

[3]  G.Y. Onoda, P.J. Steinhardt, D. DiVincenzo, and J.E.S. Socolar, *Phys. Rev. Lett.* **60**, 2653 (1988); **62**, 1210 (1989); M. Jarić and M. Ronchetti, *Phys. Rev. Lett.* **62**, 1209 (1989).

[4]  V. Elser, *Acta Cryst. A* **42**, 36 (1986).

[5]  P.A. Heiney, P.A. Bancel, P.M. Horn, L.J. Jordon, S. LaPlaca, J. Angilello, and F.W. Gayle, *Science* **238**, 660 (1987); P.M. Horn, W. Malzfedt, D.P. DiVincenzo, J. Toner, and R. Bambino, *Phys. Rev. Lett.* **57**, 1444 (1986).

[6]  P. Stephens and A. Goldman, *Phys. Rev. Lett.* **56**, 1168 (1986).

[7]  B. Dubost, J.M. Tanaka, P. Sainfort, and M. Audier, *Nature* (Lond.) **324**, 48 (1986); F.W. Gayle, *J. Mater. Res.* **2**, 1 (1987).

[8]  P.J. Steinhardt and S. Ostlund, *ibid.*, p. 314.

[9]  C.A. Gutym, A.I. Goldman, P.W. Stephens, K. Hiraga, A. Tsai, A. Inoue, and T. Masumoto, *Phys. Rev. Lett.* **62**, 2409 (1989); P. Bancel, *Phys. Rev. Lett.* **63**, 2741 (1989); A.-P. Tsai, A. Inoue, and T. Masumoto, *Jpn. J. Appl. Phys.* **26**, L1505 (1987).

[10] T.L. Ho, J.A. Jaszczak, Y.H. Li, and W.F. Saam, *Phys. Rev. Lett.* **59**, 1116 (1987).

[11] T.L. Ho, Y.H. Li, W.F. Saam, and J.A. Jaszczak, *Phys. Rev. B* **39**, 10614 (1989).

[12] R. Lipowsky and C. Henley, *Phys. Rev. Lett.* **60**, 23904 (1988).

[13] A. Garg and D. Levine, *Phys. Rev. Lett.* **59**, 1683 (1987).

[14] J.A. Jaszczak, W.F. Saam, and B. Yang, *Phys. Rev. B* **39**, 9289 (1989).

[15] A.-P. Tsai, A. Inoue, and T. Masumoto, *Jpn. J. Appl. Phys.* **26**, L1505 (1987); S. Ebalard and F. Spaepen, *J. Mater. Res.* **4**, 39 (1989).

[16] W. Ohashi and F. Spaepen, *Nature* (Lond.) **330**, 555 (1987).

[17] L. Bendersky, *Phys. Rev. Lett.* **55**, 1461 (1985); R.J. Schaefer and L. Bendersky, *Scr. Metall.* **20**, 745 (1986).

[18] K. Chattopadhyay, S. Lele, S. Ranganathan, G.N. Subbanna, and N. Thangaraj, *Curr. Sci.* **20**, 895 (1985).

[19] A.R. Kortan, F.A. Thiel, and H.S. Chen, *Phys. Rev. B* **40**, 9397 (1989); A.R. Kortan, H.S. Chen, J.M. Parsey Jr., and L.C. Kimerling, *J. Mater. Sci.* **24**, 1999 (1989).

[20] R.J. Hauy, *Essai d'une the'orie sur la structure des cristaux*, Paris, 1784; also see J.G. Burke, *Origins of the Science of Crystals*, University of California Press, Berkeley, Calif., 1966, p. 96.

[21] For an excellent review on equilibrium crystal shapes, see M. Wortis, in *Chemistry and Physics of Solid Surfaces*, Vol. VII, edited by R. Vanselow, Springer-Verlag, Berlin, 1988.

[22] A.A. Chernov, *Modern Crystallography III*, Springer-Verlag, Berlin, 1984, pp. 45–47.

[23] L.D. Landau, *Collected Papers of L.D. Landau*, edited by D. ter Haar, Gordon and Breach, New York, 1965, pp. 540–545.

[24] C. Herring, *Phys. Rev.* **82**, 87 (1951).

[25] K. Ingersent and P.J. Steinhardt, *Phys. Rev. Lett.* **60**, 2444 (1988); *Phys. Rev. B* **39**, 980 (1989).

[26] N.G. deBruijn, *Proc. Koninklijke Nederlandse Akadamie van Wetenschappen Series A* **84** (= *Indagationes Mathematics* **43**), 39 (1981).

[27] F. Gahler and J. Rhyner, *J. Phys. A* **19**, 267–277 (1986). A particular simple derivation of the equivalence between the projection method and dual method is given in D.A. Rabson, T.L. Ho, and N.D. Mermin, *Acta Cryst. A* **44**, 678 (1988).

[28] V. Elser, in *Proceedings of the 15th International Colloquium on Group Theoretical Methods in Physics, Philadelphia*, edited by R. Gilmore and D.H. Feng, World Scientific, Singapore, 1987, pp. 162–183; C. Henley, *Comm. Condens. Mater. Phys. B*, vol. XIII, 58 (1987); M. Widom, D.P. Deng, and C.L. Henley, *Phys. Rev. Lett.* **63**, 310 (1989); K. Strandburg, L.H. Tang, and M.V. Jarić, *Phys. Rev. Lett.* **63**, 314 (1989).

# 6

# Icosahedral Ordering in Supercooled Liquids and Metallic Glasses

*Subir Sachdev*

## 6.1 Introduction

A metallic glass is a solid consisting of metallic atoms arranged in a random manner with no obvious long-range correlation in the atomic positions. While such random atomic arrangements are easy to achieve in materials with covalent bonding, until 1960 the solid state of all known metals and metallic alloys consisted of regular, periodic arrangements of the atoms. The first metallic glass was produced in 1960 by Duwez and co-workers [1] by rapidly cooling a molten alloy of gold and silicon. Metallic glasses have since then been created in large numbers of simple metal, transition metal, and metalloid systems by a variety of ingenious methods. "Splatcooling" techniques have been developed to achieve cooling rates of over a million degrees per second and have created a completely new metallurgical technology. The new metallic materials so produced have proved to be of considerable technological importance for their unique magnetic, mechanical, and corrosion-resistance properties [2].

In this chapter we investigate the question of how "random" the atomic arrangements in a metallic glass really are. In particular, an attempt shall be made to identify features of the structure which are not sensitive to microscopic details like the nature of the interatomic potential, directional bonding, and local charge transfer between the atoms. Our analysis shall therefore be based on understanding the structural properties of dense and supercooled systems of atoms interacting with each other through spherically symmetric forces. We find that there are significant short-range orientational correlations between the atomic arrangements: the characterization of these orientational correlations will be the main subject of this chapter.

As constructed, the theories reviewed in this chapter are directly applicable to any dense, supercooled liquid of spheres interacting with a pairpotential which has a repulsive hardcore and a weak long-range attraction. Such systems can be easily realized in computer simulations, but there is no known bulk monoatomic metallic glass. However, amorphous films of cobalt and iron have been made by deposition of the metallic vapor on a substrate at liquid helium temperatures. We will compare our results with X-ray scattering measurements on such films. We shall in addition argue that our results can also be applied to a large class of real metallic glasses: glass forming metal-metalloid and metal-metal alloys.

A key property of the systems we shall consider is that they are frustrated. By "frustration" we mean that particles in the ground state cannot simultaneously sit in the minima presented to them by pairwise interactions with their neighbors. This leads to a large degeneracy in the ground state. In phase space, the system has large numbers of nearly equal free energy minima separated by substantial free energy barriers. If the system gets locked into one of these minima upon cooling from the liquid, a glassy or amorphous state can result. Our main objective shall be to determine the atomic correlations at a "typical" local minimum of the free energy.

## 6.2   Three-Dimensional Sphere Packings and Frustration

In this section we shall introduce several qualitative methods of characterizing the order in supercooled liquids. A quantitative approach shall be taken in the next section after the introduction of a suitably defined order parameter and associated Landau free energy. As noted above, we shall be interested in characterizing the local minima of a system of interacting particles with a Lennard–Jones-like pair potential; i.e., a potential with a strong hardcore repulsion and long attractive tail. A related problem we shall also consider is the dense random packing of identical spheres.

The sphere packing problem has a trivial solution when the spheres are constrained to move in a single plane. Three spheres will clearly lie at the vertices of an equilateral triangle. Four spheres will form two equilateral triangles sharing a common edge. It is clear that we can extend the arrangement of equilateral triangles to the triangular lattice and accommodate an infinite number of spheres. This packing is the densest possible packing and all particles sit in the minima of the potential due to their six nearest neighbors. Thus, the locally optimum arrangement of three spheres (the triangle) has a unique periodic extension and the system is clearly unfrustrated.

The physics changes dramatically when we allow the particles to move in three dimensions. The state of minimum energy for four particles is the tetrahedron, shown in Figure 6.1. Computer simulations of a Lennard–Jones system [3] show that larger numbers of particles like to arrange themselves in configurations that maximize the number of tetrahedra. In Figures 6.2 and 6.3 we show systems of 7 and 13 particles whose states of global minimum energy are the pentagonal bipyramid and the icosahedron, respectively. Both the configurations can be divided very simply

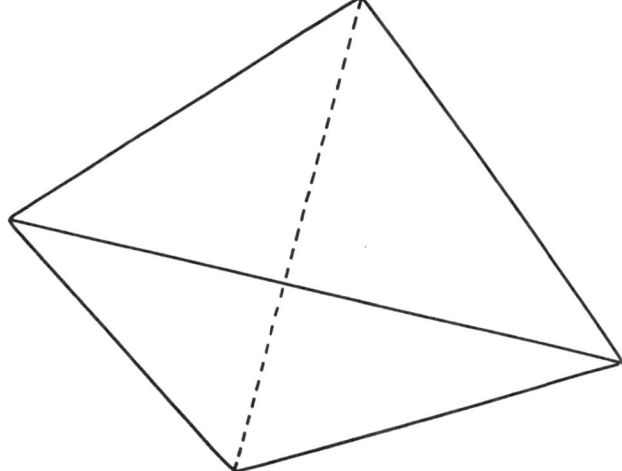

FIGURE 6.1. State of minimum energy for four particles: the tetrahedron.

into tetrahedra: the pentagonal bipyramid has five tetrahedra around the central bond. The icosahedron is, in turn, made up of 12 interpenetrating pentagonal bipyramids centered on the 12 bonds from the center to the vertices of the icosahedron. Another important local minimum is the six-particle octahedral arrangement shown in Figure 6.4; unlike the previous configurations it cannot be split into approximately equilateral tetrahedra. Note, however, that the octahedron has an appreciable "hole" in the middle, indicating that it is a poor starting point in the search for metastable minima for larger numbers of particles.

One manifestation of the frustration in the packing of these particles is the fact that the bond lengths in the pentagonal bipyramid and the icosahedron are not all equal. The bonds in the center of both solids are approximately 5% smaller than the bonds on the surface. Alternatively, if one attempted to put five perfect tetrahedra around a bond, one would be left with a gap of 7.4° between the first and last tetrahedron. See Figure 6.5. This situation stands in sharp contrast to the case in flat two-dimensional space where six equilateral triangles form a perfect hexagon.

We now turn to the analysis of simulations on the dense packing of spheres. We shall examine both computer simulations and actual experiments on the dense packing of arrays of ball bearings. An important tool in the analysis of such configurations is the Voronoi construction. The region of space closer to an atom than to any other is identified as the Voronoi polyhedron associated with that atom. Two atoms are then identified as nearest neighbors if their Voronoi polyhedra share a common face. The network of nearest-neighbor bonds obtained in this manner can be shown to consist only of tetrahedra (barring exceptional degeneracies). The decomposition of space into a tetrahedral network is physically meaningful

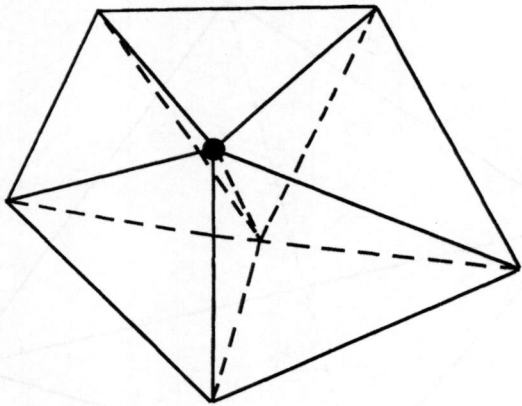

FIGURE 6.2. State of minimum energy for seven particles: fivefold bipyramid.

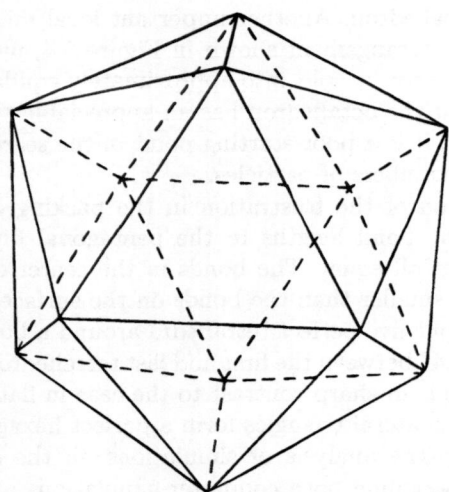

FIGURE 6.3. State of minimum energy for 13 particles: the icosahedron.

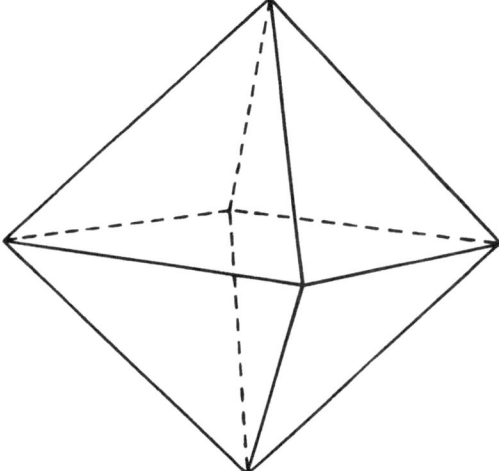

FIGURE 6.4. Local minimum of energy for six particles: the octahedron.

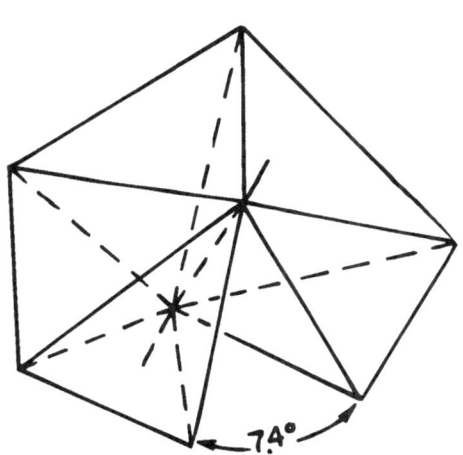

FIGURE 6.5. Five perfect tetrahedra around a bond. Note the empty space of 7.4° that is left over.

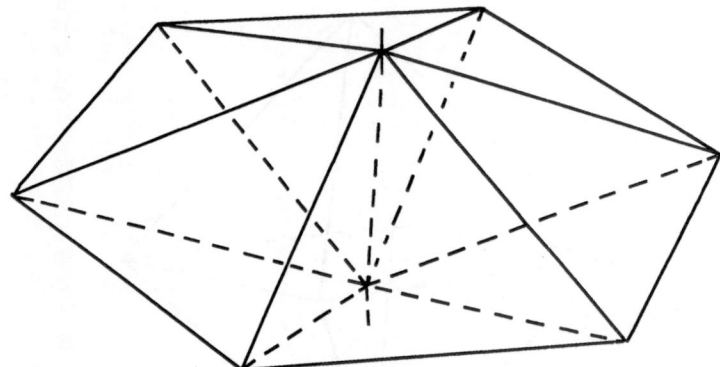

FIGURE 6.6. A sixfold bipyramid. The bond in the center is identified as a $-72°$ disclination.

only if all the tetrahedra are approximately equilateral and the number of nearly octahedral arrangements relatively small.

One of the early experiments on the packing of ball-bearing arrays was carried out by Bernal [4]. Finney [5] performed a Voronoi decomposition of the Bernal structure and found that 45% of the bonds were the centers of pentagonal bipyramids. Further dense packing experiments were performed by Bennett [6] who used an algorithm designed to give structures denser than those obtained by Bernal. A relaxation by Ichikawa [7] of the Bernal structure and a subsequent Voronoi analysis yielded pentagonal bipyramids on 52% of the bonds. Moreover, Finney and Wallace [8] found that the Bennett structure could be well described by local configurations which were either approximately equilateral tetrahedra or octahedra: over 80% of the local configurations were tetrahedra. This latter fact justifies the use of the Voronoi construction in Ichikawa's analysis of the structure.

The theory of metallic glass structure we review here attaches particular importance to the bonds with five tetrahedra around them (fivefold bonds, Figure 6.2) as a structural motif. The pentagonal bipyramid is the densest possible packing of seven particles and can thus be regarded as a locally ideal ordered configuration; this notion of ordering will be made precise in the next section. The system would clearly like to extend the fivefold bonds into all space but the geometrical properties of flat three-dimensional space make this impossible. Instead, as the dense packing simulations discussed above show, as many as 48% of the bonds turn out to be either sixfold (Figure 6.6) or fourfold (Figure 6.7). The fourfold bonds are actually distorted octahedra, and the issue of whether they are better regarded as a collection of four tetrahedra or as an octahedron can be decided by examining the bond lengths: such a criterion was used by Finney and Wallace [8] in the work discussed in the previous paragraph. The Voronoi construction will, of course, always yield only tetrahedra.

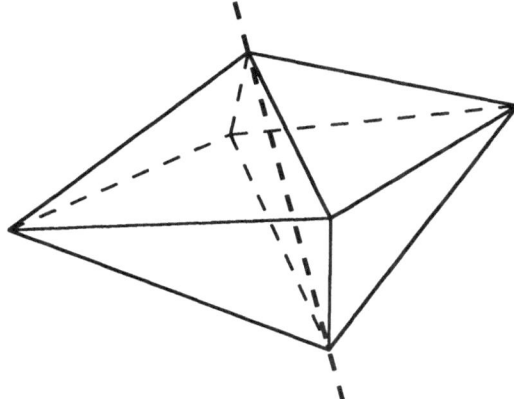

FIGURE 6.7. A fourfold bipyramid. The bond in the center is identified as a $+72°$ disclination.

Following Nelson [9], we will describe an instantaneous snapshot of a dense supercooled liquid in terms of its arrangement of fivefold bonds. If a particle has 12 fivefold bonds emerging from it, its coordination shell will form an icosahedron (Figure 6.3). Nelson [9] therefore introduced an orientational order parameter which would acquire its maximum value at a particle at the center of an icosahedron. The fourfold and sixfold bonds could then be identified as positive and negative disclination defects in the ideal icosahedral order. The connection with the bond-orientational order in two-dimensional hexatics (Chapters 2–4) is apparent: in this case, the perfectly ordered sites are particles with six neighbors and the five- and sevenfold coordinated particles form disclination defects in the orientational order. There is, however, a crucial difference between these two systems: the defects in two-dimensional hexatics are induced purely by thermal fluctuations; in contrast, the fourfold and sixfold bonds are forced into any random close packing of spheres by the frustration inherent in flat three-dimensional space. The defects are quite dense: a bond is defected with probability close to $1/2$. If we assume that the defects are randomly distributed, the probability that a particle will have exactly 12 fivefold bonds emerging from it is smaller than 1 in 1000. In their molecular dynamics simulation of 999 particles, Finney and Wallace [8] did not find a single icosahedron: this is not surprising in light of the above discussion.

We note also the work of Steinhardt et al. [10] which focused attention on the importance of fivefold bipyramids. In a computer simulation of a supercooled Lennard–Jones liquid, these investigations examined the system for the presence of bond-orientational order associated with five-fold bonds and icosahedra. They found orientational order at sufficiently low temperatures with a correlation length comparable to the size of the system.

## 6.2.1  Frank–Kasper Phases

Just prior to the work of Bernal and followers was the analysis of Frank and Kasper [11] of certain intermetallic crystalline alloys. They noted that a large number of intermetallic compounds (the Laves phases $MgCu_2$, $MgZn_2$, $CaZn_5$, the $\mu$-phase $Fe_7W_6$, $CaCu_5$, NbNi to name a few; these compounds are now widely referred to as the Frank–Kasper phases) could be understood as representations of tetrahedral close packings. The entire crystal structure was composed of approximately equilateral tetrahedra. A majority of the bonds are fivefold, with the remaining bonds sixfold (these phases have no fourfold bonds). Frank and Kasper also presented a simple and important topological argument which shows that no particle could have fivefold bonds and just a *single* sixfold bond emerging from it. In other words, the sixfold bonds must form defect lines which run through the entire crystal structure. A particle with a single defect line running through it would then have two sixfold bonds emerging from it and have a coordination shell of 14 particles [Figure 6.8(a)]. Frank and Kasper also showed that three or four defect lines could meet at a point; the particle at the intersection of the lines would then have coordination shells of 15 or 16 particles, respectively [Figures 6.8(b) and (c)]. Examples of 12, 14, 15, and 16 coordinated particles occur in the Frank–Kasper phases.

Nelson [9] argued that the defect line analysis of Frank and Kasper could be extended to provide a description of supercooled liquids and metallic glasses. It is a remarkable fact that most of the metallic alloys which form metallic glasses also form stable intermetallic compounds with a Frank–Kasper-like structure [12–14] (this point will be discussed further in Section 6.4.2). However, as the computer simulations clearly show, random close-packings also have fourfold bonds which are absent in the Frank–Kasper phases. Using modern topological methods [15], Nelson extended the arguments of Frank and Kasper to include fourfold defect lines. He showed that, like the sixfold bonds, the fourfold bonds could not end at a point and must be part of defect lines. Two, three, and four fourfold bonds can meet at a particle, making that particle eight, nine, and ten coordinated (Figure 6.9). Supercooled liquids and metallic glasses therefore form a tangled network of defect lines: an ordered arrangement of defect lines leads to the Frank–Kasper phases.

# 6.3  Structure Factor of Monoatomic Supercooled Liquids

In this section we shall review recent work on constructing a Landau free energy of a dense supercooled liquid of spheres interacting with a pair-potential with a repulsive hardcore and weak long-range attraction. A

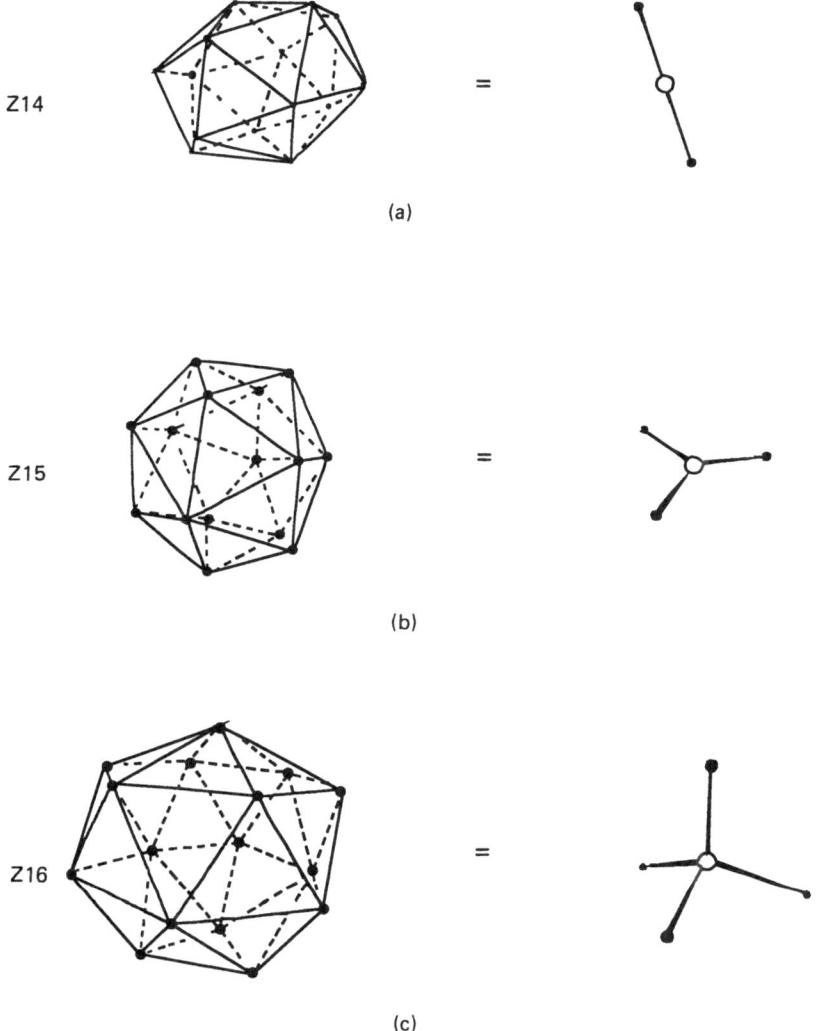

FIGURE 6.8. (a) through (c) 14, 15, and 16 coordinated particles. The figures on the right indicate the sixfold disclination lines emanating from the central atom.

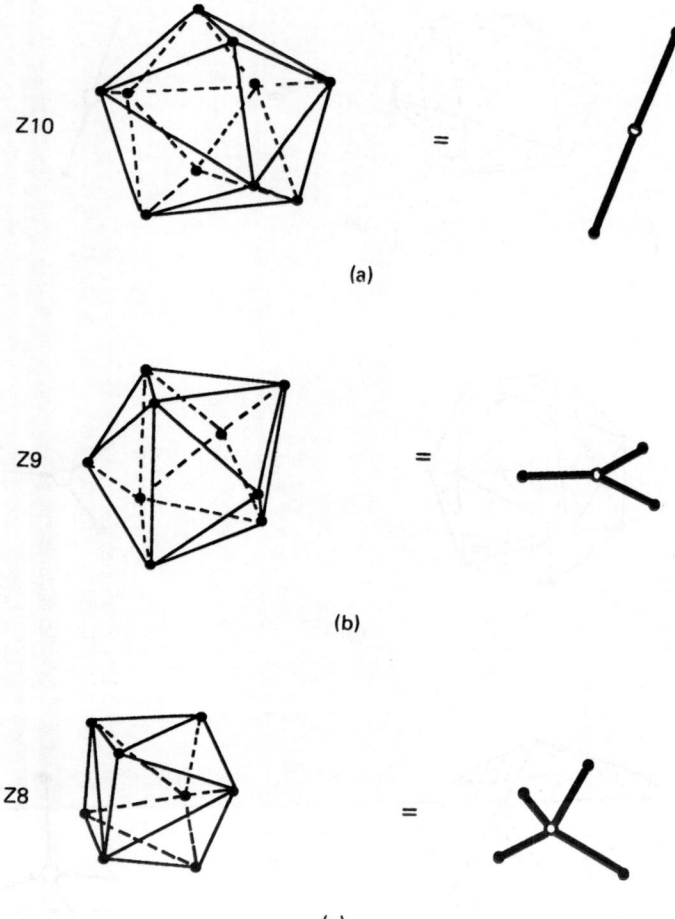

Z10

Z9

Z8

(a)

(b)

(c)

FIGURE 6.9. (a) through (c) Eight, nine, and ten coordinated particles. The figures on the right indicate the fourfold disclination lines emanating from the central atom.

Gaussian approximation to the Landau free energy will then be used to calculate the experimentally measurable structure factor.

## 6.3.1    Sphere Packings in Curved Three-Dimensional Space

A crucial first step in the analysis of any frustrated system is the understanding of the related unfrustrated system which is perfectly ordered. The unfrustrated system can then be used to define an order parameter, which is, in turn, used to characterize the unfrustrated system. The ideal, unfrustrated system associated with random close packing of spheres was first identified by Coxeter [16]. He noted that on the three-dimensional surface of a four-dimensional sphere five perfect tetrahedra can be arranged around a bond, with no gap left between the first and last tetrahedron, i.e., the gap in Figure 6.5 vanishes on this positively curved three-dimensional space. The entire surface of the sphere can be tiled with 600 perfect tetrahedra, with every tetrahedra being equivalent to any other. Every bond has exactly five perfect tetrahedra around it and every particle sits at the center of a perfect icosahedron. This packing of particles on the surface is known as polytope $\{3, 3, 5\}$. The curvature of the sphere $\kappa$, which is the inverse of the radius of the sphere, is related to the near-neighbor distance $d$ by

$$\kappa d = \frac{\pi}{5} . \tag{6.1}$$

This tiling of particles can be considered as the analog of the triangular lattice in two dimensions, with the important difference that it is of finite extent.

Motivated by his analysis of polytope $\{3, 3, 5\}$, Coxeter [16] presented a simple "mean-field" analysis of the effects of frustration in flat space. He argued that some of the properties of dense random packings in flat space could be modeled by a fictitious space-filling polytope $\{3, 3, q\}$, where $q$ is the average number of tetrahedra around a bond. On the surface of a sphere, there are five tetrahedra around every bond, so $q$ is 5. In flat space the dihedral angle of the tetrahedron is $\arccos(1/3)$, so on the average, there is space for

$$q = \frac{2\pi}{\arccos(1/3)} \tag{6.2}$$

$$\approx 5.1043$$

tetrahedra around every bond. The values of $q$ in simulations of supercooled liquids and in the Frank–Kasper phases are quite close to this value [9]. Most remarkable is the Frank–Kasper phase $Mg_{32}(AlZn)_{49}$, which approaches the ideal value of $q$ to one part in $10^4$.

The correlations in polytope $\{3, 3, 5\}$ were used as a template for the description of local order in supercooled liquids by Kléman and Sadoc [17],

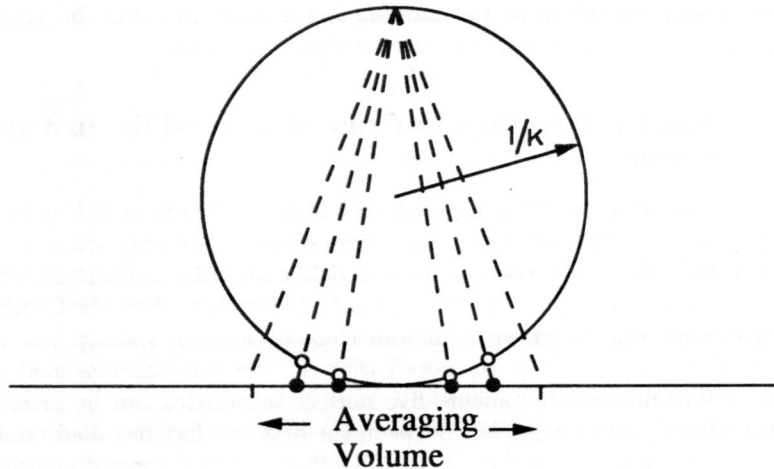

FIGURE 6.10. Schematic showing the projection of a configuration of particles onto a tangent sphere.

Sadoc [18], and Sadoc and Mosseri [19]. These investigations concentrated upon literal mappings of tetrahedra from curved unfrustrated space to a flat frustrated space.

## 6.3.2   Order Parameter

The particular approach to extending Coxeter's ideas that we shall focus on this chapter was pioneered by Nelson and Widom [20]. An order parameter will be introduced to measure the strength of the local icosahedral ordering. With *each* point in the physical flat three-dimensional space associate a tangent four-dimensional sphere. Now project the particle density in a small averaging volume $\Delta V$ on the surface of the sphere as shown in Figure 6.10. This defines a density function $\rho(\mathbf{r}, \hat{u})$ at every point $\mathbf{r}$ in three-dimensional space and every point $\hat{u}$ on the surface of the tangent sphere at $\mathbf{r}$. The physical density which is measured by X-ray scattering is clearly $\rho(\mathbf{r}, \hat{u} = -1)$, where $\hat{u} = -1$ is the "south pole" of the four-dimensional sphere. The physical question we wish to answer is: how close is the environment in the physical flat three-dimensional space around any given point $\mathbf{R}$ to the ideal ordering in polytope $\{3, 3, 5\}$? Because of the nature of the projection operation, it is clear that an equivalent question is: how close is the density $\rho(\mathbf{r} = \mathbf{R}, \hat{u})$, as a function of $\hat{u}$ near $\hat{u} = -1$, to that of polytope $\{3, 3, 5\}$? This question is most easily answered by performing an expansion of the density on the sphere in terms of the hyperspherical harmonics [21]:

$$\rho(\mathbf{r}, \hat{u}) = \sum_{n, m_a, m_b} Q_{n, m_a, m_b}(\mathbf{r}) Y^*_{n, m_a, m_b}(\hat{u}), \qquad (6.3)$$

where the $Y_{n,m_a,m_b}$ are known hyperspherical harmonics labeled by the representation index $n = 0, 1, 2, \ldots$ and the quantum numbers $m_a$, $m_b$, ranging in integer steps from $-n/2$ to $n/2$, label the basis states within each representation of $SO(4)$. The coefficients $Q_{n,m_a,m_b}(\mathbf{r})$ are defined on every point $\mathbf{r}$ of the physical space and characterize the environment in the neighborhood of the point $\mathbf{r}$. They can therefore be considered as a peculiar set of local multiparticle correlation functions. Let $\rho(\mathbf{r} = \mathbf{R}, \hat{u})$ specify a density, as a function of $\hat{u}$ on the surface of the sphere, which is identical to the density on polytope $\{3, 3, 5\}$. A simple computation [20] then shows that the order parameter $Q_{n,m_a,m_b}(\mathbf{R})$ is zero for all values of $n$, except the special values $n = 12, 20, 24, 30, 32 \ldots$. The precise values of $\mathbf{Q}_n(\mathbf{R})$ depend on location and orientation of polytope $\{3, 3, 5\}$ on the sphere, but the selection rules on $n$ are independent of these parameters. The vanishing of $\mathbf{Q}_n(\mathbf{R})$ on such a large number of $n$ values is a consequence of the high degree of symmetry of polytope $\{3, 3, 5\}$. The special set of $n$ values $12, 20, 24, 30, 32 \ldots$ thus form a "reciprocal lattice" distinguishing polytope $\{3, 3, 5\}$ from other particle arrangements on the surface of the sphere. We are now finally in a position to answer the questions posed earlier in this paragraph. The particle configuration in the neighborhood of $\mathbf{R}$ is similar to that in polytope $\{3, 3, 5\}$ if the values of $\mathbf{Q}_n(\mathbf{R})$ are largest as the following special values of $n = 12, 20, 24, 30, 32 \ldots$. The matrices $\mathbf{Q}_n(\mathbf{r})$ for these special values of $n$ are therefore the order parameters that we require. These order parameters characterize the degree of tetrahedral and fivefold bipyramidal ordering in the neighborhood of any point in space.

Before we are able to write down a Landau free energy with the order parameter introduced above, we need to understand the optimum relative position of two fragments of polytope $\{3, 3, 5\}$ at a neighboring point in space. A simple way of achieving the best relative positions which minimize the number of defect lines was suggested by Sethna [22]. Imagine "rolling" the four-dimensional sphere along a straight line in flat three-dimensional space. The particle configuration laid out by this procedure will consist of mildly distorted tetrahedra and not induce any defect lines. In fact, the tetrahedra will form a Bernal spiral shown in Figure 6.11. (Precisely such spirals exist along the axes of the Kagomé net structure found in many Frank Kasper phases.) The relative orientation of the order parameter at two neighboring points along the rolling can be easily shown to satisfy

$$\mathbf{Q}_n(\mathbf{r} + \boldsymbol{\delta}) = \exp(i\kappa \mathbf{L}_{0\mu}^n \delta_\mu) \mathbf{Q}_n(\mathbf{r}) \tag{6.4}$$

where $\mathbf{L}_{0\mu}^n$ is a $(n + 1) \times (n + 1)$ matrix which generates the rotations of $SO(4)$ performed when the sphere is rolled in the $\mu$ direction of flat three-dimensional space. One might now naively guess that it should be possible to tile all three-dimensional space with fragments of polytope $\{3, 3, 5\}$ by starting at the origin and rolling the sphere in a straight line in all directions. However, it can be shown that this procedure always introduces incompatibilities at points other than the origin. One way to illustrate this

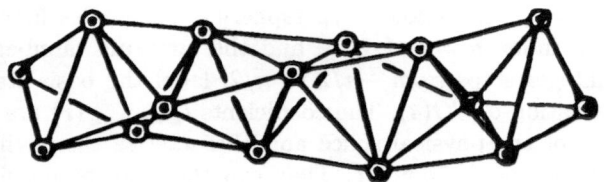

FIGURE 6.11. Tetrahedra forming a Bernel spiral.

frustration is to consider the operation of rolling the sphere in a closed path along the edges of a square in the $\mu$ and $\nu$ directions as shown in Figure 6.12. The order parameter configuration before and after the rolling operations will have the following relationship [22]

$$
\begin{aligned}
\mathbf{Q}_n(\mathbf{r})|_{\text{final}} &= \exp(-i\kappa \mathbf{L}^n_{0\mu}a)\exp(-i\kappa \mathbf{L}^n_{0\nu}a)\cdot \\
&\quad \cdot \exp(i\kappa \mathbf{L}^n_{0\mu}a)\exp(i\kappa \mathbf{L}^n_{0\mu}a)\mathbf{Q}_n(\mathbf{r})|_{\text{initial}} \\
&\approx \exp(-\kappa[\mathbf{L}^n_{0\mu},\mathbf{L}^n_{0\nu}]a^2)\mathbf{Q}_n(\mathbf{r})|_{\text{initial}} \\
&= \exp(-i\kappa \mathbf{L}^n_{\mu\nu},a^2)\mathbf{Q}_n(\mathbf{r})|_{\text{initial}}
\end{aligned} \tag{6.5}
$$

where $\mathbf{L}^n_{\mu\nu}$ is the generator of $SO(4)$ which performs rotations in the $(\mu,\nu)$ plane of flat three-dimensional space. The relative rotations between the initial and final configurations of the order parameter necessarily imply the presence of a constant density of disclination defect lines piercing every plaquette in flat three-dimensional space. These are precisely the fourfold and sixfold bonds introduced earlier; the sign of the rotation above implies an excess of sixfold bonds over fourfold bonds.

### 6.3.3   Landau Free Energy

We finally now have assembled all the ingredients necessary to write down a Landau theory of tetrahedral close packing in supercooled liquids. The Landau free energy expansion performs an expansion in gradients and powers of the order parameter retaining all terms which are consistent with the symmetry of the system. Because of the strong frustration inherent in flat three-dimensional space, we anticipate that the magnitude of $\mathbf{Q}$ will be small and that a low-order expansion will be adequate. To quadratic order the free energy takes the form [20]

$$
F = \tfrac{1}{2}\sum_n \int d^3\mathbf{r}\,\left[K_n|(\partial_\mu - i\kappa \mathbf{L}^n_{0\mu})\mathbf{Q}_n|^2 + r_n|\mathbf{Q}_n|^2\right] + \cdots, \tag{6.6}
$$

where the ellipses denote cubic and higher-order terms, and $K_n$ and $r_n$ phenomenological parameters. The gradient term has been chosen such that the system will attempt to satisfy the "rolling-sphere" relationship [Eq. (6.4)] at every point in space. Apart from this input, the form of the

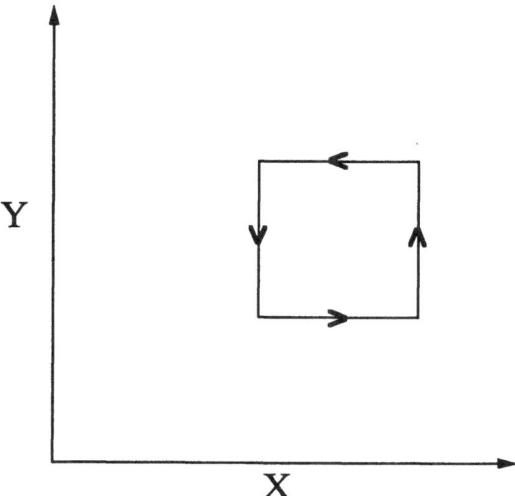

FIGURE 6.12. Rolling the sphere around a closed loop. The final state of the order parameter is related to the initial state by a rotation about an axis perpendicular to the plane of the loop.

free energy follows solely from the symmetry requirement that the energy be independent of the orientation of polytope $\{3, 3, 5\}$ defined by the order parameter $\mathbf{Q}_n$. We make the physical input of demanding local regions of polytope $\{3, 3, 5\}$ ordering by demanding that the $r_n \to +\infty$, except for the selected values $n = 12, 20, 24, 30, 32\ldots$ [23]. Fluctuations in all but these values of $n$ will be strongly suppressed.

Before turning to an analysis of the free energy $F$, it is useful at this point to recall a few essential features of a related system — an extreme type-II superconductor in a magnetic field [24] (by extreme type-II we mean that the London penetration depth is so large that we can safely ignore the changes in the extremal magnetic field due to the supercurrents). In suitable units, the Landau free energy expansion of the superconducting order parameter $\Psi$ takes the form:

$$F_{sc} = \tfrac{1}{2} \int d^3\mathbf{r} \left[ K|(\partial_\mu - iA_\mu)\Psi|^2 + r|\Psi|^2 \right] + \cdots, \qquad (6.7)$$

where $\mathbf{A}(\mathbf{r})$ is the vector potential associated with a uniform external magnetic field $\mathbf{H}$ ($\nabla \times \mathbf{A} = \mathbf{H}$), $K$ is a phenomenological stiffness, and the parameter $r$ is assumed to have the form $r = r'(T - T_c)$. The mean-field phase diagram for this system as a function of the temperature $T$ and $H$ is sketched in Figure 6.13(b). In zero field the system goes superconducting when $r = 0$, i.e., at $T = T_c$. In nonzero fields the *frustration* induced by the external field has two consequences: (1) it depresses the transition temperature: i.e., the system does not go superconducting until $r$ is sufficiently negative; and (2) the superconducting state in nonzero field is an Abrikosov flux lattice consisting of a regular array of the defect lines (vortices) at a density which relieves the frustration due to the field.

We now return to the consideration of the physics of icosahedral ordering as described by the free energy $F$. As discussed earlier, we can turn off the frustration by placing the particles on the surface of a four-dimensional sphere of radius $R$ [25]. When the curvature of the sphere $\kappa = 1/R$ satisfies the condition (6.1), the system is unfrustrated and we can replace the "covariant" derivative in Eq. (6.6) by an ordinary derivative. Thus, the unfrustrated system is the analog of the superconductor in zero field, with the low-temperature ordered state being polytope $\{3, 3, 5\}$. The parameters $r_n$ are presumably close to zero at the melting temperature of polytope $\{3, 3, 5\}$: they are not exactly zero because the presence of cubic terms in the free energy expansion drives the transition first order. As $R$ moves away from the ordered value, the melting temperature is depressed and the ordered state becomes a Frank–Kasper structure consisting of an ordered configuration of defect lines: the phase diagram is shown in Figure 6.13(a). The Frank–Kasper structures are the analog of the Abrokosov flux lattice in the superconductor. The physical flat three-dimensional space is strongly frustrated, and we presume that sluggish dynamics freeze the system at a temperature $T_g$ which is above the temperature $T_0^*$, the melting temperature of a suitable Frank–Kasper structure. We also denote in Figure 6.13(a) the melting temperature $T_m$ of the global ground state of a monoatomic metallic system which is usually a FCC crystal. An important prediction that can be made from the above considerations is that the parameters $r_n$ must be *negative* for $n = 12, 20, 24, 30, 32 \ldots$ at the glass transition temperature.

We turn finally to the use of the free energy $F$ to make quantitative experimental predictions [23,24]. We will calculate the structure factor $S(\mathbf{q})$,

$$S(\mathbf{q}) = \int d^3\mathbf{R} \, e^{i\mathbf{q}\cdot\mathbf{R}} \langle \rho(\mathbf{r} = \mathbf{R}, \hat{u} = -1)\rho(\mathbf{r} = 0, \hat{u} = -1)\rangle, \qquad (6.8)$$

which can be measured easily in electron, X-ray, or neutron scattering experiments. We will assume that the glassy configuration frozen in as the liquid is quenched can be adequately described by the equilibrium ensemble defined by $F$ with all cubic and higher-order terms omitted. With this working hypothesis, the structure factor can be easily calculated after using the relationship (6.3) between the order parameter and the density. The first step is the diagonalization of the quadratic form in $F$; details of this have been presented in Ref. 24. The diagonalization yields a spectrum of eigenvalues which depend on the wave vector $q$ and the representation index $n$. For physically reasonable values of $K_n$ and $r_n$, it is easy to see [23,24] that there is a peak in the structure factor for each of the $n$ values [12,20,24,30,32] and that this peak occurs very close to the wave vector $q$, at which the lowest eigenvalue in the representation $n$ has a minimum

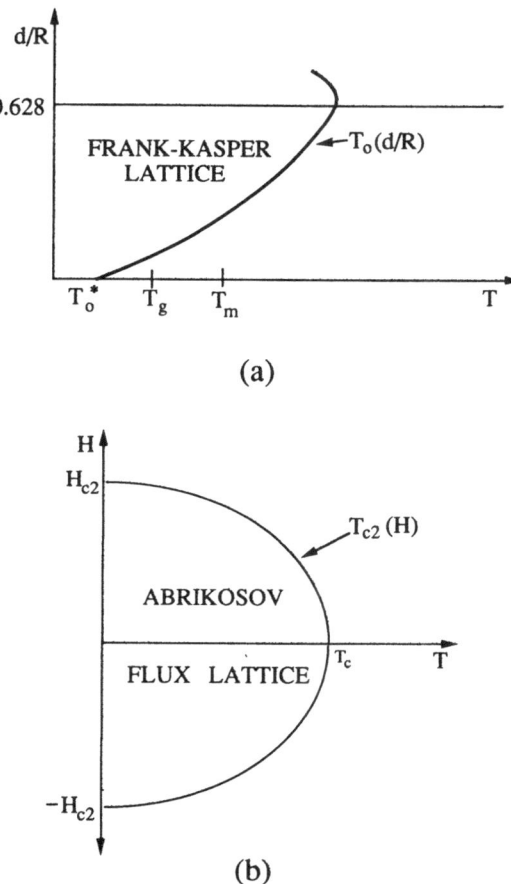

FIGURE 6.13. (a) Hypothetical phase diagram of simple fluids with short-range icosahedral order as a function of temperature and "curvature," i.e., $d/R$. $T_0(d/R)$ is the mean field instability temperature of the liquid toward a Frank–Kasper-like crystal. Its value in flat space is $T_0^*$. $T_m$ is the melting temperature of a FCC lattice. Since $N < \infty$, the transitions at finite curvature will, of course, be smeared by finite size effects. (b) Phase diagram of an extreme type-II superconductor in a magnetic field. $T_c$ is the critical temperature in the absence of a magnetic field and $T_{c2}(H)$ the locus of transitions into an Abrikosov flux lattice.

TABLE 6.1. Comparison of the theoretical predictions from the free energy $F$ for the ratio in the peak positions of the structure factor with experimental measurements on vapor-deposited iron and cobalt and computer simulations on monoatomic supercooled liquids.

|  | Theory | Amorphous Cobalt [26] | Amorphous Iron [27] | Simulations [28] |
|---|---|---|---|---|
| $q_{20}/q_{12}$ | 1.71 | 1.69 | 1.72 | 1.7 |
| $q_{24}/q_{12}$ | 2.04 | 1.97 | 1.99 | 2.0 |

as a function of $q$. Furthermore, the relative positions of the peaks are *independent* of all parameters in $F$. In particular, the strongest peaks occur at the wave vectors $q_{12}$, $q_{20}$, and $q_{24}$ and these wave vectors are predicted to have the ratios:

$$\frac{q_{20}}{q_{12}} = 1.71; \quad \frac{q_{24}}{q_{12}} = 2.04. \tag{6.9}$$

We show in Table 6.1 a comparison of these ratios to the ratios between the first, second, and third peaks in the structure factor of vapor-deposited cobalt and iron, and computer simulations of supercooled liquids: the agreement is remarkably good for $q_{20}/q_{12}$, but there are $\approx 2.5\%$ errors in $q_{24}/q_{12}$. It is argued in Ref. 24 that the effect of higher terms in the free energy must be of a form which improves the agreement in $q_{24}/q_{12}$.

Absolute predictions of the peak positions can be made once the parameter $\kappa$ is known. If we use Eq. (6.1), this involves knowledge of the hard-sphere diameter $d$. Using experimentally known peak positions, we obtain a value of $d$ which is within 10% of the interparticle spacing of crystalline ground states of the relevant atoms. Finally, we can perform a fit to the entire structure factor by using $\kappa$, $K_n$, and $r_n$ as adjustable parameters. The results of such a procedure for amorphous cobalt are shown in Figure 6.14. We find a fourth peak in the structure factor which appears to be a composite of $q_{30}$ and $q_{32}$. The values of $r_n$ can be determined from the fit, and as expected from the analysis associated with Figure 6.13, all of them are negative. We have thus obtained a consistent, phenomenological description of metallic glasses and supercooled liquids which associates peaks in the structure factor with symmetry properties of polytope $\{3,3,5\}$.

# 6.4    Application to Real Metallic Glasses

Strictly speaking, the calculation outlined above should be applicable only to monoatomic dense supercooled liquids. All of the metallic glasses manufactured so far have two or more metallic or metalloid components of differing sizes. Nevertheless, the results obtained above should be of relevance

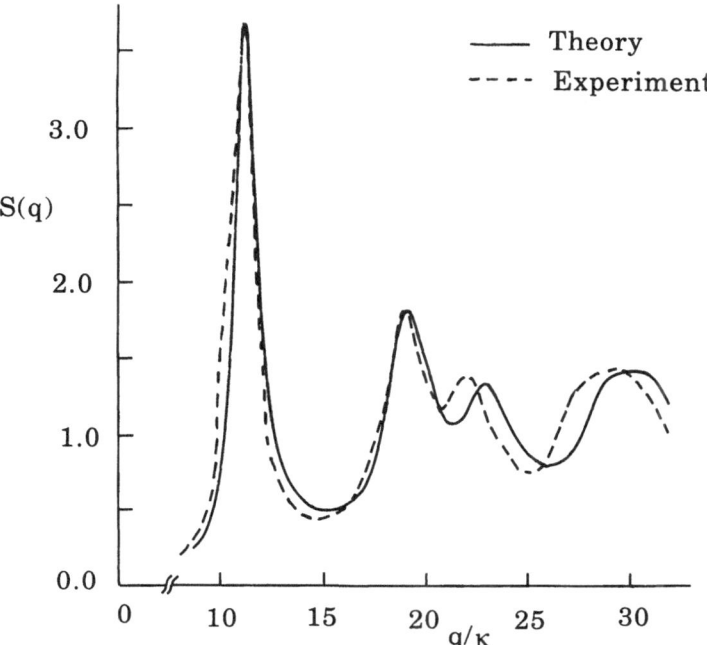

FIGURE 6.14. A comparison of the structure factor obtained by fitting the theoretical prediction to the experimental data on amorphous cobalt films obtained by Leung and Wright [26]. There are two adjustable parameters for each peak that determine its width and height. The peak positions are, however, a consequence of theory.

to real metallic glasses for reasons we shall now discuss. All metallic glasses can be broadly classified into two categories [14]: (1) the metal-metalloid glasses consisting of $M_{1-x}N_x$, where $x$ is in the range 0.15 – 0.20, $M$ is one or more of transition metals like Fe, Co, Ni, Pd, Au, and Pt, and $N$ is one or more of metalloids like P, B, Si, and C; (2) the metal-metal glasses which contain combinations of metals like Mg, Zn, Ca, Al, Cu, Ti, La, and Ce.

## 6.4.1    Metal-Metalloid Glasses

We begin with a discussion of the structure and formation of the metal-metalloid glasses. Before turning to specific materials, we present some qualitative general arguments. In the spirit of the Landau theory introduced above, we can model the effect of the metalloid component by a fluctuating impurity concentration field $c(\mathbf{r})$. The simplest way that $c(\mathbf{r})$ can couple to the icosahedral order parameter is via the replacement [23]

$$F \to F + \int d^3\mathbf{r} \left( \sum_n \gamma_n |\mathbf{Q}_n|^2 + \frac{c^2}{2\chi} - c\Delta \right), \tag{6.10}$$

where $\chi$ is the impurity concentration susceptibility, the $\gamma_n$ are coupling constants, and $\Delta$ is an impurity chemical potential. The constant $\Delta$ and $\gamma_n$ must be positive for the impurities to disrupt the local icosahedral ordering. We can now integrate out the impurity concentration and find that the only effect of the field $c(\mathbf{r})$ is to perform the replacement

$$r_n \to r_n + \frac{\gamma_n \Delta \chi}{2}. \tag{6.11}$$

From the analysis above it is clear that this replacement merely leads to a broadening of the peaks in the structure factor without significantly changing their positions. This is what is observed experimentally.

As a typical example of a metal-metalloid glass, we shall consider in detail the $Ni - P$ system. Shown in Figure 6.15 is the equilibrium phase diagram of the $Ni - P$ system. Also marked is the range of compositions at which the system forms a metallic glass. Note that this range is centered around the deepest eutectic. This is a universal feature among metal-metalloid glasses.

The stable crystal structure near the glass formability range is $Ni_3P$. A first step in understanding the structure of the metallic glass is an analysis of the structure of $Ni_3P$ [29]. The space group of the crystal is $S_4^2$, with each unit cell containing 24 $Ni$ atoms in symmetry partners of three nonequivalent positions and 8 $P$ atoms in symmetry partners of one position. We assign nearest-neighbor bonds to all $Ni - Ni$ distances which are smaller than 3.58 $A$ and to all $Ni - P$ distances which are less than 2.58 $A$. These distances are chosen so that the results are equivalent to a "weighted" [30,31] Voronoi construction. In Figure 6.16 we show the coordination shell of $P$. There are 9 $Ni$ nearest neighbors arranged so that all

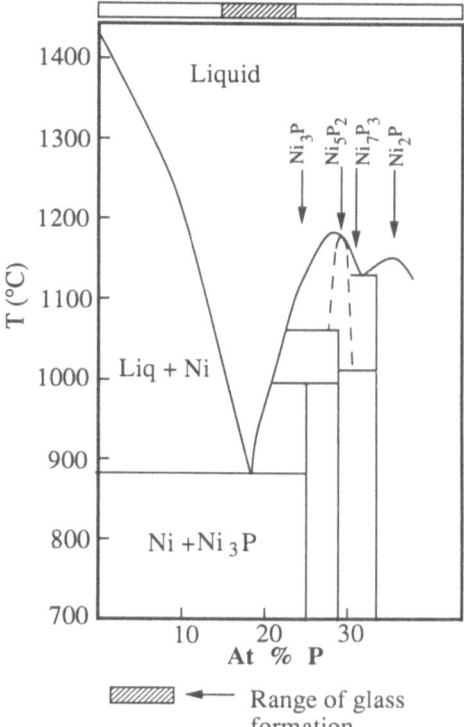

FIGURE 6.15. Phase diagram of the $Ni - P$ system. Also indicated is the range of compositions over which rapid cooling from the melt forms a glass.

TABLE 6.2. Coordination statistics for the $P$ and $Ni$ atoms in $Ni_3P$. The number of $n$-fold bonds emanating from an atom is represented by $v_n$. The total coordination number of the atom is $Z$.

| Atom | $v_4$ | $v_5$ | $v_6$ | $Z$ |
|------|-------|-------|-------|-----|
| $P$ | 3 | 6 | 0 | 9 |
| $NiI$ | 1 | 10 | 4 | 15 |
| $NiII$ | 3 | 6 | 7 | 16 |
| $NiI$ | 3 | 6 | 5 | 14 |

but three of the bonds are fivefold. These three bonds are fourfold and the entire structure forms a $Z9$ defect in the notation of Nelson [9]. The $P$ atom in $Ni_2P$ has a very similar coordination shell. It is the small size of the $P$ atom which forces in a small coordination number and fourfold coordinated bonds. The partially covalent nature of the $Ni - P$ bond is also responsible for this structure. The fourfold bonds lead to the presence of distorted octahedra. However, the shortest diagonal of the octahedron is almost 40% smaller than the other two, demonstrating that the octahedra are better described as fourfold bipyramids. The three different coordination shells of the $Ni$ atoms are more complicated. They are shown in Figure 6.17. The majority of the bonds emanating from the $Ni$ are fivefold, with the remaining bonds being either sixfold or fourfold. All the fourfold bipyramids associated with these $Ni$ atoms have at least one $P$ atom on their vertices. The diagonals of the fourfold bipyramids now differ by only 12%, so they are more nearly octahedral. However, as discussed below, these octahedra occupy only a small fraction of the space in the unit cell. In Table 6.2 we display the coordination statistics of the $Ni$ and $P$ atoms. In Table 6.3 are displayed the relative fractions of fourfold, fivefold, and sixfold bonds in crystalline $Ni_3P$. For comparison, we have also displayed the relative fractions of these bonds in the relaxed Bennett model of Ichikawa [7]. Note the remarkable similarity between the coordination topology of a dense random packing model of *identical* spheres and that of a packing of two different sizes spheres. In the case of $Ni_3P$, it is the smaller size of the $P$ atom which forces in an appreciable number of fourfold bonds. The fourfold bonds in the Ichikawa–Bennett model [7] represent a presence of octahedral "voids" which will probably decrease in number upon further relaxation.

Experiments of Cocco et. al. [32] measuring the radial distribution function of glassy $Pd_{1-x}B_x$ (which we expect to be structurally very similar to $Ni_{1-x}P_x$) showed a small peak at $\sqrt{2}$ times the $Pd$ diameter, indicating the presence of fourfold bipyramids. However, the size of the peak *decreased* with decreasing $B$ concentration [32], indicating that the small size of the $B$ atom was responsible for the fourfold bipyramids. Recognizing the importance of the special coordination topology of the $P$ atoms, Gaskell [33]

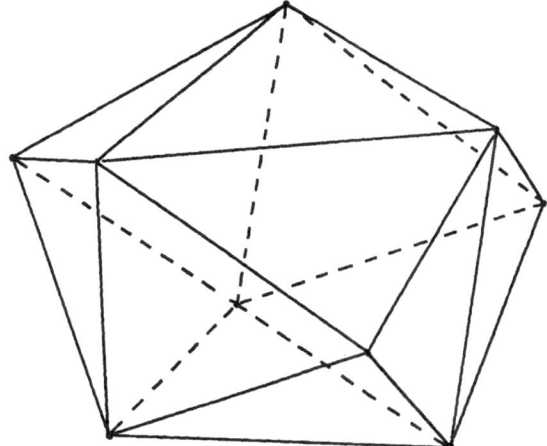

FIGURE 6.16. Coordination shell of the $P$ atom in $Ni_3P$. The topology of this shell is identical to the $Z9$ defect in Figure 6.9. We denote the number of $n$-fold bonds emanating from the central $P$ atom by $v_n$; we have $v_5 = 6$, $v_6 = 0$, and $v_4 = 3$.

TABLE 6.3. Distribution of fourfold, fivefold, and sixfold bonds in various structures. The percentage of $n$-folds is represented by $f_n$ and $q$ is the mean number of tetrahedra around a bond.

| Structure | $f_4$ | $f_5$ | $f_6$ | $q$ |
|---|---|---|---|---|
| $Ni_3P$ | 18 | 52 | 30 | 5.12 |
| Ichikawa–Bennett | 16 | 53 | 27 | 4.91 |
| Gaskell | 27 | 49 | 18 | 4.61 |

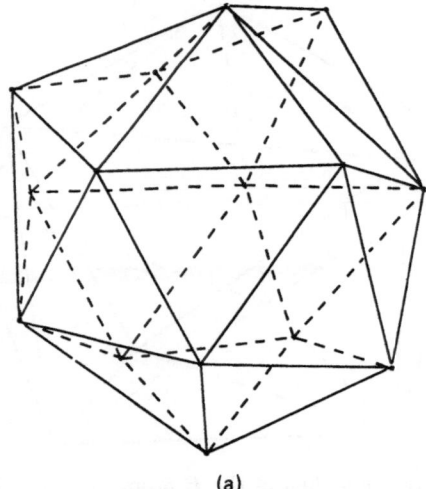

(a)

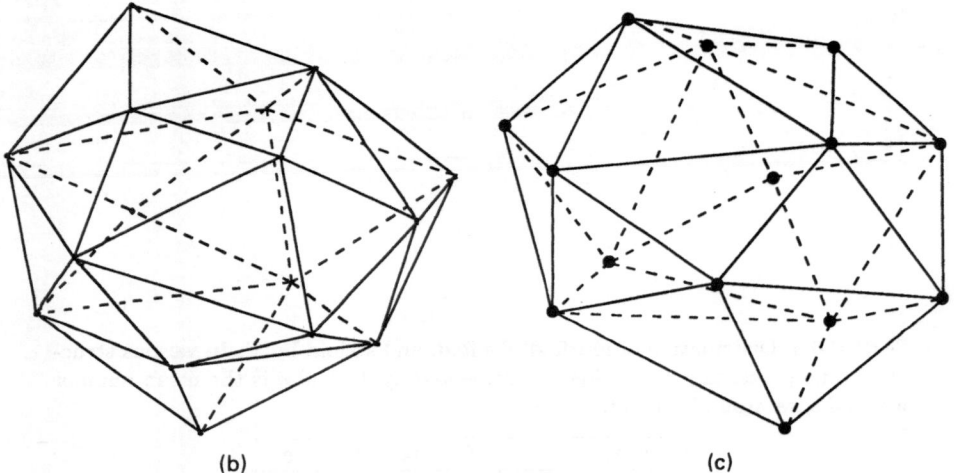

(b)                              (c)

FIGURE 6.17. Coordination shell of the (a) $NiI$ ($v_5 = 10$, $v_6 = 4$, $v_4 = 1$), (b) $NiII$ ($v_5 = 6$, $v_6 = 7$, $v_4 = 3$), and (c) $NiIII$ ($v_5 = 6$, $v_6 = 5$, $v_4 = 3$) atoms in $Ni_3P$.

has performed a simulation of the structure of *glassy* $Ni_{1-x}P_x$. The model consists essentially of a random packing of phosphorus units having a local environment similar to that of $P$ in $Ni_3P$, shown in Figure 6.16. Gellatly and Finney [30] performed a Voronoi analysis of this structure using a technique which recognized the size differences between the atoms. The results of this analysis are shown in Table 6.3. Again, almost 50% of the bonds are fivefold coordinated. There is a larger number of fourfold bonds than in the relaxed Ichikawa–Bennett model, but this is probably because the constructions constrain every phosphorus environment to be that of Figure 6.16. The main conclusion we wish to draw from all of the above arguments is that the dense random tetrahedral close packing of spheres remains a good approximation for the structure of glassy $Ni_{1-x}P_x$. With the additional broadening of the peak widths, the analysis based on polytope $\{3, 3, 5\}$ remains valid.

As noted earlier, most metal-metalloid glasses form at compositions near the eutectic [2]. This can be most simply understood by the fact that there is no *stable crystalline structure* at the eutectic composition. Increasing the nickel content in the $Ni_3P$ structure will increase the energy of the crystal drastically because the $Ni$ atoms do not prefer to sit at the smaller $P$ sites. There may well be exotic metastable crystal structures with larger unit cells at the eutectic compositions which will accommodate the tendency of the $Ni$ atoms to sit in the center of icosahedra and, at the same time, enable the $P$ atoms to maintain their environment. However, the fourfold defect lines which are forced in by the $P$ atoms act as kinetic hinderances and lock the system in a metastable glassy state. At even lower $P$ compositions, the pure crystalline $Ni$ structure is stable enough to prevent glass formation. The importance of the high energies of the crystal structure at the eutectic composition was illustrated by some experiments of Chen [34,35]. None of the thermodynamic measurements of Chen indicated any extra structural stability of the glassy structure at the eutectic composition.

## 6.4.2   Metal-Metal Glasses

Glasses are formed on the rapid cooling of many molten metal-metal alloys. These glasses may be further classified as (1) simple metal — simple metal glasses such as $Mg_{70}Zn_{30}$, $Ca_{67}Al_{33}$, (2) simple metal — transition metal glasses such as $Ca_{65}Pd_{35}$, $Ti_{60}Be_{40}$, and (3) transition metal — transition metal glasses such as $Nb_{60}Ni_{40}$, and $Fe_{55}W_{45}$. Some rare-earth alloys also form metallic glasses. A universal feature of these alloys is that at these or at differing compositions they form stable intermetallic compounds which are either Frank–Kasper or closely related structures. For the alloys mentioned above, the associated Frank–Kasper phases are $MgZn_2$, $CaAl_2$, $CaPd_2$, $Ti_2Be_{17}$, $TiBe_2$, $NbNi$, and $Fe_7W_6$. As we have already discussed, all of these crystalline phases are characterized by a tetrahedral network of near neighbor bonds, with all the tetrahedra being approximately equilateral. This supports the conjecture that the disordered glassy arrangements of these alloys are described by tetrahedral close packing. A majority of the bonds will have five tetrahedra around them, interspersed with a tangled network of lines of sixfold and some fourfold bonds.

In these systems, there are compositions at which the Frank–Kasper phases are very stable and prevent effective glass formation. The Laves phase, for instance, will form when one-third of the atoms are larger than the rest, the ratio of the radii being in the range $1.1 - 1.3$. It is easy to see how requirements of geometrical close packing of spheres in the Laves phase forces in this requirement. It is important to note, as pointed out by Hafner [14], that changes in the atomic diameter in the presence of the other element of the alloy need to be taken into account. These changes in diameter are controlled by electronegativity differences between the components: the more electropositive atom shrinks in size. At compositions in which the larger atom is in the majority, no stable crystalline structure will form, and the ease of glass formability will increase. In this regard, a calculation performed by Hafner [14] on the $Ca - Mg$ system is of particular interest. Using a self-consistent pseudopotential method, Hafner estimated the total energies of $Ca - Mg$ Laves phase and the metallic glass as modeled by a relaxed Finney structure. Remarkably, the energy of the glassy structure was a minimum at the *same* composition as the $CaMg_2$ Laves phase. This nonetheless did not include the range of glass formation. At other compositions near the eutectic, the energy of the crystalline structure was driven up considerably, leading to easier glass formation.

## 6.5   Conclusions

This chapter has reviewed a statistical theory of the packing of identical hard spheres in flat three-dimensional space. The theory begins by identifying the tetrahedral arrangement of four spheres as an important locally dense arrangement and proceeds to tile all space with the tetrahedra. Such a task is well known to be impossible for equilateral tetrahedra and distortions are inevitably introduced. Five tetrahedra can pack around a bond with a minimum of distortion (Figure 6.2). An attempt to continue this fivefold structure to all of space runs up against further geometric obstacles and one finds that sixfold (Figure 6.6) and fourfold (Figure 6.7) must be introduced. An orientational/translational order parameters was then introduced to measure the amplitude and the orientation of the fivefold bipyramids (Figure 6.2). This rather complicated multiparticle order parameter was defined by comparing the local environment in the neighborhood of any point with that of an ideal packing of spheres (polytope $\{3, 3, 5\}$) on a uniformly curved three-dimensional space. In this latter packing, the curvature of the space has been carefully chosen to allow for a tiling with identical equilateral tetrahedra, thus relieving the frustration of flat three-dimensional space. We emphasize that this curved space is used only as a mathematical tool to generate the bond-orientational order parameter which describes sphere packings in the physical flat three-dimensional space. A Landau free energy was introduced to describe fluctuations of this order parameter in a manner which incorporated the frustration inherent in flat space. The frustration induced disclinations in the order parameter which were identified with the fourfold and sixfold bipyramids. Finally, the Landau free energy was used to calculate the structure factor of the hard-sphere packing. Parameter-free predictions were made for the ratio of the peak positions of the structure factor which were in rather good agreement with experimental measurements on amorphous metallic films. To our knowledge, this is the first theory to present a direct calculation and physical interpretation of the peaks in the structure factor. All existing computer simulations have focused mainly on the peaks in the radial distribution function, which is a Fourier transform of the structure factor.

The second part of this chapter dealt with the application of these results to realistic metallic glasses which always contain combinations of metallic/metalloid atoms of different sizes. It was argued that peak position in the structure factor of these glasses should be insensitive to the presence of atoms of slightly differing radii: the main effect of the alloying should be a broadening of the peaks, an effect which is borne out experimentally. Monoatomic metallic glasses do not form because of the presence of low crystalline structures with cubic symmetry which preempt glass formation. With the introduction of atoms of different sizes, these crystalline structures are destabilized and the system is able to access a glassy state. The main shortcoming of our approach is the lack of a precise criterion for the conditions which enhance glass formability. This question has, however, been addressed with first-principles electronic structure calculations by Hafner [14].

The ideas discussed in this chapter can be applied to other properties of metallic glasses: these include electronic structure [36], phonon density of states [37], and viscous relaxation [38]. The reader is referred to the original papers for details. In the following chapter Jarić will discuss a different, but closely related, formulation of the bond-orientational order parameter and its coupling to translation symmetries and quasicrystalline order.

*Acknowledgments.* All of my work on the subject reviewed in this paper was performed in collaboration with D.R. Nelson; I am grateful to him for freely offering his insights and many useful discussions. This review has been adapted from Chapter 1 of S. Sachdev, Ph.D. Thesis, Harvard University, 1985. The work was supported, in part, by the National Science Foundation under Grants No. DMR82-07431 and DMR8857228, and by the Alfred P. Sloan Foundation.

# References

[1] W. Klement, R.H. Willens, and P. Duwez, *Nature* **187**, 809 (1960); for a historical account, see P. Duwez, *Trans. Am. Soc. Metals* **60**, 607 (1967).

[2] For a review, see *Amorphous Metallic Alloys*, edited by F.E. Luborsky, Butterworth, Boston, 1983.

[3] M.R. Hoare, *Adv. Chem. Phys.* **40**, 49 (1979).

[4] J.D. Bernal, *Proc. Roy. Soc. Lond. Ser. A* **280**, 299 (1964).

[5] J.L. Finney, *Proc. Roy. Soc. Lond. Ser. A* **319**, 479 (1979).

[6] C.H. Bennett, *J. Appl. Phys.* **43**, 2727 (1972).

[7] T. Ichikawa, *Phys. Status Solidi* **19**, 707 (1973); T. Yamamoto, H. Shituba, T. Mihara, K. Haga, and M. Doyama, in *Proceedings of International Conference on Rapidly Quenched Metals*, Vol. 1, edited by T. Masumoto and K. Suzuki, Japan Institute of Metals, Sendai, 1982.

[8] J.L. Finney and J. Wallace, *J. Non-Cryst. Solids* **43**, 165 (1981).

[9] D.R. Nelson, *Phys. Rev. B* **28**, 5515 (1983).

[10] P.J. Steinhardt, D.R. Nelson, and M. Ronchetti, *Phys. Rev. Lett.* **47**, 1297 (1981); *Phys. Rev. B* **28**, 784 (1983).

[11] F.C. Frank and J.S. Kasper, *Acta Crystrallogr.* **11**, 184 (1958), **12**, 483 (1959).

[12] D.E. Polk and B.C. Giessen, *Metallic Glasses*, American Society of Metals, Metals Park, Ohio, 1978.

[13] R. St. Amand and B.C. Giessen, *Scr. Metall.* **12**, 1021 (1978).

[14] J. Hafner, *Phys. Rev. B* **21**, 406 (1980).

[15] N.D. Mermin, *Rev. Mod. Phys.* **51**, 591 (1979).

[16] H.S.M. Coxeter, *Ill. J. Math.* **2**, 746 (1958); *Introduction to Geometry*, Wiley, New York, 1969; *Regular Polytopes*, Dover, New York, 1973.

[17] M. Kléman and J.F. Sadoc, *J. Phys. Lett.* (Paris) **40**, L569 (1979).

[18] J.F. Sadoc, *J. Phys. Colloq.* (Paris) **41**, C8-326 (1980).

[19] J.F. Sadoc and R. Mosseri, *Phil. Mag. B* **45**, 467 (1982).

[20] D.R. Nelson and M. Widom, *Nucl. Phys. B* **240**, 113 (1984).

[21] M. Bander and C. Itzykson, *Rev. Mod. Phys.* **38**, 330 (1966).

[22] J.P. Sethna, *Phys. Rev. Lett.* **50**, 2198 (1983).

[23] S. Sachdev and D.R. Nelson, *Phys. Rev. Lett.* **53**, 1947 (1984).

[24] S. Sachdev and D.R. Nelson, *Phys. Rev. B* **32**, 1480 (1985).

[25] J.P. Straley, *Phys. Rev. B* **30**, 6592 (1984).

[26] P.K. Leung and J.C. Wright, *Phil. Mag.* **30**, 185 (1974).

[27] J.-P. Lauriat, *J. Non-Cryst. Solids* **55**, 77 (1983).

[28] J.D. Weeks, *Phil. Mag.* **35**, 1345 (1977).

[29] B. Aronsson, *Acta Chem. Scand.* **9**, 137 (1955).

[30] B.J. Gellatly and J.L. Finney, *J. Non-Cryst. Solids* **50**, 313 (1982).

[31] F.M. Richards, *J. Mol. Biol.* **82**, 1 (1974).

[32] G. Cocco, S. Enzo, M. Sampoli, and L. Schiffni, *J. Non-Cryst. Solids* **61 & 62**, 577 (1984).

[33] P.H. Gaskell, *J. Non. Cryst. Solids* **32**, 207 (1979).

[34] H.S. Chen, *Acta Metall.* **24**, 153 (1976).

[35] H.S. Chen, *Mat. Sci. and Engrg.* **23**, 151 (1976).

[36] M. Widom, *Phys. Rev. B* **31**, 6456 (1985).

[37] M. Widom, *Phys. Rev. B* **34**, 756 (1986).

[38] S. Sachdev, *Phys. Rev. B* **33**, 6395 (1986).

# 7

# Orientational Order and Quasicrystals

*Marko Vukobrat Jarić*

*To the memory of my mother, Mileva Jarić.*

## 7.1 Introduction

At the end of 1984, a group of researchers at the National Institute of Standards and Technology reported a discovery of an alloy of aluminum and manganese whose electron diffraction pattern revealed a novel structure [1]. While the diffraction peaks were relatively sharp, indicating positional coherence on the order of several hundred angstroms, the pattern itself displayed a noncrystallographic, icosahedral orientational order. This discovery challenged a century-old belief that periodicity and structural order are essentially equivalent. This chapter will address the above challenge within the framework of the Landau theory of freezing [2] and from the point of view of bond-orientational order, but without a detailed introduction to quasicrystals. A more comprehensive introduction to quasicrystals can be obtained from a number of review articles [3–7], collection of reprints [8], and books [9–11].

Since the initial discovery, a number of quasicrystals have been found to possess besides icosahedral also other noncrystallographic symmetries such as octagonal, decagonal, and dodecagonal. The extent of orientational and translational order observed in quasicrystals varies. For example, translational order ranges from $\geq 10^2$ Å in i(Al-Mn-Si) to $\geq 10^4$ Å in i(Al-Cu-Fe). Different quasicrystals are also found under widely different conditions, including rapid solidification at one end of the spectrum and slow extraction from the melt at the other end, and they can be thermodynamically metastable or stable. It is likely, however, that the materials which are now commonly called quasicrystals encompass several different classes of structural order.

Theoretically, several broad classes can be identified depending on whether translational or orientational order is idealized. If the translational order is assumed to be long-ranged, but the noncrystallographic orientational order is assumed, only short-ranged, then quasicrystals are nothing but crystals with large unit cells containing many parallel, nearly icosa-hedral clusters. In this case, the icosahedral orientational order is only approximate, and the true icosahedral symmetry of the diffraction pattern could be recovered only by icosahedral multiple twinning of the crystal. Both twinning and deviations from icosahedral symmetry have been iden-tified in several phases of Al-Cu-Li and Al-Cu-Fe which were originally thought to be quasicrystalline. This class, as well as the class of structures with neither perfect translational nor icosahedral orientational order, will not be discussed here.

The remaining possibilities are either that both translational and non-crystallographic orientational order are long-ranged, or that only orienta-tional order, whether crystallographic or not, is long-ranged. Either of the two cases would represent a qualitatively new kind of structural order not observed previously. In the first case, probably applicable to materials like i(Al-Cu-Fe), quasicrystals are, in fact, quasiperiodic crystals, often exempli-fied by Penrose tiling [12]. This was one of the first proposed explanations of the quasicrystal's structural order [13], although it was applied to i(Al-Mn) which, with its very short translational correlation length, probably represents the second case, the so-called icosahedral glass, exemplified by a model of "random" packing of icosahedra [14]. Successes and failures of these two models are the subject of the chapter by A. Goldman and will not be further elaborated here.

A common feature of the two models is the presence of long-range icosa-hedral orientational order, which is the main focus of this chapter. More precisely, this chapter addresses the question of what structures with long-range icosahedral orientational order can be stabilized, irrespective of whether long-range translational order is present or not. The problem will be approached from the point of view of long-range bond-orientational or-der. The importance of icosahedral bond-orientational order has been long established in laboratory and computer experiments on metallic glasses and supercooled liquids. Even in crystals, some of which are closely related to icosahedral quasicrystals, such as $\alpha$(Al-Mn-Si) and R(Al-Cu-Li), there are clear remnants of icosahedral bond-orientational order. The topic of bond-orientational order in metallic glasses and supercooled liquids is the subject of the chapter by S. Sachdev.

Landau theories based solely on translational order parameters have not been successful in their goal to obtain a *generic* phase with long-range icosahedral translational order,[1] giving another motivation for seeking an orientational order route to icosahedral quasicrystalline phases. In fact, an explicit minimization of the bond-orientational Landau free energy results in generic stability of an icosahedral phase (icosahedratic), although oc-tahedral, uniaxial, and hexagonal phases are also present [16]. It is this

---

[1]For a review see, e.g., Ref. 15 and references therein.

stability, we argue, that is responsible for the formation of the icosahedral quasicrystalline phases [17].

In an interesting paper published in 1981, Nelson and Toner [18] first suggested the possibility of a bulk, three-dimensional bond orientationally ordered phase intermediate between a crystal and isotropic liquid. As a bond-orientational order parameter, they considered the translationally invariant $L = 4$ spherical harmonic of the bond density, which can describe an octahedratic (cubic) bond orientationally ordered phase, in contrast to the $L = 2$ harmonic which can only describe a nematic. They showed that the Landau free energy associated with this order parameter, and appropriate for the isotropic-liquid-to-octahedratic transition, contains a third-order invariant, so that according to the mean-field Landau theory the transition must be discontinuous. Similarly, since the Landau free energy appropriate for the octahedratic-to-crystal transition also contains a third-order term in the translational order parameter, they concluded that this transition must also be discontinuous. However, they noted that fluctuations may change the liquid-octahedratic transition to a continuous one, while the octahedratic-BCC crystal transition is likely to remain discontinuous. Nelson and Toner also examined physical properties of the octahedratic phase and identified its signatures in the X-ray structure function, in hydrodynamic properties, and in characteristic singularities of the thermodynamic functions.

The possibility of an icosahedratic phase was first considered in detail by Steinhardt, Nelson, and Ronchetti [19] in molecular dynamics simulations of a supercooled Lennard–Jones liquid. They used the $L = 6$ order parameter, since this is the lowest harmonic associated with icosahedral symmetry, and constructed the orientational Landau free energy to third order in this order parameter. Since the third-order term is present for $L = 6$, they concluded that the transition between a liquid and a bond orientationally ordered phase is discontinuous but might become continuous if the effect of fluctuations is taken into account. As a measure of the "icosahedrality" of the bond-orientational order, they adopted the invariant associated with the third degree term in the free energy previously used in a different context by Busse [20] and Sattinger [21]. Evaluating this invariant for many different clusters, they never observed it to exceed the value obtained for clusters with icosahedral symmetry. Steinhardt et al. also included in the free energy a translational order parameter up to third order and concluded, from its coupling with the orientational order parameter, that orientational order must be induced by the periodic order, whereas translational order is not necessarily generated by bond order.

Some of the work started in Refs. 18 and 19 was later completed and extended by the author [16,17]. First, we constructed the fourth-order terms in the orientational free energy and determined the resulting orientational phase diagram for $L = 2, 4, 6$ by an explicit minimization of the free energy [16]. Then, using the orientational translational interaction and assuming that parameters are in the icosahedratic portion of the $L = 6$ phase diagram, we determined the fundamental density waves which set up the long-range icosahedral translational order [17]. Mainly this work will be reviewed in this chapter. Some new simple illustrations will be added, and some of the calculations will be presented in a more explicit form. Also, new, more selective rotationally invariant measures of bond-orientational order parameter symmetry will be introduced.

The picture of icosahedral quasicrystals which emerges from the results of this chapter is the following. As an isotropic liquid is cooled, it develops long-range icosahedral bond-orientational order at a temperature $T_0$, above a temperature $T_t$ at which a long-range translational order would develop. Even if in a real system $T_t > T_0$, it might be possible to supercool a liquid in order to reach this isocahedral phase. Although all solidlike elastic constants vanish in an icosahedratic, it shows a resistance to torsion, distinguishing it from isotropic liquids [18]. However, it should be noted that orientational order in this phase is not necessarily connected with the intrinsic shape of the constituent molecules, like it is in the liquid crystals, but it is rather a consequence of the bond ordering.

Once the icosahedratic is formed, it may lead to quasicrystal formation by one of two routes: via quenching to an icosahedral glass, or via a coupling between the orientational and the translational order parameter which, if strong enough, may induce icosahedral long-range translational order at a temperature $T'_t$, $T_t < T'_t \leq T_0$. Since translational order necessarily implies orientational order, but not *vice-versa*, this sequence of phases — translationally ordered (quasi)crystal, orientationally ordered icosahedratic, disordered liquid — physically makes sense in view of the corresponding increases in entropy. Of course, it is also possible to break the translational symmetry of the orientational order parameter (bond density) without a reference to the translational order parameter (density), like in the quasicrystal models of a Blue Fog liquid crystal phase,[2] but this will not be discussed here.

While in the above scenario the icosahedral glass phase is necessarily metastable and preceded by an icosahedratic phase, the quasiperiodic phase can be either stable or metastable and, if the translational orientational coupling is strong enough, a separate icosahedratic phase might be preempted by a translationally ordered phase (i.e., $T'_t = T_0$). This is consistent with the conclusion that i(Al-Mn-Si), which can be obtained only by rapid solidification and which is metastable, is probably an icosahedral glass, while i(Al-Cu-Fe), which can be obtained by slow cooling and is stable all the way up to the melting, is a quasiperiodic crystal. Moreover, this scenario leads to a specific prediction that the fundamental set of reciprocal vectors associated with the peaks in the diffraction pattern is either along the fivefold

---

[2]For a review see Ref. 22 and references therein.

or along the threefold symmetry axes of an icosahedron. This prediction has been borne out by experiments. Luckily, this scenario does not predict that an icosahedral quasicrystal must occur. The easiest way around quasicrystals is for $T_t$ to be sufficiently above $T_0$. Also, the orientational order parameter might set into one of the other possible symmetries ($O_h$, $D_{\infty h}$, or $D_{6h}$), or the coupling between the orientational and the translational order parameters might be too weak, allowing for a translational order with a different symmetry. Finally, if the solid-liquid transition is strongly first order, conclusions of the Landau theory are of a limited value.

In the following section we shall define the bond-orientational order parameter, its symmetries and their measures. Then, in Section 7.3, we shall determine the bond-orientational phase diagram and use the result in Section 7.4 to determine conditions for the stability of icosahedral quasiperiodic crystals. The last section is devoted to a brief discussion of the results.

# 7.2    Bond-Orientational Order Parameter

## 7.2.1    Definition

Phenomenological theories based on the Landau theory of phase transitions are variational theories which require minimization of a free energy functional with respect to a coarse-grained order parameter. For example, the order parameter in the Landau approach to solidification [2,15,23–26] should be viewed as a coarse-grained mass density,

$$\rho(\mathbf{r}) = \int \langle \rho_\mu(\mathbf{r} + \mathbf{x}) \rangle w_t(\mathbf{x})\, d^3\mathbf{x}, \tag{7.1}$$

where

$$\rho_\mu(\mathbf{r}) = \sum_i \delta(\mathbf{r} - \mathbf{r}_i) \tag{7.2}$$

is the microscopic density, the sum is over the pointlike particles located at $\mathbf{r}_i$, $\langle \dots \rangle$ denotes a thermodynamic average, and $w_t(\mathbf{x})$ is an appropriate coarse-graining function. Such coarse graining of microscopic quantities is performed in order to avoid complexity which is irrelevant at a more macroscopic level. Coarse graining leads to a small order parameter so that a free energy expansion can be generated. However, if the expansion is also restricted to a low order in derivatives of the order parameter, some nonlocal properties will be neglected and may have to be included as additional coarse-grained order parameters. Such a nonlocal property is the bond-orientational density, which is proportional to a number of bonds emanating from $\mathbf{r}$ and pointing in a direction $\hat{\mathbf{n}}$,

$$\Omega(\mathbf{r}, \hat{\mathbf{n}}) = \int \langle \Omega_\mu(\mathbf{r} + \mathbf{x}, \hat{\mathbf{n}}) \rangle w_0(\mathbf{x})\, d^3\mathbf{x}, \tag{7.3}$$

where

$$\Omega_\mu(\mathbf{r}) = \rho_\mu(\mathbf{r}) \int [\rho_\mu(\mathbf{r}+\mathbf{n}) - \delta(\mathbf{n})] w n^2 \, dn = \sum_i \delta(\mathbf{r} - \mathbf{r}_i) \sum_{j \neq i} \delta^{(2)}(\hat{\mathbf{r}}_{ij}, \hat{\mathbf{n}}) w_{ij}$$

(7.4)

is the microscopic bond density, $\mathbf{r}_{ij} = \mathbf{r}_j - \mathbf{r}_i$, $w$ (or $w_{ij}$) is a weighting function which in the simplest case depends only on $n = |\mathbf{n}|$, and $w_0(\mathbf{x})$ is a coarse-graining function not necessarily equal to $w_t(\mathbf{x})$. The delta function on the sphere, $\delta^{(2)}(\hat{\mathbf{n}}_1, \hat{\mathbf{n}}_2)$, is defined as $\sum_{Lm} Y_{Lm}(\hat{\mathbf{n}}_1) Y_{Lm}^*(\hat{\mathbf{n}}_2)$.

It is convenient to expand the orientational order parameter in spherical harmonics with respect to $\hat{\mathbf{n}}$, and, plane waves (Fourier series/integral) with respect to $\mathbf{r}$:

$$\Omega_{Lm}(\mathbf{q}) \sim \int e^{i\mathbf{q}\cdot\mathbf{r}} \int Y_{Lm}(\hat{\mathbf{n}}) \Omega(\mathbf{r}, \hat{\mathbf{n}}) \, d^2\hat{\mathbf{n}} \, d^3\mathbf{r}.$$

(7.5)

For example, the orientational order parameter corresponding to a set of points located at $\mathbf{r}_i$ given in Eq. (4) leads to

$$\Omega_{Lm}(\mathbf{q}) \sim \sum_i e^{i\mathbf{q}\cdot\mathbf{r}_i} \sum_{j \neq i} Y_{Lm}(\mathbf{r}_{ij}) w_{ij}.$$

(7.6)

Clearly, under a rotation $g \in O(3)$, $\Omega_{Lm}(\mathbf{q})$ transforms as

$$\Omega_{Lm}(\mathbf{q}) \to D_{mm'}^L(g^{-1}) \Omega_{Lm'}(g\mathbf{q}),$$

(7.7)

where $D_{mm'}^L(g)$ is the unitary matrix representing $g$. Under a translation $\mathbf{t}$, $\Omega_L(\mathbf{q})$ transforms as

$$\Omega(\mathbf{q}) \to e^{i\mathbf{q}\cdot\mathbf{t}} \Omega(\mathbf{q}).$$

(7.8)

In this paper we shall be concerned mainly with the uniform $\mathbf{q} = 0$ orientational order parameter, in which case only even $L$ appear in Eq. (7.5). Also note that the reality of $\Omega(\mathbf{r}, \hat{\mathbf{n}})$ implies $\Omega_{Lm}^*(\mathbf{q}) = (-1)^m \Omega_{L,-m}(-\mathbf{q})$.

Steinhardt et al. [19] calculated $\Omega_{Lm}(0)$ for some icosahedral, FCC, BCC, SC, and HCP clusters and found that for the simple cubic cluster the "dominant" $L$ (in the sense to be precisely defined in Section 7.2.3) is $L = 4$, whereas for the other clusters the dominant $L$ is $L = 6$, although for BCC and FCC clusters the nonvanishing contributions arose also for $L = 4$. Therefore, we shall focus our attention on these harmonics. For completeness, we shall also analyze the case $L = 2$ which is related to nematic ordering.

## 7.2.2  Symmetries

For a given $L$, the order parameter $\Omega_{Lm}$ could be viewed as a vector in a $(2L + 1)$-dimensional space. Using the obvious vector notation, we write

$$\Omega_L = (\Omega_{L-L}, \ldots, \Omega_{L0}, \ldots, \Omega_{LL}).$$

(7.9)

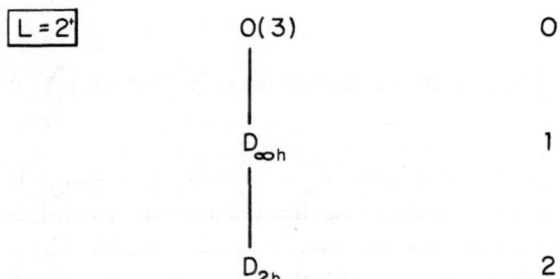

FIGURE 7.1. $O(3)$ isotropy groups for $L = 2^+$. Lines indicate group-subgroup relationships with smaller groups being at the lower levels. The integers at the right indicate dimensionalities of corresponding symmetry planes. This figure is adapted from Ref. 16.

It was found in Ref. 19 that for all clusters with octahedral symmetry

$$\Omega_2^{O_h} = 0, \tag{7.10}$$

$$\Omega_4^{O_h} \sim \left( \sqrt{\frac{5}{24}}, 0, 0, 0, \sqrt{\frac{7}{12}}, \ldots \right), \tag{7.11}$$

$$\Omega_6^{O_h} \sim \left( 0, 0, \sqrt{\frac{7}{16}}, 0, 0, 0, -\sqrt{\frac{1}{8}}, \ldots \right), \tag{7.12}$$

provided the coordinate axes were aligned with the fourfold rotation axes. Only the proportionality constant varied, depending on the specific form of the clusters (say, BCC, FCC, or SC) or on the specific form of $w_{ij}$, i.e., on the rule for selecting the bonds contributing to Eq. (6). A similar situation occurred for icosahedral clusters for which they found

$$\Omega_2^{Y_h} = 0, \tag{7.13}$$

$$\Omega_4^{Y_h} = 0, \tag{7.14}$$

$$\Omega_6^{Y_h} \sim \left( 0, i\sqrt{\frac{7}{25}}, 0, 0, 0, 0, \sqrt{\frac{11}{25}}, \ldots \right), \tag{7.15}$$

provided the $z$-axis was aligned with a fivefold rotation axis and the $x$-axis with a twofold rotation axis. Of course, as we shall show below, these observations were not coincidences.

All elements of $O(3)$ which leave a vector $\Omega_L$ invariant constitute its isotropy group. Precisely this group, which clearly depends only on the direction $\hat{\Omega}_L$ of $\Omega_L$, is the symmetry group of the phase characterized by the order parameter. Therefore, it is essential to determine the symmetries of *all* the vectors in the $(2L + 1)$-dimensional vector space $\mathcal{V}_L$

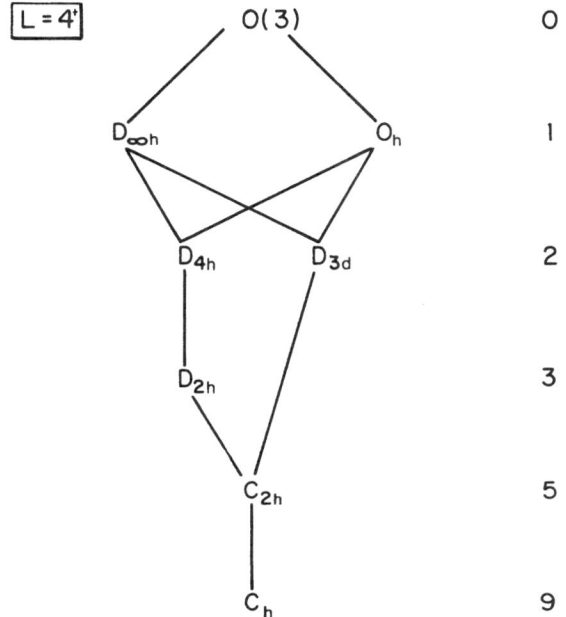

FIGURE 7.2. Isotropy groups for $L = 4^+$. This figure is adapted from Ref. 16.

spanned by $\Omega_L$. Equivalently, for a given representation $L$ one must determine all the isotropy subgroups of O(3). The isotropy subgroups of SO(3) were calculated by Michel [27] and by Ihrig and Golubitsky [28]. Since $\Omega_L \to (-1)^L \Omega_L$ under the inversion through the origin ($\mathbf{r} \to -\mathbf{r}$), the corresponding isotropy subgroups of O(3) are the same as those of SO(3) for $L$ odd, while for $L$ even they are the direct product of the isotropy subgroups of SO(3) with $C_i$.[3] The results for $L = 2$, 4, or 6 are given in Figures 7.1, 7.2, and 7.3, respectively. We note that the isotropy groups which are listed represent, in fact, the equivalence classes of isotropy groups. That is, if $G < $ O(3) is an isotropy group, then $gGg^{-1}$ is also an isotropy group for any $g \in$ O(3).

The numbers $i(G)$ listed on the right in Figures 7.1, 7.2, and 7.3 are the numbers of linearly independent vectors whose isotropy group is the group given at the left. Thus, to each isotropy group $G$ corresponds an $i(G)$-dimensional symmetry plane (linear subspace) in $\mathcal{V}_L$ generated by these vectors. This plane consists of all the vectors in $\mathcal{V}_L$ which are invariant (fixed) under the isotropy group $G$. In Tables 7.1, 7.2, and 7.3 we give explicitly parametric equations for the vectors describing all the symmetry planes. We always give only one plane representing the entire family of

---

[3]These representations of O(3) are denoted $L^+$ or $L^-$ for $L$ even or odd, respectively.

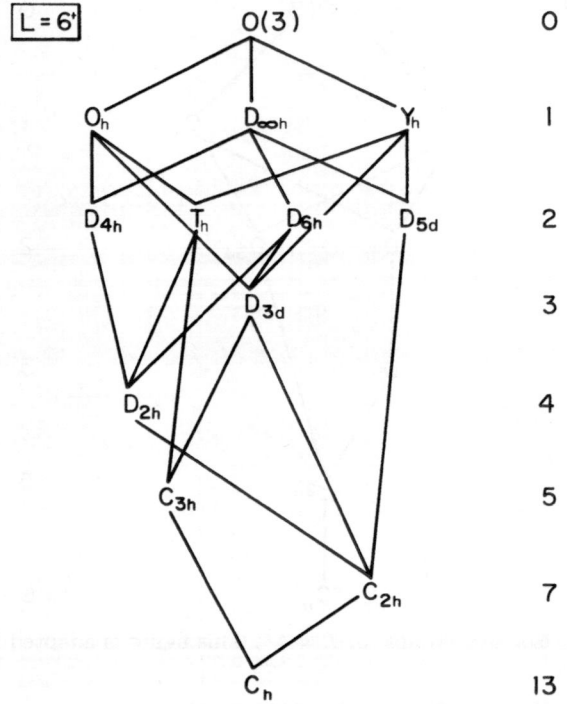

FIGURE 7.3. Isotropy groups for $L = 6^{+}$. This figure is adapted from Ref. 16.

TABLE 7.1. The isotropy groups and symmetry planes for $L = 2^+$ representation of $O(3)$.[*]

| $G$ | Dimension | Symmetry plane | Orientation |
|-----|-----------|----------------|-------------|
| $O(3)$ | 0 | 0 | |
| $D_{\infty h}$ | 1 | $(0, 0, a, \ldots)$ | $\hat{z}_\infty$ |
| $D_{2h}$ | 2 | $(1, 0, b, \ldots)$ | $\hat{z}_2, \hat{x}_2$ |

[*]The isotropy groups for $L = 2^+$ representation of $O(3)$ are given in the first column. The second column gives dimensionality of the associated symmetry plane while the third column gives the parametric form of the corresponding vectors (Latin and Greek characters are used for real and complex numbers, respectively). Only negative $m$ components are explicitly shown. For $m > 0$ we have $Q_{Lm} = (-1)^m Q_{L,-m}^*$. The last column gives the orientation of the rotation axes. For example, $\hat{n}_p$ indicates an axis of order $p$ around $\hat{n}$.

TABLE 7.2. The isotropy groups and symmetry planes for $L = 4^+$ representation of $O(3)$.[*]

| $G$ | Dimension | Symmetry plane | Orientation |
|-----|-----------|----------------|-------------|
| $O(3)$ | 0 | 0 | |
| $D_{\infty h}$ | 1 | $(0, 0, 0, 0, a, \ldots)$ | $\hat{z}_\infty$ |
| $O_h$ | 1 | $a(\sqrt{5/24}, 0, 0, 0, \sqrt{7/12}, \ldots)$ | $\hat{z}_4, \hat{x}_4$ |
| $D_{4h}$ | 2 | $(a, 0, 0, b, \ldots)$ | $\hat{z}_4, \hat{x}_2$ |
| $D_{3d}$ | 2 | $(0, ia, 0, 0, b, \ldots)$ | $\hat{z}_3, \hat{x}_2$ |
| $D_{2h}$ | 3 | $(a, 0, b, 0, c, \ldots)$ | $\hat{z}_2, \hat{x}_2$ |
| $C_{2h}$ | 5 | $(\alpha, 0, \beta, 0, c, \ldots)$ | $\hat{z}_2$ |
| $C_i$ | 9 | $(\alpha, \beta, \gamma, \delta, a, \ldots)$ | |

[*]See Table 7.1 footnote.

equivalent planes. The planes which are listed correspond to a particular setting of the group $G$ in the usual three-dimensional space whose coordinates are fixed with respect to $Y_{Lm}(\hat{n})$. For each plane, the setting of the corresponding symmetry group is also specified in the tables. The observations shown in Eqs. (7.10–15) are now an obvious consequence of the results from Tables 7.1–7.3.

It is instructive to illustrate some of the harmonics of the bond density associated with the high symmetries. In Figures 7.4–7.10 we show

$$\Omega_L(\hat{n}) \equiv \sum_{m=-L}^{L} \Omega_{Lm} Y_{Lm}^*(\hat{n}) \tag{7.16}$$

TABLE 7.3. The isotropy groups and symmetry planes for $L = 6^+$ representation of $O(3)$.*

| $G$ | Dimension | Symmetry plane | Orientation |
|---|---|---|---|
| $O(3)$ | 0 | 0 | |
| $D_{\infty h}$ | 1 | $(0,0,0,0,0,0,a,\ldots)$ | $\hat{z}_\infty$ |
| $Y_h$ | 1 | $(a0, i\frac{1}{5}\sqrt{7}, 0,0,0,0, \frac{1}{5}\sqrt{11}, \ldots)$ | $\hat{z}_5, \hat{x}_2$ |
| $O_h$ | 1 | $a(0,0, \frac{1}{4}\sqrt{7}, 0,0,0, -\frac{1}{4}\sqrt{2}, \ldots)$ | $\hat{z}_4, \hat{x}_4$ |
| $D_{6h}$ | 2 | $(a,0,0,0,0,0,b,\ldots)$ | $\hat{z}_6, \hat{x}_2$ |
| $D_{5d}$ | 2 | $(0, ia, 0,0,0,0, b, \ldots)$ | $\hat{z}_5, \hat{x}_2$ |
| $D_{4h}$ | 2 | $(0,0,a,0,0,0,b,\ldots)$ | $\hat{z}_4, \hat{x}_2$ |
| $T_h$ | 2 | $(-\sqrt{5/11}a, 0, -\sqrt{7/2}b, 0, a, 0, b, \ldots)$ | $\hat{z}_2, \hat{x}_2$ |
| $D_{3d}$ | 3 | $(a,0,0,ib,0,0,c,\ldots)$ | $\hat{z}_3, \hat{x}_2$ |
| $D_{2h}$ | 4 | $(a,0,b,0,c,0,d,\ldots)$ | $\hat{z}_2, \hat{x}_2$ |
| $C_{3h}$ | 5 | $(\alpha,0,0,\beta,0,0,a,\ldots)$ | $\hat{z}_3$ |
| $C_{2h}$ | 7 | $(\alpha,0,\beta,0,\gamma,0,a,\ldots)$ | $\hat{z}_2$ |
| $C_i$ | 13 | $(\alpha,\beta,\gamma,\delta,\nu,\mu,a,\ldots)$ | |

*See Table 7.1 for footnote.

for icosahedral $Y_h$, octahedral $O_h$, tetrahedral $T_h$, uniaxial $D_{\infty h}$, hexagonal $D_{6h}$, pentagonal $D_{5d}$, and tetragonal $D_{4h}$ symmetries, respectively. According to Tables 7.1–7.3, the only nonvanishing icosahedrally symmetric contribution to the bond density comes from the $L = 6$ harmonic. By substituting Eq. (7.15) into Eq. (7.16), we obtain

$$\Omega_6^{Y_h}(\hat{n}) \sim Y_{60}(\hat{n}) - i\sqrt{\frac{7}{11}} [Y_{65}(\hat{n}) + Y_{6-5}(\hat{n})] \sim \text{const} + I_{Y_h}^6(\hat{n}), \quad (7.17)$$

where $I_{Y_h}^6(\hat{n}) = \prod_{i=1}^6 \hat{v}_i \cdot \hat{n}$ and $\hat{v}_i$ point along the six fivefold rotation axes of the icosahedron. As can be seen from Figure 7.4, depending on the overall sign, the corresponding $\Omega(\hat{n})$ is maximal either in the directions of icosahedral vertices (fivefold symmetry axes) or faces (threefold symmetry axes).

Similarly, we see from Tables 7.1–7.3 that the first two nonvanishing harmonics for $O_h$ are $L = 4$ and 6. The octahedral bond density for $L = 4$ obtained by substituting Eq. (7.11) into Eq. (7.16) is

$$\Omega_4^{O_h}(\hat{n}) \sim Y_{40}(\hat{n}) + \sqrt{\frac{5}{14}} [Y_{44}(\hat{n}) + Y_{4-4}(\hat{n})] \sim \text{const} + (n_1^4 + n_2^4 + n_3^4). \quad (7.18)$$

This density, the so-called fourth-order cubic (octahedral) harmonic, has maxima either in the directions of the vertices or faces of a cube, as can be seen in Figure 7.5(a). The octahedral density associated with $L = 6$,

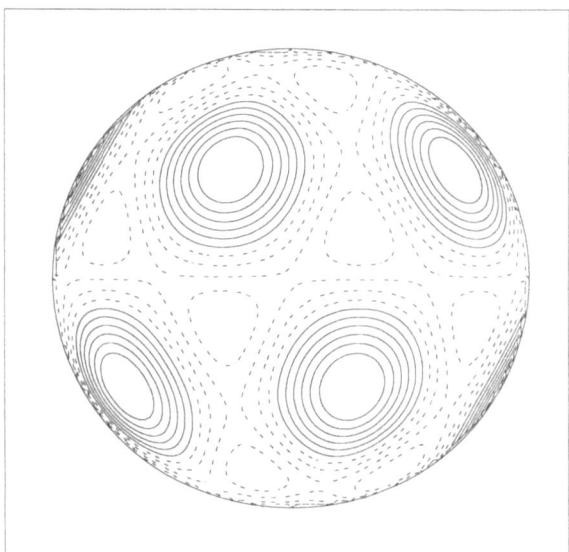

FIGURE 7.4. The $Y_h$ bond density $\Omega(\hat{n}) = \sum_m \hat{\Omega}_{Lm} Y^*_{Lm}(\hat{n})$ for $L = 6$. Contour diagram of the bond density is shown on a sphere viewed from the positive $x$-direction, with the $y$-axis directed to the right and the $z$-axis to the top. Negative contours are drawn by dashed lines. Note that by changing the sign of $\hat{\Omega}$, all minima would become maxima and all maxima would become minima. The same conventions are used in Figs. 7.5 through 7.10.

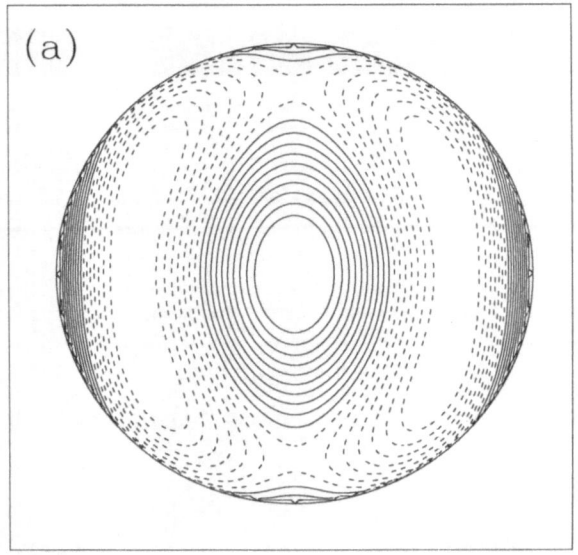

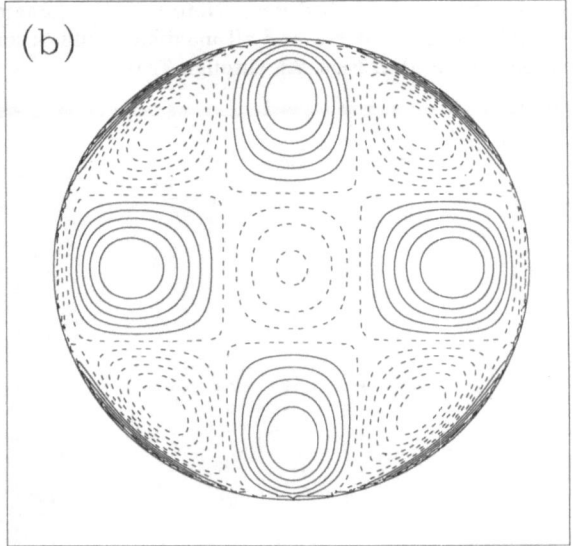

FIGURE 7.5. The $O_h$ bond density for (a) $L = 4$ and (b) $L = 6$.

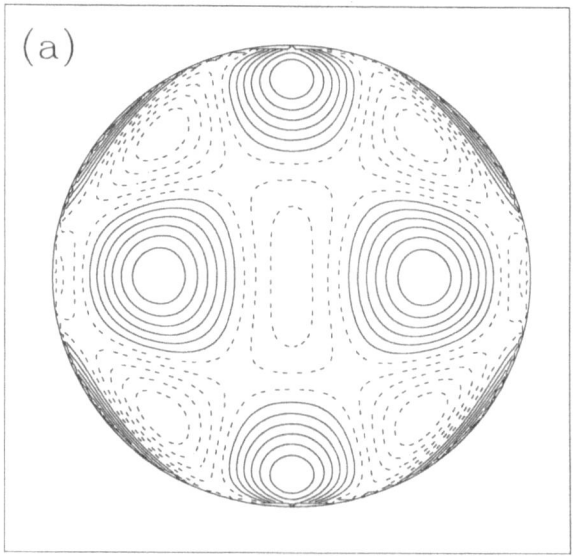

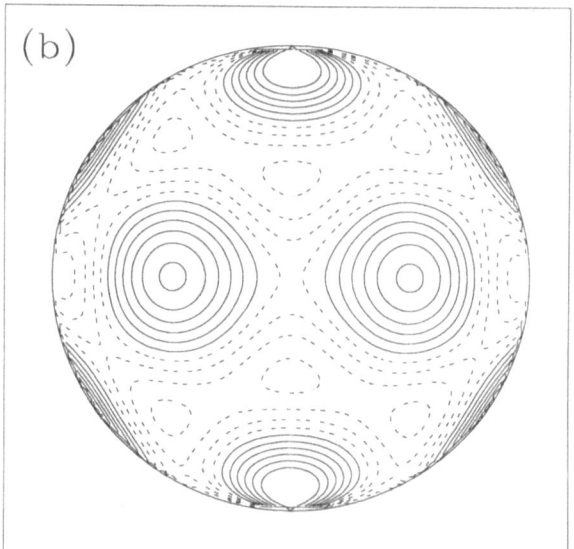

FIGURE 7.6. The $T_h$ bond density for $L = 6$. The three values of the angle parametrizing the densities are (a) $\alpha = 30°$, (b) $60°$, and (c) $90°$.

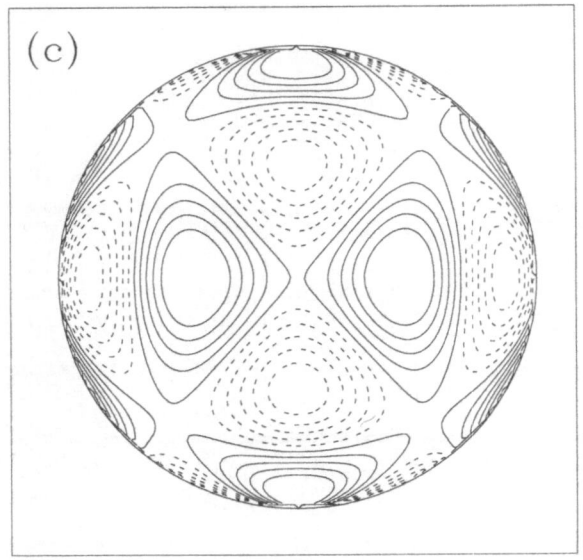

FIGURE 7.6. (*continued*)

obtained by substituting Eq. (7.12) into Eq. (7.16),

$$\Omega_6^{O_h}(\hat{\mathbf{n}}) \sim Y_{60}(\hat{\mathbf{n}}) - \sqrt{\frac{7}{2}}[Y_{64}(\hat{\mathbf{n}}) + Y_{6-4}(\hat{\mathbf{n}})]$$

$$\sim \text{const} - (n_1^4 + n_2^4 + n_3^4) + \frac{15}{11}(n_1^6 + n_2^6 + n_3^6),$$

(7.19)

is the sixth-order octahedral harmonic. It has maxima in the directions of either the vertices or edges of a cube, as shown in Figure 7.5(b).

Again, Tables 7.1–7.3 show that for tetrahedral symmetry $T_h$, like for $Y_h$, the first nonvanishing contribution comes from $L = 6$. However, the difference is that the $T_h$-symmetry plane is two-dimensional,

$$\Omega_6^{T_h} = \left(-\sqrt{\frac{5}{11}}a, 0, -\sqrt{\frac{7}{2}}, b, 0, a, 0, b, \dots\right),$$

(7.20)

so that, aside from an overall factor, the tetrahedral bond density can be parametrized by introducing an "angle" $\alpha$ so that

$$a = \sqrt{\frac{11}{32}}\sin\alpha$$

(7.21)

and

$$b = -\sqrt{\frac{1}{8}}\cos\alpha.$$

(7.22)

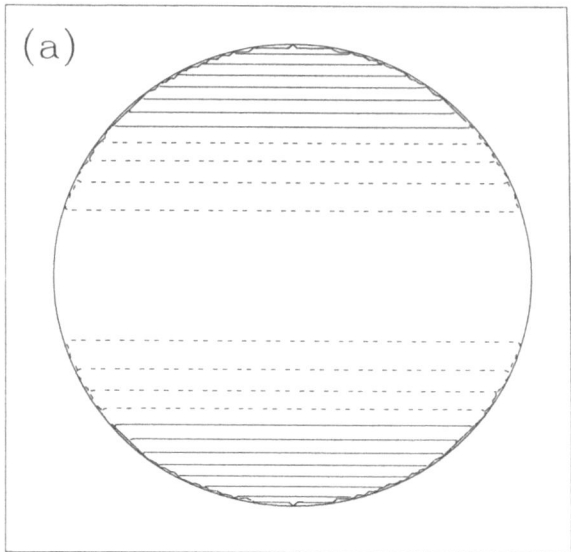

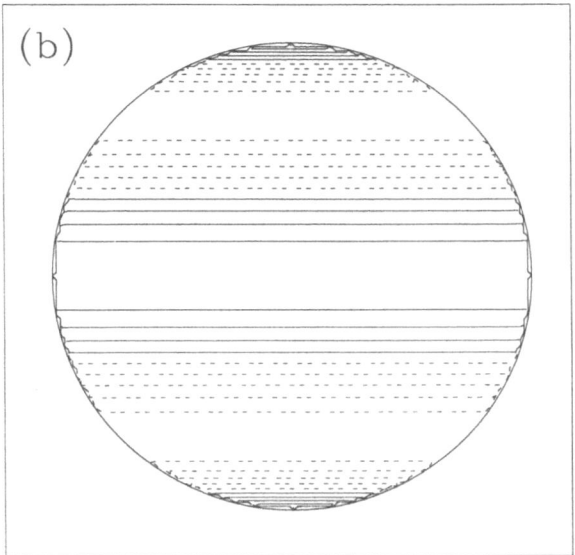

FIGURE 7.7. The $D_{\infty h}$ bond density for (a) $L = 2$, (b) $L = 4$, and (c) $L = 6$.

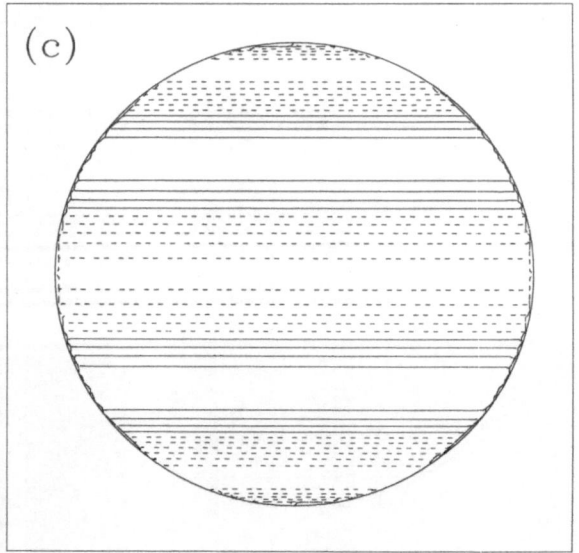

FIGURE 7.7. (*continued*)

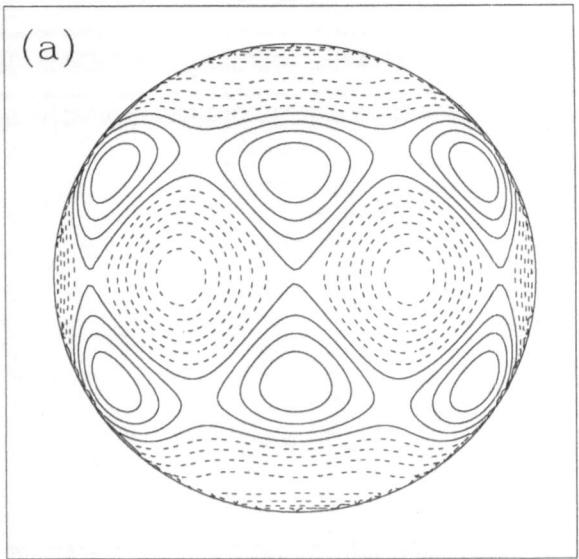

FIGURE 7.8. The $D_{6h}$ bond density for $L = 6$. The three values of the angle parametrizing the densities are (a) $\alpha = 30°$, (b) $62.89846° = \alpha_2$, and (c) $90°$.

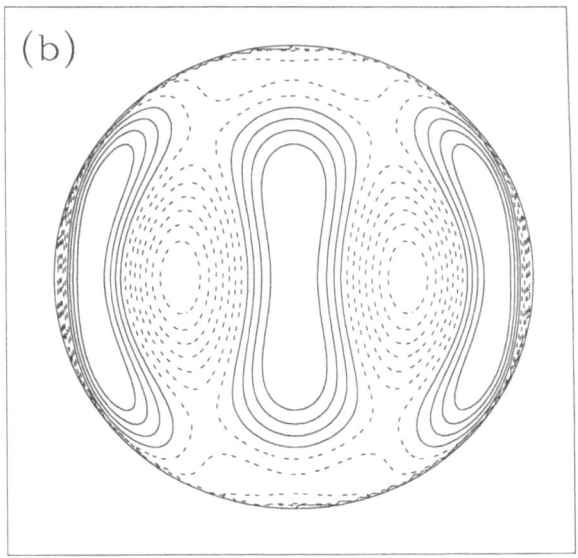

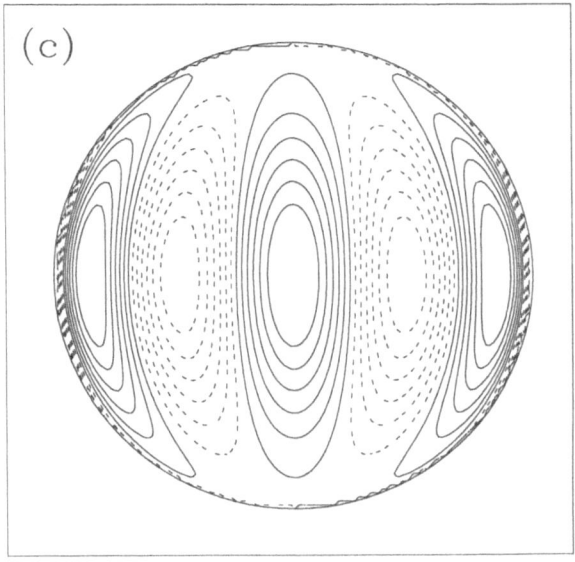

FIGURE 7.8. (*continued*)

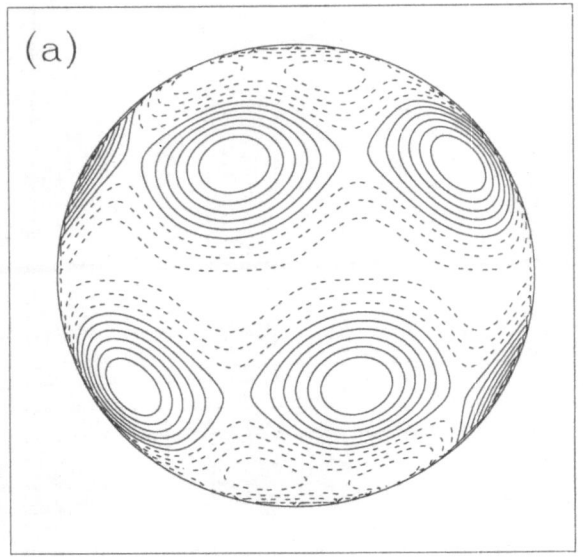

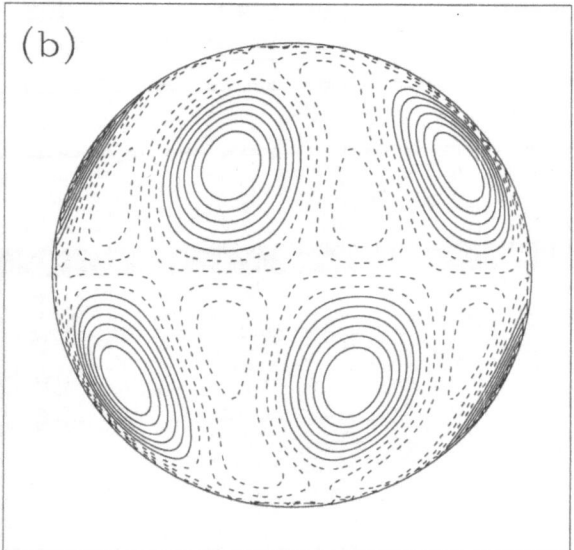

FIGURE 7.9. The $D_{5d}$ bond density for $L = 6$. The three values of the angle parametrizing the densities are (a) $\alpha = 30°$, (b) $60°$, and (c) $90°$.

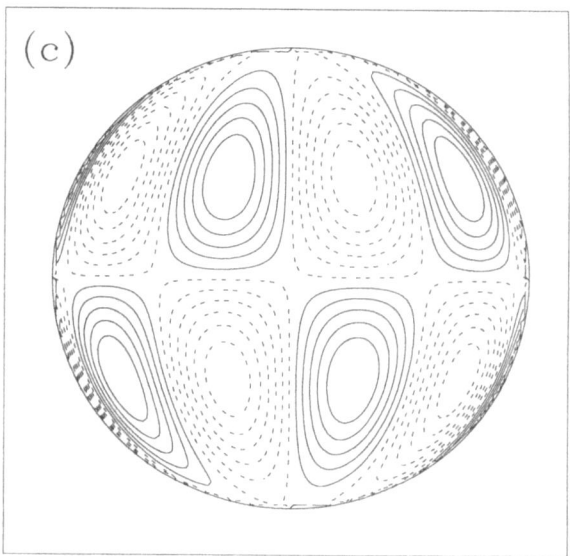

FIGURE 7.9. (*continued*)

For special values of $\alpha$ one recovers either $Y_h$ ($\tan \alpha = -4\sqrt{\frac{21}{11}}$, Figure 7.4)[4] or $O_h$ ($\alpha = 0$, Figure 7.5) densities. Bond densities for two other values of $\alpha$ are shown in Figure 7.6.

The bond densities with uniaxial $D_{\infty h}$ symmetry have a contribution for each $L = 2$, 4, and 6. The bond densities with this symmetry can be calculated using Eq. (7.16) and the appropriate $D_{\infty h}$ order parameters $\Omega_L$ given in Tables 7.1–7.3. Explicitly, we find for $L = 2$

$$\Omega_2^{D_{\infty h}}(\hat{\mathbf{n}}) \sim Y_{20}(\hat{\mathbf{n}}) \sim (3n_3^2 - 1). \tag{7.23}$$

Therefore, depending on the sign of the prefactor, this bond density is maximal either in the $z$-direction or along the equator in the $xy$-plane. We illustrate this density in Figure 7.7(a). The bond density with $D_{\infty h}$ symmetry for $L = 4$,

$$\Omega_4^{D_{\infty h}}(\hat{\mathbf{n}}) \sim Y_{40}(\hat{\mathbf{n}}) \sim 35n_3^4 - 30n_3^2 + 3, \tag{7.24}$$

is significantly different from the one for $L = 2$. The absolute minimum (maximum) along the equator changes to a local maximum (minimum), while the absolute minimum (maximum) moves to a cone around the $z$-axis at a definitive angle

$$\cos^2 \theta = \frac{3}{7}. \tag{7.25}$$

---

[4]The orientation of $Y_h$ obtained by taking this particular $\alpha$ is different from the orientation in Figure 7.4. See the discussion following Eq. (7.34) below.

This is shown in Figure 7.7(b). The density for $L = 6$,

$$\Omega_6^{D_{\infty h}}(\hat{n}) \sim Y_{60}(\hat{n}) \sim 231n_3^6 - 315n_3^4 + 105n_3^2 - 5, \qquad (7.26)$$

has a richer structure than in the $L = 4$ case. Additional extrema appear and the angle of the cone of absolute minima (maxima) reduces to

$$\cos^2 \theta = \frac{1}{33}(15 + \sqrt{60}), \qquad (7.27)$$

as shown in Figure 7.7(c).

As seen from Tables 7.1–7.3, the bond density associated with the hexagonal $D_{6h}$ symmetry is nonzero only for $L = 6$. It is also parametrized with one parameter,

$$\Omega_6^{D_{6h}} \sim \left( \sqrt{\frac{1}{2}} \sin \alpha, 0, 0, 0, 0, 0, \cos \alpha, \dots \right), \qquad (7.28)$$

and the corresponding density is

$$\Omega_6^{D_{6h}}(\hat{n}) \sim Y_{60}(\hat{n}) \cos \alpha + \sqrt{\frac{1}{2}}[Y_{66}(\hat{n}) + Y_{6-6}(\hat{n})] \sin \alpha$$

$$\sim (231n_3^6 - 315n_3^4 + 105n_3^2 - 5) \cos \alpha \qquad (7.29)$$

$$+ \sqrt{\frac{231}{2}}(n_1^6 - 15n_1^4 n_2^2 + 15n_1^2 n_2^4 - n_2^6) \sin \alpha.$$

For $\cos \alpha = 1$, the $D_{\infty h}$ symmetry [Figure 7.7.(c)] is recovered, while for $\cos \alpha = 0$, the absolute maxima are in the equatorial plane either in the directions of the edges or faces of a hexagon. As $\cos^2 \alpha$ is increased, while the possibility of the maxima in the directions of vertices of the hexagon remains, the maxima in the directions of hexagon faces are moved out of the equatorial plane. This is illustrated in Figure 7.8 for three representative values of $\alpha$.

A similar situation occurs for the $D_{5d}$ symmetry for which

$$\Omega_6^{D_{5d}} \sim \left( 0, i\sqrt{\frac{1}{2}} \sin \alpha, 0, 0, 0, 0, \cos \alpha, \dots \right). \qquad (7.30)$$

Clearly, for $\alpha = 0$ the $D_{\infty h}$ symmetry [(Figure 7.7(c)] is recovered, while for $\tan \alpha = \sqrt{\frac{14}{11}}$ the $Y_h$ symmetry is recovered (Figure 7.4). The $D_{5d}$ densities for other values of $\alpha$ are shown in Figure 7.9.

For tetragonal symmetry $D_{4h}$, the bond density has a contribution from $L = 4$ and 6. For $L = 4$ Table 7.2 gives

$$\Omega_4^{D_{4h}} \sim \left( \sqrt{\frac{1}{2}} \sin \alpha, 0, 0, 0, \cos \alpha, \dots \right), \qquad (7.31)$$

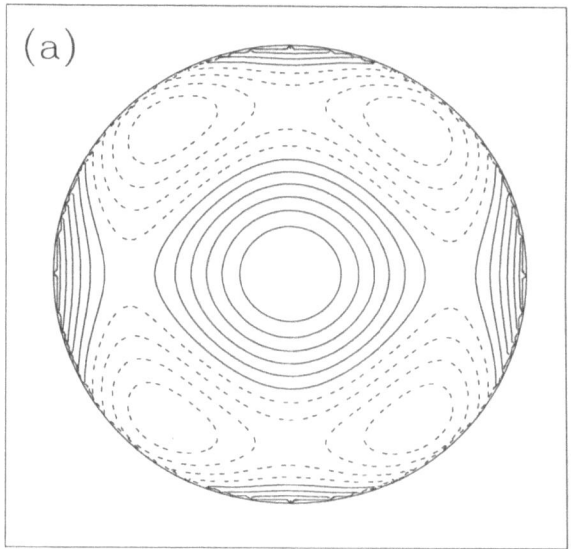

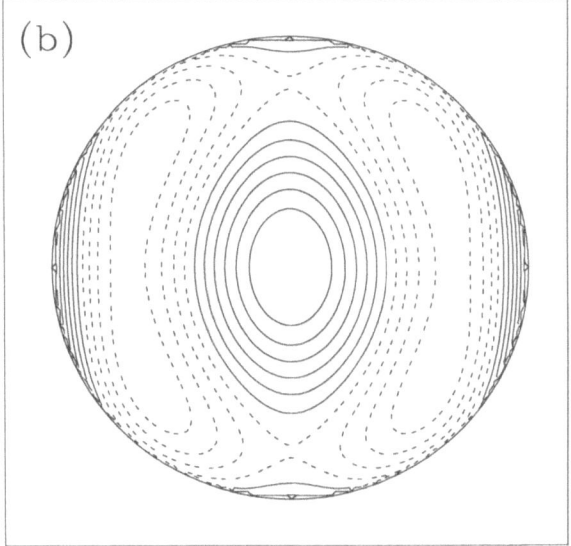

FIGURE 7.10. The $D_{4h}$ bond density for (a–c) $L = 4$ and (d–f) $L = 6$. The three values of the angle parametrizing the densities for each $L$ are (a,d) $\alpha = 30°$, (b,e) $60°$, and (c,f) $90°$.

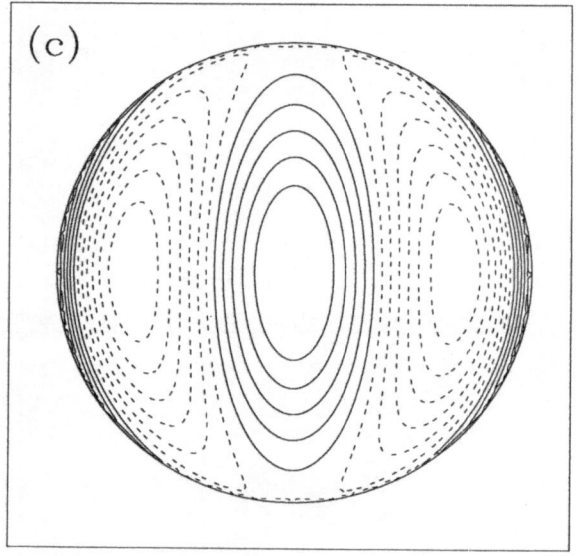

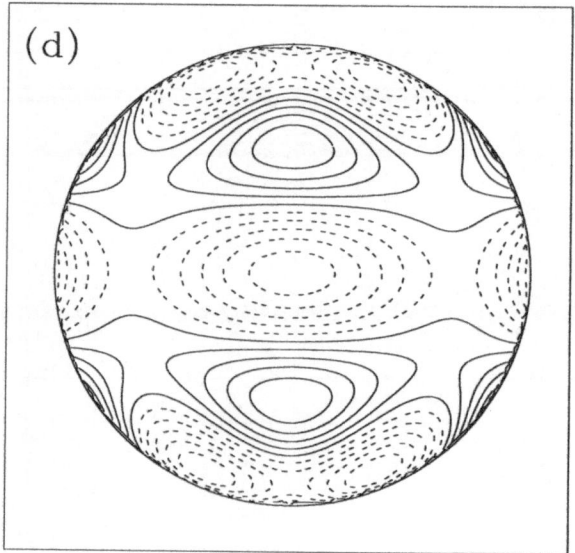

FIGURE 7.10. (*continued*)

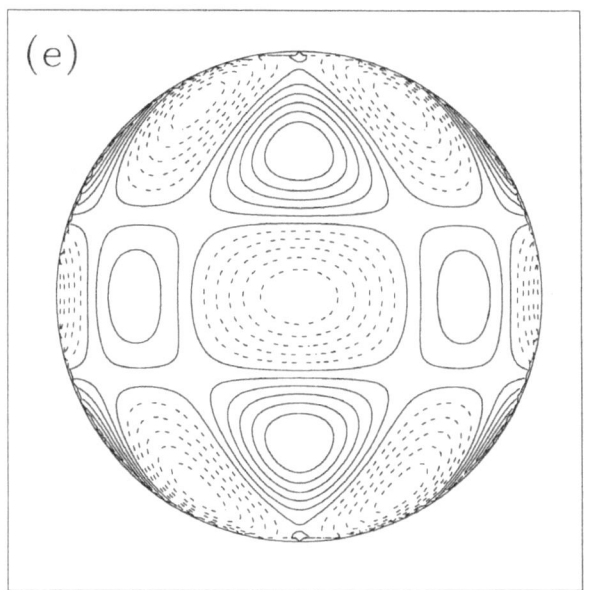

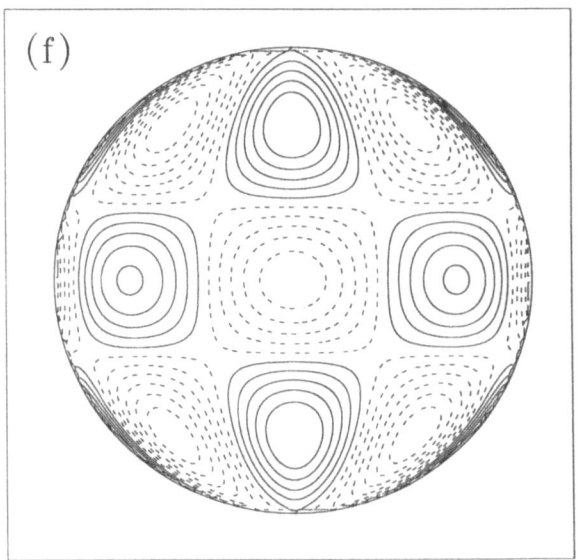

FIGURE 7.10. (*continued*)

so that the density reduces to $D_{\infty h}$ at $\alpha = 0$ or to $O_h$ at $\tan \alpha = \sqrt{\frac{5}{7}}$. As we shall see, an important point in the phase diagram is associated with $\alpha = \frac{\pi}{2}$, in which case the bond density is

$$\Omega_4^{D_{4h}}(\hat{n}) \sim Y_{44}(\hat{n}) + Y_{4-4}(\hat{n}) \sim n_1^4 - 6n_1^2 n_2^2 + n_2^4 \qquad (7.32)$$

and it has maxima in the meridian planes associated with either vertices or midedges of a square in the equatorial plane, as shown in Figure 7.10. For $L = 6$ one finds from Table 7.3 that

$$\Omega_6^{D_{4h}} \sim \left( 0, 0, \sqrt{\frac{1}{2}} \sin \alpha, 0, 0, 0, \cos \alpha, \dots \right). \qquad (7.33)$$

Therefore, the corresponding density reduces to $D_{\infty h}$ for $\alpha = 0$ and to $O_h$ for $\tan \alpha = -\sqrt{7}$. The density for other values of $\alpha$ is shown in Figure 7.10.

## 7.2.3  Measures

As can be seen from Table 7.3, the $L = 6$ harmonic of any bond-orientational order parameter with icosahedral symmetry must always have the form Eq. (7.17), provided that the coordinate system in which the order parameter is expressed is appropriately chosen (in this case, a fivefold axis is the $z$-axis, while a twofold axis is the $x$-axis). In particular, the result will be independent of the specific icosahedral cluster considered, or of details of a definition of the order parameter in Eqs. (7.3) and (7.4) (i.e., of $w$ or $w_0$). Obviously, this is not an accident, but a consequence of the fact that in $V_L$ the $Y_h$-symmetry plane is one-dimensional. Other symmetries of the bond-orientational order parameter can be analyzed in a similar fashion.

However, if one is given a bond density and wants to determine its symmetry, the above observation is useless, since without knowing the symmetry, one also does not know how to orient the coordinate system to recover the "canonical" form of the order parameter. For example, one might have

$$\Omega_6^{Y_h} \sim \tau^3 + 21(\tau^2 n_1^2 - n_3^2)(\tau^2 n_2^2 - n_1^2)(\tau^2 n_3^2 - n_2^2), \quad \tau \equiv \frac{(1 + \sqrt{5})}{2}, \quad (7.34)$$

which has icosahedral symmetry but is quite different from Eq. (7.17).[5] Here, the $z$-axis is a twofold axis, while it is a fivefold axis in Eq. (7.17), whereas a fivefold axis is here in the $(\tau \, 0 \, 1)$ direction (the $x$-axis is a two-fold axis in both cases). Clearly, the symmetry of a bond density is equivalent to the symmetry of the same bond density rotated by any rotation in $O(3)$. Therefore, it is necessary to find a coordinate independent, i.e., $O(3)$-invariant measure of the bond density. In addition, the measure must be

---

[5]It is this orientation of $Y_h$ that is recovered for $\tan \alpha = -4\sqrt{\frac{21}{11}}$ in Eqs. (7.20) and (7.21).

able to resolve different symmetries, such as these listed in Tables 7.1–7.3 for $L = 2$, 4, and 6. If only a limited resolution is required, e.g., only between the highest symmetries, only a few components of the measure might suffice.

Therefore, for a given $L$, each component of the required measure $I(\Omega_L)$ is an O(3)-invariant function of $\Omega_{Lm}$. It is natural to first consider possible polynomial functions, which leads us to a study of the ring of invariant polynomials (it is a ring since sums and products of invariant polynomials are also invariant polynomials). A zero-order polynomial, a constant, is an invariant, but obviously has no resolution. A first-order polynomial cannot be O(3)-invariant since the O(3) representation labeled by $L \neq 0$ is irreducible and different from the identity representation. For the same reason, there is only one second-order invariant, the magnitude square of $\Omega_L$,

$$I_L^2(\Omega_L) = \Omega_L \cdot \Omega_L \equiv \|\Omega_L\|^2 = \sum_{m=-L}^{L} (-1)^m \Omega_{Lm} \Omega_{L,-m}. \tag{7.35}$$

However, since $D^L$ acts linearly on $\mathcal{V}_L$ (the representation is unitary), the symmetry of $\Omega_L$ depends only on its direction, not on its magnitude, so that $I_L^2$ cannot resolve the symmetry of $\Omega_L$. Therefore, we must consider third- and higher-degree polynomials in $\hat{\Omega}_L \equiv \Omega_L / \|\Omega_L\|$.

A third-degree invariant measure of the bond-orientational order was previously used [16–21]. It can be constructed using Wigner's $3j$ symbol,

$$I_L^3(\Omega_L) = \begin{pmatrix} L & L & L \\ m_1 & m_2 & m_3 \end{pmatrix} \Omega_{Lm_1} \Omega_{Lm_2} \Omega_{Lm_3}, \quad L \text{ even}, \tag{7.36}$$

where

$$\begin{pmatrix} L & L & L \\ m_1 & m_2 & m_3 \end{pmatrix} \tag{7.37}$$

is a Wigner's $3j$ symbol. However, while this measure can resolve the $O_h$ symmetry for $L = 4$, or the $Y_h$ symmetry for $L = 6$, it cannot resolve any other symmetry. Therefore, it is necessary to determine other invariant polynomials.

The total number $N_L^p$ of linearly independent invariants of a given degree $p$ is given by

$$N_L^p = \frac{1}{\pi} \int_0^\pi \chi_L^{[p]}(\phi)(1 - \cos\phi) \, d\phi, \tag{7.38}$$

where $\chi_L^{[p]}$ is a character of the $p$th symmetrized power of $D^L$ and can be calculated using the standard formulas for the characters of the symmetrized power of a representation. In particular, for $N_L^3$ and $N_L^4$ we have

$$N_L^3 = \frac{1}{\pi} \int_0^\pi \left[ \frac{1}{6}\chi_L(\phi)^3 + \frac{1}{2}\chi_L(\phi)\chi_L(2\phi) + \frac{1}{3}\chi_L(3\pi) \right] (1 - \cos\phi)\, d\phi$$

$$= \begin{cases} 1, & L \text{ even} \\ 0, & L \text{ odd} \end{cases} \tag{7.39}$$

and

$$N_L^4 = \frac{1}{\pi} \int_0^\pi \left[ \frac{1}{24}\chi_L(\phi)^4 + \frac{1}{4}\chi_L(\phi)^2\chi_L(2\phi) + \frac{1}{8}\chi_L(2\phi)^2 + \frac{1}{3}\chi_L(\phi)\chi_L(3\phi) \right.$$

$$\left. + \frac{1}{4}\chi_L(4\phi) \right] (1 - \cos\phi)\, d\phi = \frac{1}{2}L, \quad L = 2, 4, 6, \tag{7.40}$$

where the character for the $L$th irreducible representation of SO(3) is

$$\chi_L(\phi) = \frac{\sin\left(L + \frac{1}{2}\right)\phi}{\sin\frac{1}{2}\phi} \tag{7.41}$$

[it suffices to consider SO(3) rather than O(3)]. As we mentioned previously, $N_L^2 = 1$, since the representations are irreducible over the reals. Alternatively, one may determine these numbers from the Molien generating function[6] $M_L(t)$ which has a power series expansion

$$M_L(t) = \sum_{p=0}^\infty N_L^p t^p \tag{7.42}$$

and which can be evaluated using

$$M_L(t) = \frac{1}{\pi} \int_0^\pi \left[ \prod_{m=-L}^L (1 - te^{im\phi})^{-1} \right] (1 - \cos\phi)\, d\phi, \tag{7.43}$$

where $e^{im\phi}$ are eigenvalues of $D^L$ for a rotation by $\phi$ around any axis. The Molien function has been calculated [30] for irreducible representations of SO(3) with $L \leq \frac{13}{2}$. For example,

$$M_6(t) = \frac{(1 + t^{33})(1 + t^4 + t^5 + \cdots + t^{48} + t^{52})}{\prod_{i=2}^{11}(1 - t^i)} = 1 + t^2 + t^3 + 3t^4 + \cdots, \tag{7.44}$$

and one obtains exactly the same result as by Eqs. (7.39) and (7.40).

The actual invariants can be expressed in terms of Wigner's 3$j$ symbols. They are shown graphically in Figure 7.11. In addition to Eqs. (7.35) and (7.36), these invariants are

---

[6]See, e.g., the introduction in Ref. 29 and references therein.

$$I_L^2 \qquad I_L^3 \qquad I_L^{4,0} \qquad \tilde{I}_L^{4,1} \qquad \tilde{I}_L^{4,2} \qquad \cdots$$

FIGURE 7.11. Graphic representation of the invariants given in Eqs. (7.35), (7.36), and (7.44–7.46). This figure is adapted from Ref. 16.

$$I_L^{4,0}(\Omega_L) = I_L^2(\Omega_L)^2, \tag{7.45}$$

$$\tilde{I}_L^{4,1}(\Omega_L) = (-1)^n \begin{pmatrix} L & L & L \\ m_1 & m_2 & n \end{pmatrix} \begin{pmatrix} L & L & L \\ -n & m_3 & m_4 \end{pmatrix} \times$$

$$\times\; \Omega_{Lm_1}\Omega_{Lm_2}\Omega_{Lm_3}\Omega_{Lm_4}, \qquad L = 4, 6 \tag{7.46}$$

and

$$\tilde{I}_L^{4,2}(\Omega_L) = (-1)^{\Sigma n_i} \begin{pmatrix} L & L & L \\ -n_1 & m_1 & n_2 \end{pmatrix} \begin{pmatrix} L & L & L \\ -n_2 & m_2 & n_3 \end{pmatrix} \begin{pmatrix} L & L & L \\ -n_3 & m_3 & n_4 \end{pmatrix}$$

$$\times \begin{pmatrix} L & L & L \\ -n_4 & m_4 & n_1 \end{pmatrix} \Omega_{Lm_1}\Omega_{Lm_2}\Omega_{Lm_3}\Omega_{Lm_4}, \qquad L = 6, \tag{7.47}$$

where the sum over repeated lowercase Latin indices is understood. In our actual calculations we used instead of $\tilde{I}_L^{4,1}$ and $\tilde{I}_L^{4,2}$ the two linearly independent quartic invariants,

$$I_L^{4,1}(\Omega_L) = \frac{\partial I_L^3}{\partial \Omega_L} \cdot \frac{\partial I_L^3}{\partial \Omega_L}, \qquad L = 4, 6 \tag{7.48}$$

and

$$I_L^{4,2}(\Omega_L) = \frac{\partial^2 I_L^{4,1}}{\partial \Omega_L \partial \Omega_L} : \frac{\partial^2 I_L^{4,1}}{\partial \Omega_L \partial \Omega_L}, \qquad L = 6, \tag{7.49}$$

which are linear combinations of the original invariants [the dot product is defined in Eq. (7.35)]. These invariants were easier to construct and they have the advantage of being nonnegative.

The third- and fourth-degree invariants are sufficient to distinguish between the highest symmetries. Moreover, since the symmetries do not depend on the sign of $I_L^3$, we shall consider its square, so that our measure of the symmetry of the bond-orientational order parameter is $(Z, X_1, X_2, \ldots)$, where

$$Z = I_L^3(\hat{\Omega}_L)^2 \tag{7.50}$$

and

$$X_i = I_L^{4,i}(\hat{\Omega}_L), \quad i = 1, 2, \ldots \ . \tag{7.51}$$

The domain $\mathcal{D}_L$ of $(Z, X_1, X_2, \ldots)$ is explicitly given by allowing $\hat{\Omega}_L$ to take all possible values on a $2L$-dimensional unit sphere. Since the unit sphere is compact and connected, and the map is polynomial, the domain $\mathcal{D}_L$ is also compact and connected. This domain is called the orbit space since entire orbits of vectors $\hat{\Omega}_L$ generated by the action of O(3) are mapped into single points in $\mathcal{D}_L$. Clearly, at best, only the symmetries which occur within the boundary of the domain $\mathcal{D}_L$ can be resolved, and we must determine the boundary.

We shall investigate the domain $\mathcal{D}_L$ analytically and numerically.[7] With the help of the explicit equations for the symmetry planes given in Tables 7.1–7.3, we shall construct $\mathcal{D}_L$ (and its singularities) by direct evaluations of Eqs. (7.50) and (7.51). However, we shall first recall several important general results applicable to such domains [31–33]. Just as the action of SO(3) on the $(2L+1)$-dimensional representation space introduces certain geometrical features in that space, namely, the symmetry planes, so it does in $\mathcal{D}_L$, namely, the images of the intersections of the unit sphere with the symmetry planes. Clearly, all equivalent planes will have the same image. It is also known that the image of the symmetry plane associated with an isotropy group will contain in its boundaries as singularities (cusps) the images of the symmetry planes associated with larger isotropy groups. It follows from the above that the higher singularities will be associated with the higher isotropy groups and that the boundary of the domain $\mathcal{D}_L$ will generally contain some singularities. In particular, the symmetry planes associated with the maximal isotropy groups, which are typically one-dimensional, will be typically mapped into singular points (point cusps) at the surface of $\mathcal{D}_L$.

We shall determine $\mathcal{D}_L$ for $L = 2$, 4, and 6. Starting with the simplest case $L = 2$, we see that the only nontrivial invariant is $I_2^3$ and the orbit space is one-dimensional. Therefore, $\mathcal{D}_2$ is a line segment $0 \le Z \le Z_{\max}$, where $Z_{\max}$ is the image of the symmetry plane associated with the maximal isotropy group $D_{\infty h}$ (see Figure 7.12).

The situation is more complex for $L = 4$. In this case, there are two nontrivial invariants $I_4^3$ and $I_4^{4,1}$ so that the orbit space is two-dimensional. There are two maximal isotropy groups, $D_{\infty h}$ and $O_h$, whose symmetry planes are one-dimensional. They are mapped into two convex cusps in the boundary of $\mathcal{D}_4$ (see Figure 7.13). Their coordinates are

$$D_{\infty h} : (Z, X_1) = \left( \frac{18}{1001}, \frac{162}{1001} \right) \tag{7.52}$$

---

[7]Most calculations, numerical as well as analytical, were performed using the programs SMP, MATHEMATICA, or MACSYMA. However, we found a problem in Wigner's $3j$ symbols in a version of SMP.

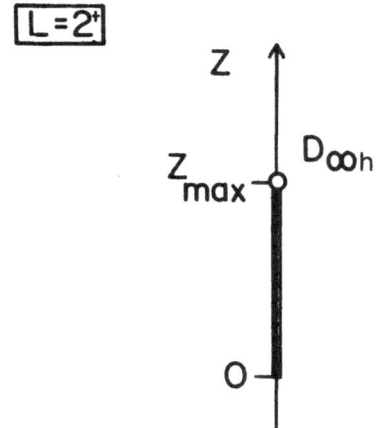

FIGURE 7.12. The orbit space for $L = 2^+$. The thick line is the image of the $D_{2h}$ symmetry plane. This figure is adapted from Ref. 16.

and

$$O_h : (Z, X_1) = \left( \frac{98}{3861}, \frac{98}{429} \right). \tag{7.53}$$

Similarly, the symmetry 2-planes associated with the isotropy groups $D_{3d}$ and $D_{4h}$ are mapped into one-dimensional boundaries of the domain $\mathcal{D}_4$. Since these singularities are in the boundaries of those with lower symmetries and they close a finite region (note that $Z \geq 0$), it follows that the points of lower symmetry are in the interior of this region. We verified this by evaluating the image of 250 points of symmetry $D_{2h}$. In the representation space, the points were uniformly distributed over a 10-turn spiral going from the north to the south pole of the two-sphere parametrizing $\hat{\Omega}_4$ of symmetry $D_{2h}$ (see Table 7.2).

The most complicated of the three cases is the $L = 6$ case. Since for $L = 6$ there are three nontrivial invariants, $I_6^3$, $I_6^{4,1}$, and $I_6^{4,2}$, the orbit space is three-dimensional. We show in Figures 7.14–7.16 the projections of $\mathcal{D}_6$ onto the three coordinate planes. The three maximal isotropy groups, $D_{\infty h}$, $Y_h$, and $O_h$, are associated with the three convex cusps seen in the figures. The $(X_1, X_2, Z)$ coordinates of these cusps are

$$D_{\infty h} : (0.0779406, 6.5385, 0.00866007), \tag{7.54}$$

$$Y_h : (0.259347, 12.9744, 0.0288164), \tag{7.55}$$

and

$$O_h : (0.0015581, 5.01628, 0.000173201). \tag{7.56}$$

Simimlarly to the $L = 4$ case, there is an additional cusp associated with $I_6^3(\hat{\Omega}_6) = 0$. In this case, the cusp is the end of a cusp line of symmetry $D_{6h}$

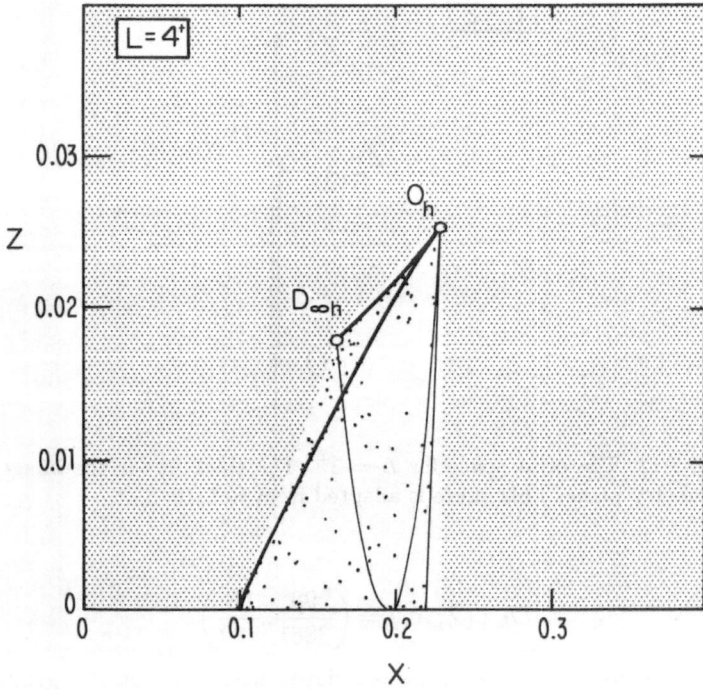

FIGURE 7.13. The orbit space for $L = 4^+$. The thick curve corresponds to the $D_{4h}$ symmetry planes, while the thin curve corresponds to the $D_{3d}$ symmetry planes. The points are images of points from the $D_{2h}$ symmetry planes. The shading outlines the convex hull of the orbit space. This figure is adapted from Ref. 16.

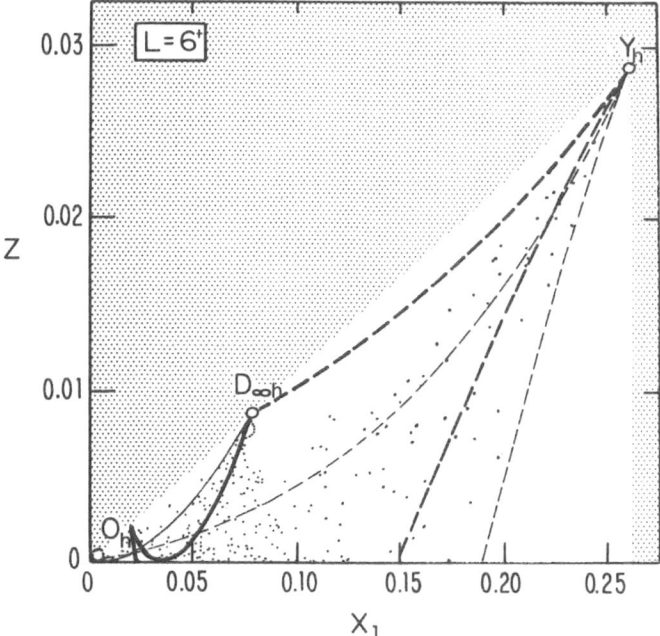

FIGURE 7.14. The orbit space for $L = 6^+$. Projection on $X_1 Z$ plane. Thick curve is the image of $D_{6h}$ symmetry planes, thick dashed curve is the image of $D_{5d}$ symmetry planes, thin curve is the image of $D_{4h}$ symmetry planes, and thin dashed curve is the image of $T_h$ symmetry planes. The points are images of points from $D_{3d}$ and $D_{2h}$ symmetry planes. Unshaded region is the projection of the convex hull of the orbit space. This figure is adapted from Ref. 16.

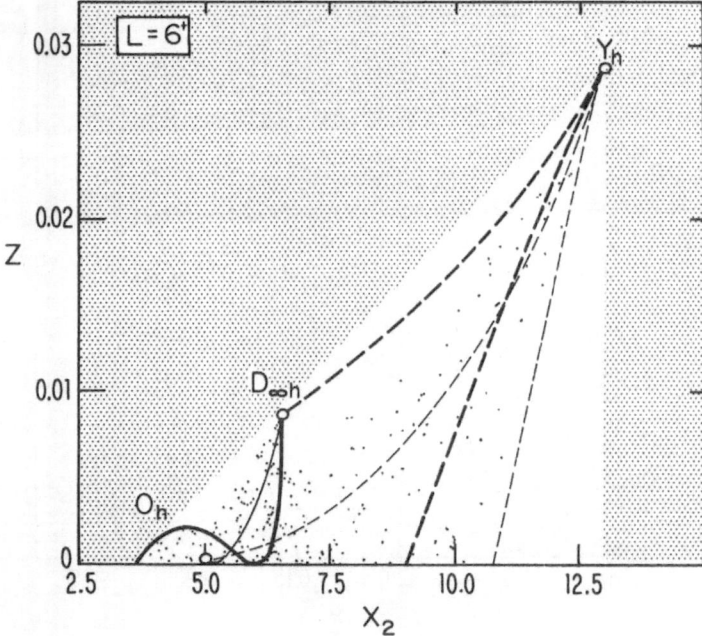

FIGURE 7.15. The orbit space for $L = 6^+$. Projection on $X_2 Z$ plane. The symbols are the same as in Fig. 7.14. This figure is adapted from Ref. 16.

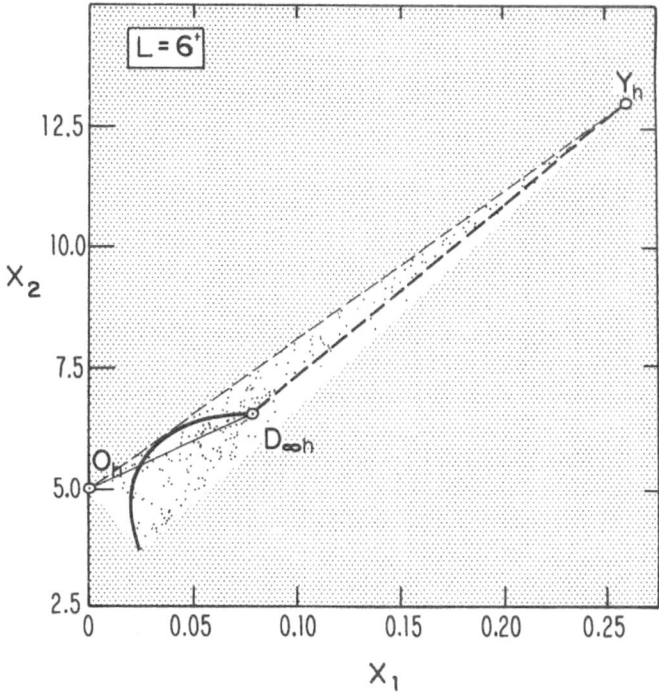

FIGURE 7.16. The orbit space for $L = 6^+$. Projection on $X_1 X_2$ plane. The symbols are the same as in Fig. 7.14. This figure is adapted from Ref. 16.

parametrized by an angle $\alpha$ as in Eq. (7.28). Explicitly, the line is given by

$$D_{6h}(\alpha) : \left\{ \begin{array}{l} X_1(\alpha) = 0.023577 - 0.0385806\cos^2\alpha + 0.0929442\cos^4\alpha \\ X_2(\alpha) = 3.63222 + 5.56319\cos^2\alpha - 2.65691\cos^4\alpha \\ Z(\alpha) = \cos^2\alpha(0.153548 - 0.2466075\cos^2\alpha)^2 \end{array} \right\}.$$

$$(7.57)$$

The end-point cusp occurs for $\alpha = \frac{1}{2}\pi$, at which

$$\hat{\Omega}_6^{D_{6h}} = \sqrt{\frac{1}{2}}(1,0,0,0,0,0,0,\ldots). \tag{7.58}$$

We have also analyzed the symmetry 2-planes associated with the symmetries $D_{4h}$, $D_{5d}$, and $T_h$. Their images are the lines shown in Figures 7.14–7.16. Next, we evaluated the images of 200 points from the $D_{3d}$ symmetry 3-plane and of the same number of points from the $D_{2h}$ symmetry 4-plane. The points were selected in a similar way as in the case of $D_{2h}$ symmetry described previously for $L = 4$. These points are also shown in Figures 7.14–7.16. Since the symmetries which we just analyzed apparently close a region in the orbit space, we conclude that the lower symmetries must belong to the interior of the region. A much more extensive evaluation [34] (involving $\sim 10^9$ points) confirms this conclusion.

# 7.3   Bond-Orientational Phase Diagram

## 7.3.1   Summary

Our basic assumption in this section is the applicability of the Landau theory to our problem. It is valid only for continuous or weakly discontinuous transitions, and then only to the extent that fluctuations can be neglected. The full treatment of fluctuations is not completed at the present. Moreover, we shall make the assumption that a particular $L = 2$, 4, or 6 is the primary (dominant) order parameter so that the interaction terms between different harmonics can be neglected to lowest order. This assumption can be rigorously satisfied only sufficiently near continuous transitions and it is approximately satisfied for weakly discontinuous transitions.

For $L = 2$ harmonics and a quartic free energy, we shall find that there is only one low symmetry phase accessible either through a first-order or through a multicritical continuous transition. This is a "uniaxial," $D_{\infty h}$ phase in which the bond density $\Omega(\hat{n})$ is maximal either in the $z$-direction or in the equatorial $xy$-plane, depending on the sign of the cubic term in the free energy. This is in agreement with previously known results. In particular, a biaxial $D_{2h}$ phase can be stabilized only by inclusion of sixth-degree terms in the free energy [35].

For $L = 4$ harmonics, we shall see that a cubic (octahedral) and a uniaxial phase can be accessed through a first-order transition. The octahedral and a $D_{4h}$ phase can be also accessed through a multicritical continuous transition. The bond density in the octahedral phase is maximal either in the directions of the vertices or the faces of a cube. The density in the $D_{4h}$ phase is maximal in the equatorial plane in the directions of either the vertices or the faces of a square. The density associated with the uniaxial phase has the maxima either in the direction of the $z$-axis *or* in a cone around the $z$-axis. The second possibility does not seem to have been emphasized in the past. The results of a partial analysis of a third-degree free energy, performed by Nelson and Toner [18], are confirmed in the present calculations where they appear as a special case.

For $L = 6$ harmonics, we shall find that icosahedral, uniaxial, octahedral, and $D_{6h}$ phases can be accessed through a first-order transition. The octahedral, icosahedral, and $D_{6h}$ phases can also be accessed through a multicritical continuous transition. The bond density in the icosahedral phase is maximal either in the directions of the vertices or the faces of an icosahedron. The bond density in the cubic phase is maximal either in the directions of the edges or the vertices of a cube. The uniaxial phase is characterized by bond density with much richer structure than the $L = 2$ or 4 harmonics. However, similarly to the $L = 4$ case, the absolute maximum of the density is either in the $z$-direction or the maxima are in the directions forming a cone around the $z$-axis.

As pointed out in Ref. 19, the density associated with $D_{6h}$ symmetry is not unique but can be parametrized with a single variable. These densities range from the case where the absolute maxima are in the equatorial plane to the case where they can be outside of this plane. However, we shall find that the density associated with the hexagonal close-packed structure is not stable. This is in agreement with a recent molecular-field, self-consistent calculation by Haymet [36] who found that the HCP phase has higher energy than either the octahedral or icosahedral phase. Although he analyzed octahedral and icosahedral ordering for $L = 4$ and 6, respectively, he did not evaluate the stability of these structures relative to other low symmetry structures. Our conclusions for $L = 6$ are also consistent with the partial analysis of Steinhardt et al. [19]. We recover their results as a special case. In particular, we proved their ansatz for the directional maxima of the third-degree term in the free energy. Therefore, the third-degree term can indeed be used as a sensitive quantitative measure of icosahedral symmetry as suggested in Ref. 19.

## 7.3.2   Free Energy

Orientational free energy for a given harmonic $L$ must be rotationally invariant. This means that terms of a given degree in the free energy expansion must be separately invariant under all rotations. However, we have already determined the necessary O(3)-invariant polynomials in the order parameter. These are the components $[(I^3)^2, I^{4,1}, I^{4,2}, \ldots]$ of our measure

of the order parameter's symmetry. Therefore, our orientational free energy
for a given harmonic $L = 2$, 4, and 6 is

$$F_L(\Omega_L) = F_L^2(\Omega_L) + F_L^3(\Omega_L) + F_L^4(\Omega_L)$$

$$= \tfrac{1}{2} a_L I_L^2(\Omega_L) + \tfrac{1}{3} b_L I_L^3(\Omega_L) + \tfrac{1}{4} \sum_{i=0}^{L/2-1} c_L^i I_L^{4,i}(\Omega_L), \qquad (7.59)$$

where $a_L$, $b_L$, and $c_L^i$ are (phenomenological) functions of thermodynamic
variables. The equilibrium value of the order parameter $\Omega_L$ minimizes the
free energy Eq. (7.59), while its symmetry (isotropy group) determines the
symmetry of the ordered phase. Although the minimization of Eq. (7.59)
is a formidable task, it is not an impossible one. Namely, several group
theoretical techniques have been developed in the last ten years designed
to deal precisely with the problem of such complexity [37].

### 7.3.3    Minimization

The importance of the symmetry planes for the minimization of the free
energy comes from the fact that the free energy at a symmetry plane is
necessarily extremal in the directions perpendicular to the symmetry plane
[38]. Therefore, when extremizing the free energy *within* a symmetry plane,
one is automatically extremizing it on the whole space $\mathcal{V}_L$. While these
observations could be used in a direct minimization [38] in $\mathcal{V}_L$, they are
also useful in another approach which we shall employ here.

The most suitable method for the present minimization problem is the
orbit space approach originally introduced by Kim [31] in the context of
the Higgs symmetry-breaking in gauge theories. When combined with the
above symmetry considerations, this method offers a powerful means for
analytical and numerical analysis [32]. The resulting orbit space approach,
which we shall describe below, is essentially a linear programming method.

By choosing a line in a direction $\hat{\Omega}_L$ parametrized by $\omega = \|\Omega_L\|$, the free
energy Eq. (7.59) can be written as

$$F_L(\Omega_L) = \omega^2 F_L^2(\hat{\Omega}_L) + \omega^3 F_L^3(\hat{\Omega}_L) + \omega^4 F_L^4(\hat{\Omega}_L) \qquad (7.60)$$

and the minimization can be split into a minimization with respect to $\omega$
and $\hat{\Omega}_L$. By minimizing with respect to $\omega$, one finds that transition occurs
when

$$H(\hat{\Omega}_L) \equiv F_L^3(\hat{\Omega}_L)^2 - 2 a_L F_L^4(\hat{\Omega}_L) = 0. \qquad (7.61)$$

For $H(\hat{\Omega}_L) \leq 0$, the stable equilibrium corresponds to $\omega = 0$, while for
$H(\hat{\Omega}_L) \geq 0$, it corresponds to $\omega \neq 0$. Therefore, a transition occurs when

$$\max H(\hat{\Omega}_L) = 0. \qquad (7.62)$$

This condition must be consistent with the requirement that the free energy
be bounded from below, i.e.,

$$F_L^4(\hat{\Omega}_L) > 0 \qquad (7.63)$$

for all $\hat{\Omega}_L$. The conditions Eqs. (7.62) and (7.63) determine the transition
surface (in the space of parameters $a_L$, $b_L$, $c_L^i$) as well as the associated
direction $\hat{\Omega}_L^0$.

If $F_L^3(\hat{\Omega}_L) \equiv 0$, then Eqs. (7.62) and (7.63) imply that a *continuous*
transition occurs at the critical surface $a_L = 0$, the low-symmetry $\hat{\Omega}_L^0$
*minimizes* $F_L^4(\hat{\Omega}_L)$, and the constant part in $F_L^4(\hat{\Omega}_L)$ can be always chosen
so that Eq. (7.63) is satisfied. Thus, the usual result for simple continuous
transitions is recovered.

If $F_L^3(\hat{\Omega}_L)$ is not identically equal to zero but $F_L^4(\hat{\Omega}_L) = \text{const} = \frac{1}{4}c_L^0$,
then the *first-order* transition surface is given by

$$\frac{1}{2}a_L c_L^0 = \max F_L^3(\hat{\Omega}_L)^2 \qquad (7.64)$$

and the low-symmetry $\hat{\Omega}_L^0$ *maximizes* $F_L^3(\hat{\Omega}_L)^2$ (or $|F_L^3(\hat{\Omega}_L)|$). Since it is
assumed that $F_L^3(\hat{\Omega}_L)$ is not identically zero, it follows that $\max F_L^3(\hat{\Omega}_L)^2 >$
0 and the condition Eq. (7.63) can be satisfied. Therefore, the first-order
transition surface has co-dimension one. In this way, we see that the usual
analysis [18,19] which deals exclusively with $F_L^3(\hat{\Omega}_L)$ is applicable when
$F_L^4(\hat{\Omega}_L) = \text{const}$.

When neither $F_L^4(\hat{\Omega}_L) = \text{const}$ nor $F_L^3(\hat{\Omega}_L) \equiv 0$, the situation is slightly
more complicated. Clearly, $\hat{\Omega}_L^0$ which maximizes Eq. (7.61) does not depend
on an additive constant in $F_L^4(\hat{\Omega}_L)$. Providing that $F_L^3(\hat{\Omega}_L^0)$ is nonzero, it
is clear that this additive constant can be chosen so that the condition
Eq. (7.62) and now the weaker condition Eq. (7.63) are both satisfied. The
first-order transition surface has co-dimension one and is given explicitly
by

$$2a_L F_L^4(\hat{\Omega}_L^0) = F_L^3(\hat{\Omega}_L^0)^2, \qquad (7.65)$$

where $\hat{\Omega}_L^0$ depends implicitly only on the parameters $a_L c_L^i/b_L^2$, $i =$
$1,2,\ldots$ .

If for some values of the parameters there exists a maximum of Eq. (7.61)
such that $F_L^3(\hat{\Omega}_L^0) = 0$, then Eqs. (7.62) and (7.63) force the critical surface
at $a_L = 0$ which, in turn, forces $b_L = 0$. This corresponds to a continuous
phase transition across a critical surface of co-dimension two.

In summary, when $F_L^3(\hat{\Omega}_L)$ is not identically equal to zero, then the absolute *maximum* of $H(\hat{\Omega}_L)$, Eq. (7.61), for which $F_L^3(\hat{\Omega}_L) \neq 0$ corresponds
to a first-order transition surface of co-dimension one, while the absolute
*minimum* of $F_L^4(\hat{\Omega}_L)$ corresponds to a second-order transition surface of
co-dimension two.

Our task is, then, to maximize Eq. (7.61) for all values of the parameters $a_L$, $b_L$, and $c_L^i$. Since $H(\hat{\Omega}_L)$ is linear in these parameters, we can alternatively maximize the *linear* form

$$2b_L^2 Z - 9a_L c_L^1 X_1 - 9a_L c_L^2 X_2 - \ldots, \qquad (7.66)$$

with respect to the orbit space variables $(Z, X_1, X_2, \ldots)$. The absolute maxima of a linear form occur at the boundaries of its domain. Therefore, one reduces the problem of maximizing $H(\hat{\Omega}_L)$ to a problem of investigating the boundary of the domain $\mathcal{D}_L$. More precisely, one needs to determine the minimal convex envelope of $\mathcal{D}_L$.

It is obvious that a maximum $\hat{\Omega}_L^0$ of $H(\hat{\Omega}_L)$ leads to an orbit of equivalent maxima: they are generated by acting on $\hat{\Omega}_L^0$ with all the rotations in $O(3)$. However, the new variables $Z, X_1, X_2, \ldots$ are invariant under $O(3)$ so that entire orbits are mapped into single points in $\mathcal{D}_L$. This removes an unnecessary redundancy. The importance of the singularities at the surface of $\mathcal{D}_L$ is in the following. By fixing the ratios between the parameters $b_L^2$, $a_L c_L^1$, $a_L c_L^2, \ldots$, one selects in the orbit space $(Z, X_1, X_2, \ldots)$ a plane whose normal is parallel to $(2b_L^2, -9a_L c_L^1, -9a_L c_L^2, \ldots)$. The point $(Z^0, X_1^0, X_2^0, \ldots)$ at the surface of $\mathcal{D}_L$ at which such a plane touches $\mathcal{D}_L$, but nowhere crosses it, corresponds to the maximum of $H$. The family of such planes, generated by varying the parameters, generates the convex hull of $\mathcal{D}_L$. Since the family of these planes is dual to the family of their normals, we see that the convex hull is dual to the phase diagram. Clearly, only the points on the surface of $\mathcal{D}_L$ which are contained in the convex hull can be directly accessed in a phase transition from the $O(3)$ phase. Now, the singular *points* (cusps) at the surface of $\mathcal{D}_L$ (which are associated with maximal isotropy groups) are typically convex so that they are most likely associated with a maximum. All of these observations become clearer as we explicitly construct the convex hull of $\mathcal{D}_L$ for $L = 2, 4$, and 6.

For $L = 2$ the orbit space is one-dimensional with a single "cusp," the end of the line segment at $Z = Z_{\max}$, associated with the uniaxial $D_{\infty h}$ symmetry.

For $L = 4$ the orbit space has cusps associated with $O_h$, $D_{\infty h}$, and $D_{4h}$ symmetries. Their coordinates are given in Eqs. (7.52) and (7.53) for $D_{\infty h}$ and $O_h$, and

$$D_{4h} : (Z, X_1) = \left(0, \frac{14}{143}\right), \qquad (7.67)$$

for $D_{4h}$ with

$$\hat{\Omega}_4^{D_{4h}} = \left(\sqrt{\frac{1}{2}}, 0, 0, 0, 0, 0, 0, 0, \sqrt{\frac{1}{2}}\right). \qquad (7.68)$$

Clearly, the convex hull of $\mathcal{D}_4$ is in the present case a convex polygon connecting the points Eqs. (7.67), (7.52), (7.53) and the point $(0, \frac{98}{429})$, as shown in Figure 7.13. Therefore, the three cusps, Eqs. (7.52), (7.53), and (7.67), necessarily correspond to extrema of the free energy.

Next, we construct the complex hull of $\mathcal{D}_6$. If we ignore for the moment all other symmetries, a careful analysis of Eqs. (7.54) to (7.57) leads to the convex hull consisting of two planar sections and three conical sections (besides the obvious vertical planes and the $Z = 0$ plane). The two planes are

$$\pi_1 : D_{\infty h} Y_h O_h D_{6h}(\alpha_2), \tag{7.69}$$

$$\pi_2 : D_{\infty h} Y_h D_{6h}(\alpha_1), \tag{7.70}$$

and the three cones are

$$\kappa_1 : O_h D_{6h}(\alpha), \quad \cos^2 \alpha \leq \cos^2 \alpha_2, \tag{7.71}$$

$$\kappa_2 : Y_h D_{6h}(\alpha), \quad \cos^2 \alpha \leq \cos^2 \alpha_1, \tag{7.72}$$

$$\kappa_3 : D_{\infty h} D_{6h}(\alpha), \quad \cos^2 \alpha_1 \leq \cos^2 \alpha \leq \cos^2 \alpha_2, \tag{7.73}$$

where

$$\cos^2 \alpha_1 = 0.120815, \tag{7.74}$$

$$\cos^2 \alpha_2 = 0.207543. \tag{7.75}$$

The projections of the convex hull are shown in Figures 7.14–7.16 and its top view in Figure 7.17.

### 7.3.4  Phase Diagram

For $L = 2$ there is a single low-symmetry phase, the uniaxial $D_{\infty h}$ phase. This phase can be accessed either through a discontinuous or a multicritical continuous transition. The transition surface is given by

$$9 a_2 c_2^0 = 2 Z_{max} b_2{}^2, \quad c_2^0 > 0. \tag{7.76}$$

Generally, for discontinuous transitions the overall sign of $\Omega_L(\hat{n})$ is determined by $b_L$, while for continuous transitions the densities of either sign have equal energy.

From Figure 7.13 we see that for $L = 4$ the low-symmetry phases are $O_h$, $D_{\infty h}$, and $D_{4h}$. Since $Z = 0$ for $D_{4h}$, this symmetry can be accessed only in a continuous transition through the surface

$$a_4 = b_4 = 0, \tag{7.77}$$

providing that

$$c_4^0 > -\frac{14}{143} c_4^1 < 0. \tag{7.78}$$

For

$$c_4^0 > -\frac{98}{429} c_4^1 > 0 \tag{7.79}$$

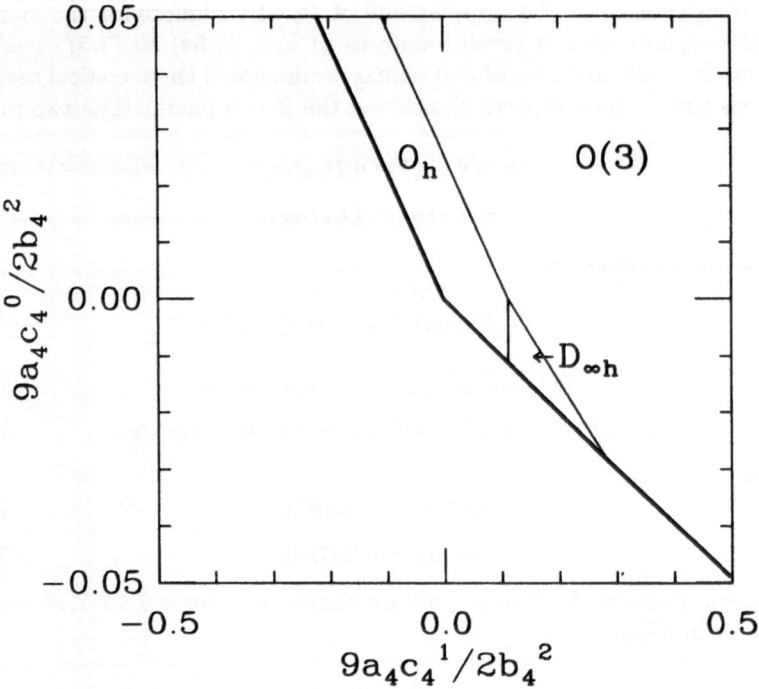

FIGURE 7.17. The $L = 4^+$ phase diagram. The thick line outlines the region of stability of free energy.

the continuous transition leads to the symmetry $O_h$. The phase with symmetry $O_h$ could be also reached by a discontinuous transition through the surface

$$9a_4c_4^0 = 2b_4^2 I_4^3(O_h)^2 - 9a_4c_4^1 I_4^{4,1}(O_h) = \frac{196}{3861}b_4^2 - \frac{196}{143}a_4c_4^1, \quad (7.80)$$

providing that

$$-\infty < \frac{9a_4c_4^1}{2b_4^2} < \frac{I_4^3(O_h)^2 - I_4^3(D_{\infty h})^2}{I_4^{4,1}(O_h) - I_4^{4,1}(D_{\infty h})} = \frac{1}{9}. \quad (7.81)$$

Similarly, the phase with the symmetry $D_{\infty h}$ can be reached by a first-order transition through the surface

$$9a_4c_4^0 = 2b_4^2 I_4^3(D_{\infty h})^2 - 9a_4c_4^1 I_4^{4,1}(D_{\infty h}) = \frac{36}{1001}b_4^2 - \frac{1458}{1001}a_4c_4^1, \quad (7.82)$$

where

$$\frac{1}{9} < \frac{9a_4c_4^1}{2b_4^2} < \frac{I_4^3(D_{\infty h})^2 - I_4^3(D_{4h})^2}{I_4^{4,1}(D_{\infty h}) - I_4^{4,1}(D_{4h})} = \frac{9}{32}. \quad (7.83)$$

The resulting phase diagram is shown in Figure 7.17. Clearly, the obtained phase diagram is completely equivalent ("dual") to the convex hull of $\mathcal{D}_4$.

The phase diagram for the $L = 6$ case is completely determined by the convex hull given in Eqs. (7.54–7.57) and (7.69–7.75), and shown in Figures 7.14–7.16 and in Figure 7.18. For example, the face $D_{\infty h}Y_h D_{6h}(\alpha_1)$ corresponds to a triple point where phases $D_{\infty h}$, $Y_h$, and $D_{6h}$ can coexist. The edge $D_{\infty h}Y_h$ corresponds to coexistence line along which phases $D_{\infty h}$ and $Y_h$ coexist. The vertex $D_{\infty h}$ corresponds to the $D_{\infty h}$ phase. Similarly, the (curved) edge $D_{6h}(\alpha)$, $\alpha_2 < \alpha < \frac{\pi}{2}$, corresponds to the $D_{6h}$ phase. This illustrates in which sense the convex hull is dual to the phase diagram. Thus, in a given problem, it is sufficient to specify the elements of the convex envelope as we have done here.

Because of its complexity and the above-mentioned duality, we shall not explicitly construct the phase diagram for $L = 6$ for which we constructed the convex hull in the previous section. As can be seen from Figure 7.18, the phase diagram is dominated in this case by the icosahedral phase which can be accessed either through a first-order transition or through a multicritical continuous one. A cubic phase is also accessible through either a first-order or a multicritical continuous transition. The uniaxial phase can be accessed only through a first-order phase transition.

The density in a plane with $D_{6h}$ symmetry can be parametrized with one parameter, the angle $\alpha$, as shown earlier in Eq. (7.28). Therefore, unlike the previous cases, the form of the density is expected to vary with temperature within the $D_{6h}$ phase, although the symmetry of the density is fixed. The states with $\alpha_2 < \alpha < \pi/2$, and only these states, can be reached through a first-order transition. The only other accessible state, with $\alpha = \pi/2$, can

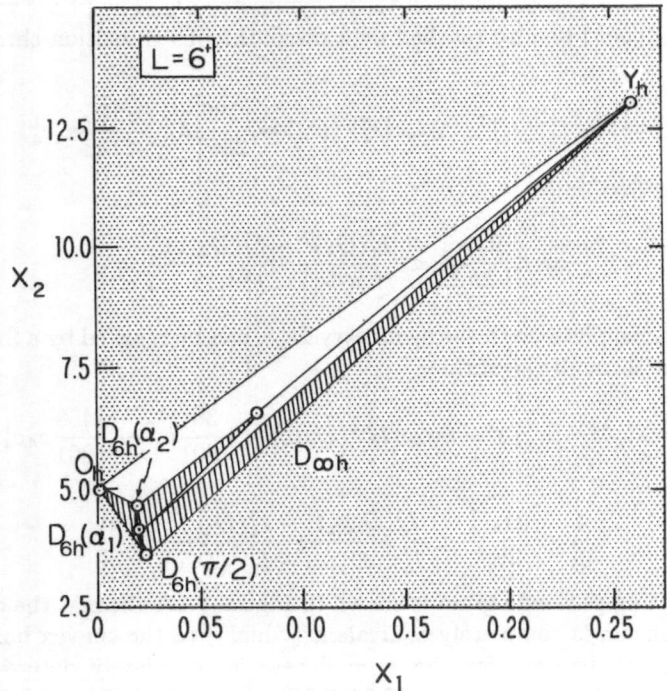

FIGURE 7.18. The convex hull for $L = 6^+$ as seen from the positive $Z$-direction. Conical portions are shaded. See also Fig. 7.16. This figure is adapted from Ref. 16.

be reached through a multicritical continuous transition. We emphasize that the $D_{6h}$ state associated with the hexagonal close-packed structure corresponding to $\cos^2 \alpha = \frac{91}{173} = 0.52601$ is not accessible by a simple transition from the $O(3)$ phase.

# 7.4   Icosahedral Quasicrystals

## 7.4.1   Summary

Using the result that the bond-orientational phase diagram is dominated by an icosahedratic phase, we recently showed that a coupling between the orientational and positional order parameters stabilizes an icosahedral quasicrystalline phase characterized by the fundamental wave vectors pointing into the vertices or faces of an icosahedron [17]. This route to quasicrystals will be the focal point of this section, but we shall also discuss results of some other approaches.

One of the first problems presented to theorists by the discovery of quasicrystals was understanding their thermodynamic stability. This task was undertaken in several recent investigations using phenomenological approaches and two-dimensional microscopic models. Widom et al. [39] found in Monte Carlo simulations that a two-dimensional, two-component Lennard–Jones system has an equilibrium pentagonal quasicrystal state corresponding to a disordered Penrose lattice which, like crystal states in two dimensions, preserves quasi-long-range order. On the other hand, using a density functional theory of freezing, Sachdev and Nelson [40] found that a single component hard-sphere system in three dimensions would yield only a metastable icosahedral quasicrystalline state.

A different phenomenological approach was taken by Bak [23], Mermin and Troian [24], and Kalugin, Kitaev, and Levitov [41], who based their investigations on the Landau theory of solidifcation as formulated by Baym et al. [25] and by Alexander and McTague [26]. In order to bypass the conclusion that a body-centered-cubic crystalline structure is favored [25,26], they either included higher-order terms in the Landau expansion [23], or they introduced an additional component to the density order parameter [24,41]. These approaches, based exclusively on an analysis of the translational ordering, explain the stability of the icosahedral quasicrystalline ordering by requiring a careful balance between infinitely many couplings. Although this can be perhaps justified on phenomenological grounds, the conditions for the quasicrystalline phase are not sufficiently generic, and there is a need for a different approach.

The main idea of the approach of Ref. 17 is to include into a theory an orientational order parameter which may stabilize the quasicrystalline phase. This does not seem unnatural when one recalls the well-known fact that many supercooled liquids and metallic glasses show a short-range icosahedral bond-orientational order [19] and that crystal structures of many alloys contain characteristic (nearly) icosahedral clusters.[8] It is perhaps not an accident that many crystalline alloys related to quasicrystalline ones contain such clusters.[9] Therefore, we shall adopt a view of solidification as an interplay between orientational and translational order parameters [18]. Indeed, the main result of the present approach is that the icosahedral quasicrystalline order can be stabilized and, in fact, triggered by the long-range icosahedral bond-orientational order. While translational ordering will always trigger simultaneous onset of the orientational ordering, the onset of the orientational ordering may (because the transition is first-order!) but need not trigger a simultaneous onset of the translational ordering: A transition into a phase with long-range orientational order but without any long-range translational order may precede the complete translational ordering. If the coupling is not sufficiently strong to induce icosahedral translational order, a quasicrystal phase can still be obtained by quenching the icosahedratic. This quasicrystalline phase, the icosahedral glass phase, has long-range icosahedral orientational order, but only short-range icosahedral translational order. On the other hand, if the coupling is strong enough, it will induce long-range icosahedral translational order characterizing an equilibrium quasiperiodic quasicrystal phase. Moreover, it is a direct result of the theory that the fundamental icosahedral wave vectors of the so-obtained quasiperiodic or icosahedral glass quasicrystals point in the directions of the 12 vertices of an icosahedron or in the directions of its 20 faces. This is in agreement with the experimental observations which definitely rule out the edge model and seem to be best-fitted by either the vertex model or the face model.

These results were obtained within the context of the Landau theory of solidification, which will be summarized and some of its features highlighted in the following section. In Section 7.4.3, we shall introduce a coupling between orientational and translational order parameters and we shall deduce the main consequences of such a coupling.

## 7.4.2   Landau Theory of Freezing

The order parameter in the Landau theory of freezing is the Fourier transform $\rho(\mathbf{q})$ of the coarse-grained density [see Eq. (7.1)]. In the isotropic, translationally disordered phase, the only nonzero component of the order parameter is the $\mathbf{q} = 0$ component. In a translationally ordered phase, components associated with a fundamental set of wave vectors $\{\mathbf{q}\}$ also become nonzero. If we assume that the coupling between components with different $q$'s, associated with different representations of the Euclidean group, can

---

[8]For a review see Ref. 42.
[9]For a review see Ref. 43.

be neglected, the problem of translational ordering becomes mathematically analogous to the problem of orientational ordering studied in Section 7.3. Near the transition, the translational free energy can be expanded in analogy with Eq. (7.59),

$$F_q[\rho_q(\hat{\mathbf{q}})] = F_q^2[\rho_q(\hat{\mathbf{q}})] + F_q^3[\rho_q(\hat{\mathbf{q}})] + F_q^4[\rho_q(\hat{\mathbf{q}})], \tag{7.84}$$

where $F_q^p[\rho_q(\hat{\mathbf{q}})]$ is of order $p$ in $\rho_q(\hat{\mathbf{q}})$. Each of these terms must be invariant under all rotations and translations. The discrete version of these terms is

$$F_q^2[\rho_q(\hat{\mathbf{q}})] = 12A_q \sum_{\hat{\mathbf{q}}} |\rho_q(\hat{\mathbf{q}})|^2 \equiv \tfrac{1}{2} A_q |\rho_q|^2, \tag{7.85}$$

$$F_q^3[\rho_q(\hat{\mathbf{q}})] = \tfrac{1}{3} B_q \sum_{\triangle} \rho_q(\hat{\mathbf{q}}_1)\rho_q(\hat{\mathbf{q}}_2)\rho_q(\hat{\mathbf{q}}_3), \tag{7.86}$$

and

$$F_q^4 = \tfrac{1}{4} \sum_{\diamond} C_q(\hat{\mathbf{q}}_q \cdot \hat{\mathbf{q}}_2, \hat{\mathbf{q}}_2 \cdot \hat{\mathbf{q}}_3)\rho_q(\hat{\mathbf{q}}_1)\rho_q(\hat{\mathbf{q}}_3)\rho_q(\hat{\mathbf{q}}_4), \tag{7.87}$$

where $\sum_{\triangle}$ is the sum over all triples $\hat{\mathbf{q}}_1, \hat{\mathbf{q}}_2, \hat{\mathbf{q}}_3 \in \{\hat{\mathbf{q}}\}$ such that $\hat{\mathbf{q}}_1 + \hat{\mathbf{q}}_2 + \hat{\mathbf{q}}_3 = 0$, while $\sum_{\diamond}$ is the sum over all quadruplets $\hat{\mathbf{q}}_1, \hat{\mathbf{q}}_2, \hat{\mathbf{q}}_3, \hat{\mathbf{q}}_4 \in \{\hat{\mathbf{q}}\}$ such that $\hat{\mathbf{q}}_1 + \hat{\mathbf{q}}_2 + \hat{\mathbf{q}}_3 + \hat{\mathbf{q}}_4 = 0$.

Since the symmetry of a translationally ordered phase depends only on the direction

$$\hat{\rho}_q(\hat{\mathbf{q}}) \equiv \frac{\rho_q(\hat{\mathbf{q}})}{|\rho_q|} \tag{7.88}$$

of the order parameter, and not on its magnitude, it will be determined by the cubic and quartic terms. The quadratic term and its square in the quartic term only play a role in determining $q$ and $|\rho_q|$. In particular, which fundamental set of the directions $\{\hat{\mathbf{q}}\}$ will be chosen depends on the higher than second-order terms. More explicitly, the transition surface is given by

$$\max_{\hat{\rho}_q(\hat{\mathbf{q}})} \{F_q^3[\hat{\rho}_q(\hat{\mathbf{q}})]^2 - 2A_q F_q^4[\hat{\rho}_q(\hat{\mathbf{q}})]\} = 0, \tag{7.89}$$

in analogy with Eq. (7.62). The maximization in Eq. (7.89) also implies a maximization over $q$ and over the sets $\{\hat{\mathbf{q}}\}$ of the directions of the fundamental wave vectors. If the $q$-dependence is neglected in $B_q$ and $C_q$, one concludes that $q$ is determined by the minimum of $A_q$.

While the rotational invariance leaves only one independent cubic coupling constant $B_q$, there is a two-dimensional continuum of quartic coupling constants $C_q$. The coupling $C_q$ has an obvious discrete symmetry arising from permutations $p$ of $\hat{\mathbf{q}}_1$, $\hat{\mathbf{q}}_2$, $\hat{\mathbf{q}}_3$, and $\hat{\mathbf{q}}_4$,

$$C_q(\hat{\mathbf{q}}_1 \cdot \hat{\mathbf{q}}_2, \hat{\mathbf{q}}_2 \cdot \hat{\mathbf{q}}_3) = C_q(\hat{\mathbf{q}}_{p(1)} \cdot \hat{\mathbf{q}}_{p(2)}, \hat{\mathbf{q}}_{p(2)} \cdot \hat{\mathbf{q}}_{p(3)}). \tag{7.90}$$

Explicit relations can be obtained using the fact that $\hat{\mathbf{q}}_1 + \hat{\mathbf{q}}_2 + \hat{\mathbf{q}}_3 + \hat{\mathbf{q}}_4 = 0$, which leads to

$$\hat{\mathbf{q}}_1 \cdot \hat{\mathbf{q}}_3 = -1 - x - y, \tag{7.91}$$

$$\hat{\mathbf{q}}_1 \cdot \hat{\mathbf{q}}_4 = y, \tag{7.92}$$

$$\hat{\mathbf{q}}_2 \cdot \hat{\mathbf{q}}_4 = -1 - x - y \tag{7.93}$$

and

$$\hat{\mathbf{q}}_3 \cdot \hat{\mathbf{q}}_4 = x, \tag{7.94}$$

where $x \equiv \hat{\mathbf{q}}_1 \cdot \hat{\mathbf{q}}_2$ and $y \equiv \hat{\mathbf{q}}_2 \cdot \hat{\mathbf{q}}_3$. Therefore, the symmetries of $C_q(x, y)$ are generated by

$$C_q(x, y) = C_q(y, x) \tag{7.95}$$

and

$$C_q(x, y) = C_q(x, -1 - x - y). \tag{7.96}$$

Clearly, for any set $\{\hat{\mathbf{q}}\}$ there will be quartic contributions with $y = -x$ (or equivalently, $y = 0$) associated with $\hat{\mathbf{q}}_3 = -\hat{\mathbf{q}}_1$ and $\hat{\mathbf{q}}_4 = -\hat{\mathbf{q}}_2$. In particular, a contribution with $x = -y = 1$ associated with $\hat{\mathbf{q}}_1 = \hat{\mathbf{q}}_2 = -\hat{\mathbf{q}}_3 = -\hat{\mathbf{q}}_4$, will always be present. However, the precise functional form of $C_q(x, y)$ can depend on specific features, such as bond angles, packing considerations, and band structure, so that very little can be said regarding the universal trends in solidification. Namely, two (inequivalent) sets of fundamental wave vectors will generally give rise to nonzero terms with *different* $(x, y)$'s and, consequently, if associated couplings are suitably chosen, either set could be assumed to have lower energy.[10]

A simple illustration of different $(x, y)$'s can be made for octahedral face, edge, and vertex sets $\{\hat{\mathbf{q}}\}$. The directions associated with these sets are the following:

$$\begin{cases} \hat{\mathbf{f}}_1^{O_h} = \dfrac{1}{\sqrt{3}}(1, 1, 1), \quad \hat{\mathbf{f}}_2^{O_h} = \dfrac{1}{\sqrt{3}}(1, 1, -1), \\[2mm] \hat{\mathbf{f}}_3^{O_h} = \dfrac{1}{\sqrt{3}}(1, -1, -1), \quad \hat{\mathbf{f}}_4^{O_h} = \dfrac{1}{\sqrt{3}}(1, -1, 1), \end{cases} \tag{7.97}$$

$$\begin{cases} \hat{\mathbf{e}}_1^{O_h} = \dfrac{1}{\sqrt{2}}(1, 0, 1), \quad \hat{\mathbf{e}}_2^{O_h} = \dfrac{1}{\sqrt{2}}(1, 1, 0), \quad \hat{\mathbf{e}}_3^{O_h} = \dfrac{1}{\sqrt{2}}(0, 1, 1), \\[2mm] \hat{\mathbf{e}}_4^{O_h} = \dfrac{1}{\sqrt{2}}(-1, 0, 1), \quad \hat{\mathbf{e}}_5^{O_h} = \dfrac{1}{\sqrt{2}}(1, -1, 0), \quad \hat{\mathbf{e}}_6^{O_h} = \dfrac{1}{\sqrt{2}}(0, 1, -1), \end{cases} \tag{7.98}$$

and

$$\hat{\mathbf{v}}_1^{O_h} = (1, 0, 0), \quad \hat{\mathbf{v}}_2^{O_h} = (0, 1, 0), \quad \hat{\mathbf{v}}_3^{O_h} = (0, 0, 1). \tag{7.99}$$

---

[10]This difficulty is amplified as one includes into the expansion Eq. (7.84) higher and higher terms. In fact, one can show rigorously [44] that if the degree of the expansion is not limited, then for *every* subgroup of SO(3) there will exist a set $\{\hat{\mathbf{q}}\}$ of that particular symmetry and a set of coupling constants such that $\{\rho_q(\hat{\mathbf{q}})\}$ minimizes the free energy.

These sets generate reciprocal lattices for FCC, BCC, and SC crystal structures, respectively. Besides $(x, y) = (1, -1)$, the other symmetry inequivalent possibilities are

$$(x, -x); \quad x = \tfrac{1}{3} = \hat{\mathbf{f}}_1^{O_h} \cdot \hat{\mathbf{f}}_2^{O_h} \tag{7.100}$$

and

$$\left(-\tfrac{1}{3}, -\tfrac{1}{3}\right); \quad (\hat{\mathbf{q}}_1 = \hat{\mathbf{f}}_1^{O_h}, \hat{\mathbf{q}}_2 = -\hat{\mathbf{f}}_2^{O_h}, \hat{\mathbf{q}}_3 = \hat{\mathbf{f}}_3^{O_h}, \hat{\mathbf{q}}_4 = \hat{\mathbf{f}}_4^{O_h}) \tag{7.101}$$

for the face set,

$$(x, -x); \quad x = \begin{cases} 0 = \hat{\mathbf{e}}_1^{O_h} \cdot \hat{\mathbf{e}}_4^{O_h} \\[2mm] \tfrac{1}{2} = \hat{\mathbf{e}}_1^{O_h} \cdot \hat{\mathbf{e}}_2^{O_h} \end{cases} \tag{7.102}$$

and

$$\left(0, -\tfrac{1}{2}\right); \quad (\hat{\mathbf{q}}_1 = \hat{\mathbf{e}}_1^{O_h}, \hat{\mathbf{q}}_2 = -\hat{\mathbf{e}}_4^{O_h}, \hat{\mathbf{q}}_3 = -\hat{\mathbf{e}}_5^{O_h}, \hat{\mathbf{q}}_4 = -\hat{\mathbf{e}}_2^{O_h}) \tag{7.103}$$

for the edge set, and

$$(x, -x); \quad x = 0 = \hat{\mathbf{v}}_1^{O_h} \cdot \hat{\mathbf{v}}_2^{O_h} \tag{7.104}$$

for the vertex set.

Different specific assumptions about the fundamental dependence of $C_q(x, y)$ will lead to different, nonuniversal conclusions. An implicit assumption made by most authors is that only the isotropic (in $\rho$ space) component of $C_q$ is nonzero. That is, the quartic term is $\overline{C}_q |\rho_q|^4$, where $C_q(x, -x) \equiv \overline{C}_q$ for all $x$. In this case, the free energy $F_q$ is minimized by a set $\{\rho_q(\hat{\mathbf{q}})\}$ which *maximizes* $|F_q^3[\hat{\rho}_q(\hat{\mathbf{q}})]|^2$.

If it is further assumed that all $\rho_q(\hat{\mathbf{q}})$ have the same amplitudes and phases, then

$$|F_q^3| = \tfrac{1}{3} B_q \frac{n_3}{n^{3/2}}, \tag{7.105}$$

where $n$ is the number of elements in $\{\hat{\mathbf{q}}\}$ (note that if $\hat{\mathbf{q}} \in \{\hat{\mathbf{q}}\}$, then $-\hat{\mathbf{q}} \in \{\hat{\mathbf{q}}\}$), and $n_3$ is the number of distinct zero-sum triplets in the set. This automatically eliminates all the sets for which $n_3 = 0$,[11] such as the sets for FCC and SC structures given in Eqs. (7.97) and (7.99). Several other sets have been considered [17,25,26] and it was found that the BCC set, Eq. (7.98) always wins with

$$\left(\frac{n_3}{n^{3/2}}\right)_{\text{BCC}} = \frac{8}{12^{3/2}} \approx 0.19245. \tag{7.106}$$

---

[11]In fact, this conclusion is independent of the assumptions about the phases and amplitudes.

In particular, this value was found to be considerably higher than the value

$$\left(\frac{n_3}{n^{3/2}}\right)_{\text{IE}} = \frac{20}{30^{3/2}} \approx 0.12172 \qquad (7.107)$$

obtained for the icosahedral edge (IE) set. The IE set consists of $n = 30$ vectors which point along the twofold axes of an icosahedron. For example, they can be obtained by all sign combinations and all cyclic permutations in $\pm(1, 0, 0)$ and $(\pm\tau, \pm1, \pm\tau^{-1})$. This set gives rise to $n_3 = 20$ zero-sum triplets. It can be verified that $n_3 = 0$ for the icosahedral vertex set (along the fivefold axes) and for the icosahedral face set (along the threefold axes). Therefore, it was concluded that BCC translational order is most stable. This conclusion was unchanged even with a relaxation of the assumptions regarding the amplitudes and phases of $\rho_q(\hat{\mathbf{q}})$ [24].

It is interesting to observe, however, that if one assumes 20 ideal tetrahedra to pack into an ideal icosahedron, and if one identifies the edges of the tetrahedra with the set $\{\hat{\mathbf{q}}\}$, then the resulting value

$$\left(\frac{n_3}{n^{3/2}}\right)_{\text{TCP}} = \frac{50}{42^{3/2}} \approx 0.18369 \qquad (7.108)$$

is much closer to the BCC value. This assumption is equivalent to neglecting the 5% difference between the edge and center-to-vertex distances in an ideal icosahedron.

In a related approach, Kalugin, Kitaev, and Levitov [41] argued that the icosahedral edge components $\rho_k(\hat{\mathbf{k}})$, which are second harmonics of the icosahedral vertex components $\rho_q(\hat{\mathbf{q}})$, must be included into the analysis since, in contrast to the ordinary crystals, for the icosahedral vertex model $k/q \approx 1$, so that the minimum of $A_q$ need not differentiate between $k$ and $q$. They then concluded that an icosahedral structure can be stabilized in this way. However, a subsequent, more detailed analysis showed this not to be the case [15].

Mermin and Troian [24] attempted to stabilize icosahedral translational ordering in another way.[12] Instead of considering only one density order parameter $\rho$, they included another density $\rho'$, viewing $\rho$ and $\rho'$ as densities of two different atomic species. Then, they assumed that $\rho'$ is high above its ordering temperature so that its ordering is induced by the ordering of $\rho$. Consequently, only the lowest-order terms in $\rho'$ need be considered. They are the interaction term,

$$\sum_{\mathbf{k}=-\mathbf{q}_1-\mathbf{q}_2} D_q(\hat{\mathbf{q}}_1 \cdot \hat{\mathbf{q}}_2)\rho'_k(\hat{\mathbf{k}})\rho_q(\hat{\mathbf{q}}_1)\rho_q(\hat{\mathbf{q}}_2), \qquad (7.109)$$

---

[12]The view of Ref. 24 presented here, although different from the original presentation, serves to emphasize some silent features of that work.

and the associated quadratic term,

$$\frac{1}{2} \sum_{\mathbf{k}=-\mathbf{q}_1-\mathbf{q}_2} A'_k |\rho'_k|^2, \tag{7.110}$$

where the sum is over all $\hat{\mathbf{q}}_1, \hat{\mathbf{q}}_2 \in \{\hat{\mathbf{q}}\}$ so that $k$ cannot exceed $2q$. Therefore, a minimization with respect to $\rho'_k$ can be easily performed, and one recovers Eq. (7.84), but with an additional quartic coupling for $\rho$,

$$-\frac{1}{2} \sum_{\hat{\mathbf{q}}_1, \hat{\mathbf{q}}_2 \in \{\hat{\mathbf{q}}\}} \frac{D_q(\hat{\mathbf{q}}_1 \cdot \hat{\mathbf{q}}_2)^2}{A'_{|\mathbf{q}+\mathbf{q}_2|}} |\rho_q(\hat{\mathbf{q}}_1)|^2 |\rho_q(\hat{\mathbf{q}}_2)|^2. \tag{7.111}$$

In other words, the coupling $C_q(x, -x)$ is transformed into

$$C_q^{eff}(x, -x) = C_q(x, -x) - \frac{1}{2} \frac{D_q(x)^2}{A'_{q\sqrt{2(1+x)}}}. \tag{7.112}$$

Since it is assumed that $A' > 0$, and since $D_q(x)^2 > 0$, the effective quartic coupling is reduced with respect to the bare coupling. In fact, since Mermin and Troian assumed that $C_q(x, y) = \text{const}$, we see that the quartic interaction becomes dominated by the coupling in Eq. (7.112) and the favored structure will be determined by the minimum of Eq. (7.112). Furthermore, if $A'$ can be assumed a slowly varying function of $k$, the minimum of Eq. (7.112) corresponds to the maximum of $D_q(x)^2$. Assuming that the maximum occurs at $x = \pm\frac{1}{\sqrt{5}}$, Mermin and Troian found that the icosahedral vertex set is selected for $\rho$ (inducing the icosahedral vertex set for $\rho'$). However, since $n_3 = 0$ for the icosahedral vertex set, the corresponding structure, although selected by the quartic terms, could be stabilized only far below another translationally ordered phase characterized by a set with $n_3 > 0$. This presents a serious problem, since one of the underlying assumptions of the Landau expansion is that it is an expansion near the isotropic phase. Moreover, the assumptions that $C_q(x, y) = \text{const}$ and that $D_q(x)^2$ reaches a sharp maximum at $x = \pm\frac{1}{\sqrt{5}}$ are not sufficiently generic.

In order to stabilize the icosahedral structure, Bak [23] extended previous assumptions by adding a fifth-degree term to the expansion. He then considered only a contribution arising from those $\mathbf{q}$'s, which form a regular pentagon. There are no such $\mathbf{q}$'s for the BCC set, but they exist in the icosahedral edge set. Therefore, provided that the free energy is bounded from below by a positive and isotropic in $\rho$-space sixth-degree term, this fifth-order coupling can be chosen to stabilize an icosahedral phase relative to the BCC phase. However, this is a highly nongeneric result and near the isotropic phase this icosahedral phase would occupy a volume of co-dimension greater than or equal to one (rather than a volume of co-dimension zero).

In fact, as we already mentioned in our discussion of the Mermin–Troian approach, similar nongeneric assumptions which can stabilize an icosahedral phase can be introduced already at the level of the fourth-degree free energy. For example, if one assigns a large negative value to $C_q(x, y)$ for those $(x, y)$ which appear for the icosahedral edge set but do not arise for the BCC set, the icosahedral symmetry could win. For the icosahedral edge set we find

$$(x, -x); \quad x = 0, \frac{1}{2\tau}, \frac{1}{2}, \frac{\tau}{2}, 1 \tag{7.113}$$

and

$$(x, y) = \left(-\frac{1}{2}, -\frac{1}{2\tau}\right). \tag{7.114}$$

Comparing this with the results Eqs. (7.102) and (7.103) for the BCC set, we see that an icosahedral edge structure can be stabilized by sufficiently negative $C_q\left(-\frac{1}{2}, -\frac{1}{2\tau}\right)$, $C_q\left(\frac{1}{2\tau}, -\frac{1}{2\tau}\right)$, or $C_q\left(\frac{\tau}{2}, -\frac{\tau}{2}\right)$, irrespective of the cubic terms.[13]

In conclusion, we have seen that a generic icosahedral phase could not be found by considering only a translational order parameter. Moreover, even at the lowest level of nongenericity, when an icosahedral phase is stabilized with special assumptions about the quartic coupling $C_q(x, y)$, only the icosahedral edge set $\{\hat{\mathbf{q}}\}$ can be selected, contrary to all current experimental observations.

## 7.4.3   Orientational-Translational Coupling

In this section we shall investigate whether a phase with icosahedral translational order can be stabilized by a coupling between orientational and translational order parameters. Therefore, we shall consider a free energy of the form

$$F_{L,q}[\rho_q(\hat{\mathbf{q}}), \Omega_{Lm}] = F_q[\rho_q(\hat{\mathbf{q}})] + F_L(\Omega_{Lm}) + F_{L,q}^{\text{int}}[\rho_q(\hat{\mathbf{q}}), \Omega_{Lm}]. \tag{7.115}$$

To lowest order, the coupling $F_{L,q}^{\text{int}}$ can be determined simply by the requirement of translational and rotational invariance. This results in an interaction free energy density in which $\Omega_{Lm}$ couples directly to the structure factor $|\rho_q(\hat{\mathbf{q}})|^2$,

$$\alpha_{L,q}\Omega_{Lm}Y_{Lm}^*(\hat{\mathbf{q}})|\rho_q(\hat{\mathbf{q}})|^2. \tag{7.116}$$

An underlying physical argument for this form is that the equilibrium $\rho$ need not have the symmetry of the equilibrium $\Omega$ even though the structure factor must have this symmetry.[14] Nevertheless, from a physical point of

---

[13]The BCC structure is favored when these couplings are positive and $C_q\left(0, -\frac{1}{2}\right)$ is negative.

[14]For example, a *general* Ammann icosahedral tiling does not have a center of *exact* icosahedral symmetry, although its structure factor does.

view, the coupling Eq. (7.116) seems somewhat confusing: The structure factor at $\mathbf{q}$ in the *reciprocal* space is coupled to the density of bonds pointing along $\hat{\mathbf{q}}$ in the *physical* space. Therefore, we shall first present a physically motivated derivation of the coupling given in Eq. (7.116).

The coupling between the orientational and translational order parameters should be such as to lock the bond density given in Eqs. (7.3) and (7.4) to a bond density

$$\overline{\Omega}(\mathbf{x}, \hat{\mathbf{n}}) = \rho(\mathbf{x}) \int [\rho(\mathbf{x} + \mathbf{n}) - \delta(\mathbf{n})] w(n) n^2 \, dn \qquad (7.117)$$

associated with the translational order parameter. In this equation, $w(n)$ is a positive bond-weighting function which should decrease to zero as $n$ is increased. The required locking can be achieved by adding to the free energy density a term

$$\lambda_{L,q} |\Omega_{Lm}(\mathbf{q}) - \overline{\Omega}_{Lm}(\mathbf{q})|^2, \qquad (7.118)$$

and assuming $\lambda_{L,q}$ to be large. The coupling given in Eq. (7.116) now follows directly from Eq. (7.118), and one finds

$$\alpha_{L,q} = -2\lambda_{L,0} w_{L,q}^*, \qquad (7.119)$$

where

$$w_{L,q} \equiv 4\pi(-1)^{L/2} \int_0^\infty j_L(qn) w(n) n^2 dn, \quad (L \text{ even}), \qquad (7.120)$$

and $j_L$ is the spherical Bessel function. Moreover, this implies that to the lowest order in $q$

$$\alpha_{L,q} = \alpha_L q^L, \qquad (7.121)$$

where

$$\alpha_L = \lambda_{L,0} \frac{8\pi(-1)^{\frac{L}{2}+1}}{(2L+1)!!} \int_0^\infty w(n) n^{L+2} \, dn. \qquad (7.122)$$

Consequences of the coupling Eq. (7.116) are easy to deduce. If the translational ordering temperature $T_t$ given by Eq. (7.89) is greater than the orientational ordering temperature $T_0$ given by Eq. (7.62) then, because the interaction Eq. (7.116) is linear in $\Omega$, the ordering of $\rho$ at $T_t$ will necessarily induce an ordering in $\Omega$. On the other hand, if $T_0 > T_t$, then since Eq. (7.116) is quadratic in $\rho$, the ordering of $\Omega$ at $T_0$ will generate an effective translational quadratic coupling $A_q^{\mathrm{eff}}(\hat{\mathbf{q}})$ without necessarily inducing an ordering of $\rho$. However, when the transition at $T_0$ is discontinuous, like in the case of the icosahedral orientational ordering, the minimum of $A_q^{\mathrm{eff}}(\hat{\mathbf{q}})$ might be sufficiently reduced relative to the minimum of $A_q$ so that $\rho$ orders. Indeed, the effective coupling is

$$A_q^{\mathrm{eff}}(\hat{\mathbf{q}}) = A_q + \alpha_{6,q} \sum_{m=-6}^{6} \Omega_{6m}^{Y_h} Y_{6m}^*(\hat{\mathbf{q}}). \qquad (7.123)$$

Owing to Eq. (7.121), $A_q^{\text{eff}}(\hat{\mathbf{q}})$ is an analytic function of $\mathbf{q}$. It is easy to verify that the absolute maximum and minimum of $\sum_m \hat{\Omega}_{6m}^{Y_h} Y_{6m}^*(\hat{\mathbf{q}})$ are $(143/4\pi)^{1/2}/5 > 0$ and $-(143/4\pi)^{1/2}/9 < 0$, and that they occur in the directions of icosahedral vertices and faces, respectively. Thus, $\min A_q^{\text{eff}}(\hat{\mathbf{q}}) < \min A_q$, and if the coupling $\alpha_6$ is sufficiently strong, $\rho$ might order at $T_0$ or at some slightly lower temperature. Moreover, depending on the sign of $B_q$, which fixes the sign of $\sum_m \Omega_{6m}^{Y_h} Y_{6m}^*(\hat{\mathbf{q}})$, the minimum of $A_q^{\text{eff}}(\hat{\mathbf{q}})$ will correspond to $\hat{\mathbf{q}}$'s pointing either along three- or fivefold axes of an icosahedron. This is in agreement with the experimental observations and contrasts the result obtained with a purely translational order parameter which can only select the icosahedral edge set. If the translational ordering is induced below $T_0$, this transition would, in principle, be continuous since there are no icosahedral vertex or face invariants of the third degree in $\rho$.

Note that in the case of the icosahedral face set selected by orientational ordering, corresponding icosahedral translational ordering would compete with other symmetries. In particular, an octahedral face set, which would induce a FCC structure, would compete since this set is contained in the icosahedral face set. The selection between such different possibilities is now completely due to the relevant quartic couplings $C_q$. However, the icosahedral phase, which is associated with a maximal subgroup of the mathematical symmetry group of the resulting free energy, will occupy in the phase space a volume of co-dimension zero.

Once the fundamental order parameters set in, they will generate in the usual manner the "higher harmonics." In particular, through a linear coupling with $\rho$, the orientational order parameter will develop Fourier components at the wave vectors $\{\mathbf{q}\}$ and their harmonics.

# 7.5    Conclusion

Three-dimensional bond-orientational order and its relevance to icosahedral quasicrystals were considered in this chapter within a mean-field theory. Quartic Landau free energies were minimized for order parameters which transform like $L = 2$, 4, and 6 irreducible representations of SO(3) and all low-symmetry phases were determined. For $L = 4$, although the phase diagram is dominated by octahedral symmetry, we find that a uniaxial as well as a $D_{4h}$ phase can be stabilized. For $L = 6$ the phase diagram is dominated by the icosahedral phase, but uniaxial, octahedral, and hexagonal phases can be also stabilized. All of these orientationally ordered phases can be interpreted as analogs of the hexatic phase [45] (i.e., bond orientationally ordered) or, alternatively, as mesomorphic phases of aligned icosahedral (octahedral, hexagonal, etc.) "molecules."

By considering an *independent* translational order parameter, which is usually assumed to be dominant in driving solidification, it was shown that an icosahedral translationally ordered phase cannot be generically stable. However, in real systems, the orientational order parameter will be coupled to the positional order parameter and solidification can be driven by the orientational ordering. Namely, as the orientational ordering sets in, it effectively renormalizes the quadratic coupling in the positional free energy by adding a term proportional to the orientational density. It was demonstrated that in this way a phase with long-range icosahedral orientational and translational order can be generically stable. The induced positional ordering then depends on the precise form of the orientational density. For example, such an analysis for $L = 6$ leads to the conclusion that an icosahedral quasicrystal phase can be stabilized with fundamental wave vectors pointing into faces or vertices of an icosahedron. Similarly, for $L = 4$ octahedral orientational ordering stabilizes SC or FCC structurs which are otherwise suppressed by a BCC phase.

Two pictures of quasicrystals emerge. In one, they are quenched icosahedratics, thus metastable, where icosahedral orientational order is long-ranged but the translational order is only short-ranged. In the other, they are quasiperiodic, with long-range icosahedral translational order, and they can be a genuine equilibrium phase.

The conclusion that two different types of quasicrystal states could exist is consistent with experimental observations. The first type of the quasicrystalline state, the icosahedral glass, is probably represented by i(Al-Mn), which is metastable and has broad peak widths. The second type, the quasiperiodic crystal, is most likely seen in i(Al-Cu-Fe), which has resolution-limited peak widths and is an equilibrium phase. The conclusion that the icosahedral edge model is not favored compared to the face or vertex models is also in good agreement with the experimental evidence. However, direct experimental evidence in support of a specific mechanism for quasicrystalline ordering outlined here is not available at present. It seems plausible that experiments analyzing a supercooled melt near a transition into a quasicrystal phase could detect the icosahedratic phase. In principle, this phase could be identified by observing icosahedrally modulated rings in the structure factor, or by observing icosahedral anisotropy in higher-order response functions.

On the theoretical side, the most important question that needs to be resolved concerns the effect of fluctuations on the nature of transitions between various phases discussed here. The conclusion that a direct transition from isotropic-liquid to icosahedral quasiperiodic phase remains first-order even with the inclusion of fluctuations is based on general arguments [47]. However, the effects of fluctuations on transitions between isotropic-liquid and icosahedratic, and between icosahedratic and a quasiperiodic crystal, remain to be fully understood. Also, other physical properties of icosahedratics need to be investigated in order to identify the most promising methods for experimental verification of this new phase.

It is a pleasure to thank the Physics Departments at the University of

California, Santa Cruz; Università degli Studi di Trento, Povo; and Universidade Federal de Pernambuco, Recife, where parts of this chapter were completed. During different stages of this work, I had valuable assistance from, or discussions with, M.D. Coutinho-Filho, B. Halperin, S. Johnson, M. Kardar, W. Kinnersley, D.R. Nelson, S.-Y. Qiu, M. Ronchetti, S. Sachdev, P.J. Steinhardt, M. Widom, and P. Young. This work was supported in part through NSF Grants DMR8821802 and DMR8721673.

# References

[1] D. Shechtman, I. Blech, D. Gratias, and J.W. Cahn, *Phys. Rev. Lett.* **53**, 1951 (1984).

[2] L.D. Landau and E.M. Lifshitz, *Statistical Physics*, Pergamon, London, 1969, Chap. 14; L.D. Landau, in *Collected Papers of L.D. Landau*, edited D. ter Haar, Gordon and Breach Pergamon, New York, 1965, p. 193.

[3] C.L. Henley, *Comm. Cond. Matt. Phys.* **13**, 59 (1987).

[4] D. Gratias, *Contemp. Phys.* **28**, 219 (1987).

[5] D.R. Nelson, *Sci. Am.* **255**, 42 (1986).

[6] M.V. Jarić, *Encyclopedia of Modern Physics*, edited by R.A. Meyers, Academic Press, San Diego, Calif., 1990, p. 551.

[7] C. Janot and J.M. Dubois, *J. Phys. F* **18**, 2303 (1988).

[8] P.J. Steinhardt and S. Ostlund, eds., *The Physics of Quasicrystals*, World Scientific, Singapore, 1987.

[9] M.V. Jarić, ed., *Introduction to Quasicrystals*, Academic Press, Boston, 1988.

[10] M.V. Jarić, ed., *Introduction to the Mathematics of Quasicrystals*, Academic Press, Boston, 1989.

[11] M.V. Jarić and D. Gratias, eds., *Extended Icosahedral Structures*, Academic Press, Boston, 1989.

[12] R. Penrose, *Bull. Inst. Math. Appl.* **10**, 266 (1974); *Math. Intelligencer* **2**, 32 (1979).

[13] D. Levine and P.J. Steinhardt, *Phys. Rev. Lett.* **53**, 2477 (1984).

[14] D. Shechtman and I. Blech, *Metall. Trans. A* **16**, 1005 (1985).

[15] O. Biham, D. Mukamel, and S. Shtrikman, in *Introduction to Quasicrystals*, edited by M.V. Jarić, Academic Press, Boston, 1989, p. 171.

[16] M.V. Jarić, *Nucl. Phys. B* [FS15] **265**, 647 (1986).

[17] M.V. Jarić, *Phys. Rev. Lett.* **55**, 607 (1985).

[18] D.R. Nelson and J. Toner, *Phys. Rev. B* **24**, 363 (1981).

[19] P.J. Seinhardt, D.R. Nelson, and M. Ronchetti, *Phys. Rev. B* **28**, 784 (1983).

[20] F.H. Busse, *J. Fluid Mech.* **72**, 67 (1975).

[21] D.H. Sattinger, *J. Math. Phys.* **19**, 1720 (1978).

[22] R. Hornreich, in *Extended Icosahedral Structures*, edited by M.V. Jarić and D. Gratias, Academic Press, Boston, 1989, p. 221.

[23] P. Bak, *Phys. Rev. Lett.* **54**, 1517 (1985).

[24] N.D. Mermin and S. Troian, *Phys. Rev. Lett.* **54**, 1524 (1985); S. Troian, *J. Phys.* (Paris) **47**, C3-271 (1986); S. Troian, Ph.D. thesis, Cornell University, Ithaca, N.Y., 1986, unpublished.

[25] G. Baym, H.A. Bethe, and C.I. Pethick, *Nucl. Phys. A* **175**, 225 (1971).

[26] S. Alexander and J. McTague, *Phys. Rev. Lett.* **41**, 702 (1978).

[27] L. Michel, *Rev. Mod. Phys.* **52**, 617 (1980).

[28] E. Ihrig and M. Golubitsky, *Physica* **13D**, 1 and references therein (1984).

[29] M.V. Jarić, L. Michel, and R.T. Sharp, *J. Phys.* (Paris) **45**, 1 (1984).

[30] J. Bistricky, R. Gaskell, J. Patera, and R.T. Sharp, *J. Math. Phys.* **23**, 1560 (1982).

[31] J.S. Kim, *Nucl. Phys. B* **196**, 285 (1982); **197**, 174 (1982); S. Frautschi and J.S. Kim, *Nucl. Phys. B* **196**, 301 (1982).

[32] M.V. Jarić, *J. Math. Phys.* **24**, 917 (1983).

[33] L. Michel, *Lecture Notes in Physics* **6**, Springer-Verlag, New York, 1970.

[34] W. Kinnersley, unpublished.

[35] R. Alben, *Phys. Rev. Lett.* **30**, 778 (1973).

[36] A.D.J. Haymet, unpublished; *Phys. Rev. B* **27**, 1725 (1983).

[37] M.V. Jarić, *Lec. Notes Phys.* **201**, 397 (1984).

[38] M.V. Jarić, *Phys. Rev. Lett.* **48**, 1641 (1982).

[39] M. Widom, K.J. Strandburg, and R.H. Swendsen, *Phys. Rev. Lett.* **58**, 706 (1987).

[40] D.R. Nelson and S. Sachdev, *Phys. Rev. B* **32**, 4592 (1985).

[41] A. Kalugin, A. Yu. Kitaev, and L.S. Levitov, *Pis'ma Zh. Eksp. Teor. Fiz.* **41**, 119 (1985) [*JETP Lett.* **41**, 145 (1985)]; L. Gronlund and N.D. Mermin, *Phys. Rev. B* **38**, 3699 (1988).

[42] M. Widom, in *Introduction to Quasicrystals*, edited by M.V. Jarić, Academic, Boston, 1988, p. 59.

[43] D. Shoemaker and C. Shoemaker, in *Introduction to Quasicrystals*, edited by M.V. Jarić, Academic, Boston, 1988, p. 1.

[44] L. Michel, unpublished.

[45] B.I. Halperin and D.R. Nelson, *Phys. Rev. Lett.* **41**, 121 (1978); (E) **41**, 519 (1978); D.R. Nelson and B.I. Halperin, *Phys. Rev. B* **19**, 2456 (1979).

[46] P.A. Bancel et al., *Phys. Rev. Lett.* **54**, 2422 (1985).

[47] S.A. Brazovskii, *Zh. Eksp. Teor. Fiz.* **68**, 42 (1975) [*Sov. Phys. JETP* **41**, 85 (1975)]; S.A. Brazovskii and S.G. Dimitriev, *Zh. Eksp. Teor. Fiz.* **69**, 979 (1975) [*Sov. Phys. JETP* **42**, 497 (1976)].

# 8

# Icosahedral Glass Models for Quasicrystals

*A.I. Goldman*

## 8.1 Introduction

The discovery of icosahedral phase alloys by Shechtman et al. [1] has provided us with an opportunity to reevaluate many of our long-held ideas and prejudices about the relationship between positional order, bond-orientational order, and periodic translational order in condensed matter systems. Traditionally, we have grouped solids into two categories. Glasses are viewed as solids which, at best, can be characterized as having short-range chemical order. Crystals, at the other extreme, are described as a periodic stacking of well-defined unit cells, identically decorated with atoms, into structures which have long-range periodic translational order.

The constraint of periodicity limits the possible rotational symmetries allowed for crystals to two-, three-, four-, and sixfold rotation axes. As illustrated nicely by Kittel [2], one cannot tile the plane completely by unit cells with fivefold symmetry, and so such tilings in two dimensions, and the related icosahedral symmetry in three dimensions, are excluded from the realm of crystallography. This does not mean that the unit cells themselves cannot contain clusters of atoms with icosahedral or other "forbidden symmetries." There are, in fact, many examples of periodic crystalline structures with unit cells which contain such arrangements [3], and they assume an important role in the description of icosahedral quasicrystalline structures.

The difference between crystalline and glassy structures is most clearly illustrated by their diffraction patterns. The diffuse halos which are observed in the diffraction patterns from glassy materials reflect the short-range nature of the positional correlations between atoms. In contrast, the sharp diffraction spots characteristic of crystalline diffraction patterns indicate the presence of long-range order. Since we have identified sharp diffraction peaks with crystallinity, and crystallinity with periodic translational order for so long, the observation of sharp diffraction peaks arranged in a pattern with the forbidden icosahedral symmetry caused quite a stir.

It is important to realize that this observation does not, in fact, challenge the paradigms of crystallography. It remains true that rotational symme-

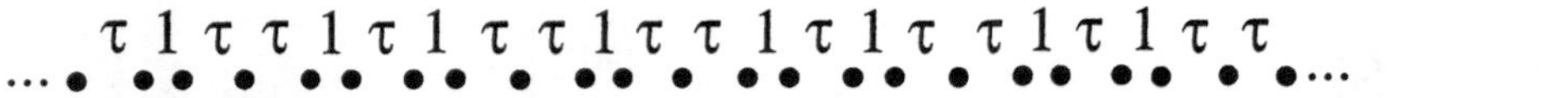

$$\ldots \overset{\tau}{\bullet} \overset{1}{\bullet} \overset{\tau}{\bullet} \overset{\tau}{\bullet} \overset{1}{\bullet} \overset{\tau}{\bullet} \overset{1}{\bullet} \overset{\tau}{\bullet} \overset{\tau}{\bullet} \overset{1}{\bullet} \overset{\tau}{\bullet} \overset{\tau}{\bullet} \overset{1}{\bullet} \overset{\tau}{\bullet} \overset{1}{\bullet} \overset{\tau}{\bullet} \overset{\tau}{\bullet} \overset{1}{\bullet} \overset{\tau}{\bullet} \overset{1}{\bullet} \overset{\tau}{\bullet} \overset{\tau}{\bullet} \ldots$$

FIGURE 8.1. A Fibonacci sequence of atoms. The distances 1 and $\tau$ that separate atoms are arranged quasiperiodically.

tries beyond those stated above are inconsistent with long-range translational periodicity. However, it is now well established that the 230 space groups on which we have traditionally based all structural identifications are but a subset of the allowed space groups if we also consider structures which have aperiodic translational order.

## 8.2    Icosahedral Crystals and Quasiperiodicity

One simple example of a system with long-range positional order that is aperiodic is shown in Figure 8.1. Atoms along the one-dimensional chain are arranged in a Fibonacci sequence of long $(L = \tau)$ and short $(S = 1)$ interatomic displacements according to the rule:

$$R_n = \frac{n}{1 + \frac{1}{\tau^2}} - \frac{1}{\tau}\{n\tau\}.$$

Here, $\tau = (\sqrt{5} + 1)/2$ is known as the golden mean, and the curly brackets denote the fractional part of the enclosed expression. The very existence of a rule specifying the atomic positions means that the structure is well-ordered, albeit aperiodic. The calculated diffraction pattern from this chain (as $L \to \infty$) is shown in Figure 8.2 and consists of a set of $\delta$-functions at positions:

$$G = \frac{2\pi}{\tau^2 + 1}(m + n\tau),$$

where $m$ and $n$ are positive and negative integers. The sequence of strong diffraction peaks in this pattern is also aperiodic and exhibits interesting scaling behavior; the pattern is invariant under inflations or deflations by the golden mean.

Structures which are characterized by long-range positional order which is aperiodic and exhibit icosahedral or other "forbidden" rotational symmetries have been termed "quasicrystals." Electron diffraction patterns of the high-symmetry twofold and fivefold planes of representative simple icosahedral (SI) and face-centered icosahedral (FCI) structures are shown in Figures 8.3 and 8.4, respectively [4]. All of the diffraction spots in these two patterns can be indexed by a set of six miller indices $(n_1, n_2, n_3, n_4, n_5, n_6)$ such that:

$$\mathbf{G} = G_0 \sum_i n_i \hat{e}_i^{\parallel}.$$

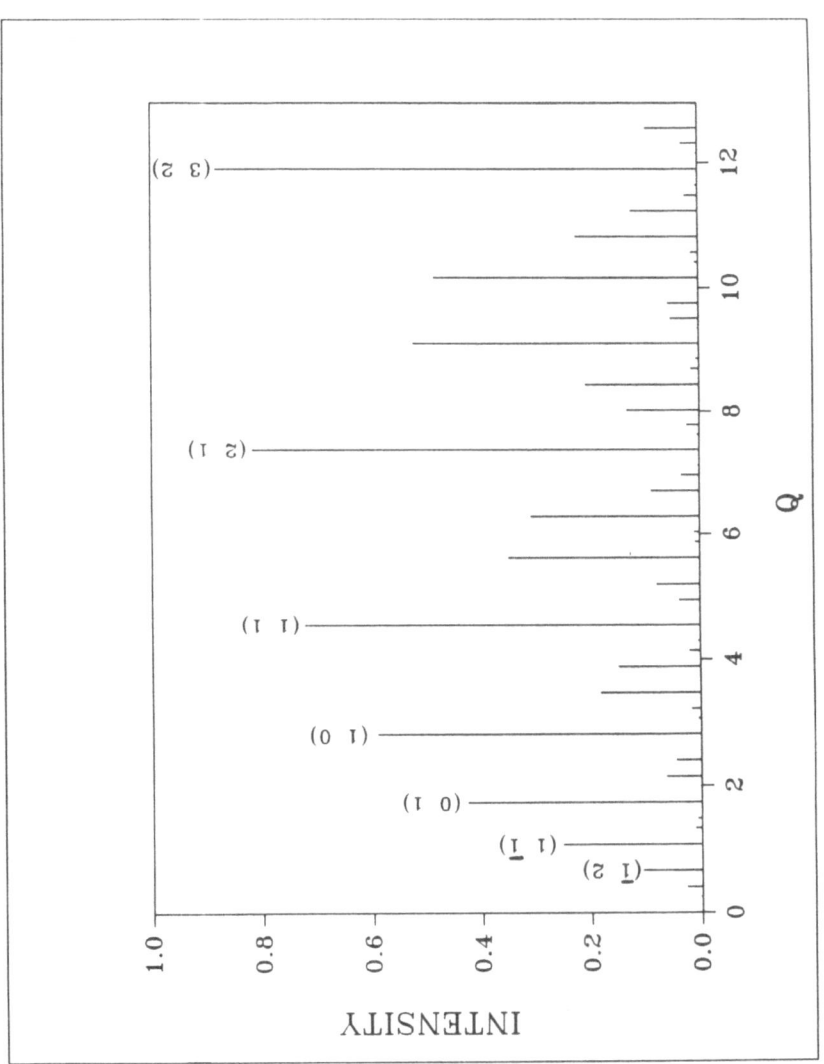

FIGURE 8.2. The Fourier transform of an infinitely long Fibonacci sequence of displacements illustrated in Fig. 8.1. The peaks are labeled by a set of two indices $(m, n)$. The strong peaks in the diffraction pattern are related by inflations of the golden mean.

The $\hat{e}_i^{\parallel}$ are six unit vectors which point from the center to the vertices of an icosahedron as shown in Figure 8.5:

$$\hat{e}_1^{\parallel} = (\tau^2 + 1)^{-1/2}(1,\tau,0), \qquad \hat{e}_2^{\parallel} = (\tau^2 + 1)^{-1/2}(\tau,0,1),$$

$$\hat{e}_3^{\parallel} = (\tau^2 + 1)^{-1/2}(\tau,0,-1), \qquad \hat{e}_4^{\parallel} = (\tau^2 + 1)^{-1/2}(0,1,-\tau),$$

$$\hat{e}_5^{\parallel} = (\tau^2 + 1)^{-1/2}(-1,\tau,0), \qquad \hat{e}_6^{\parallel} = (\tau^2 + 1)^{-1/2}(0,1,\tau).$$

The meaning of the superscript on the unit vectors will be discussed in Section 8.3.3.

The SI and FCI patterns are distinguished by the selection rules for allowed reflections. For the SI pattern, all combinations of integer $n_i$ are permissible, while for the FCI structure, all indices for a particular reflection must have the same parity. The fivefold plane of both the SI and FCI patterns contains five pairs of twofold axes 72° apart. Along these twofold axes there is a quasiperiodic sequence of bright spots related by inflations of $\tau$. The golden mean enters here because of the geometry of icosahedra. The twofold plane deserves careful inspection because it contains all three high-symmetry axes as shown. If we focus on the fivefold axes in Figures 8.3 and 8.4, we see a clear difference between the SI and FCI patterns. The bright spots along this direction in the SI pattern are related by an inflation of $\tau^3$, while those in the FCI pattern are related by $\tau$ inflations. The SI diffraction patterns are invariant under inflations or deflations by $\tau^3$, while the FCI patterns are invariant under inflations or deflations by $\tau$.

One of the interesting implications of the inflation/deflation symmetry of these patterns is that the fundamental length scale, or "quasilattice constant," of the structure cannot be determined from diffraction measurements within factors of $\tau$ or $\tau^3$, as the case may be. One choice of a fundamental reciprocal lattice vector may yield a length scale that corresponds to interatomic distances, while another may be more appropriate to distances between clusters of atoms.

# 8.3    Survey of Quasicrystalline Models for the Icosahedral Phase

There are two categories of structural models for the icosahedral phase alloys. The quasicrystalline models, described in this section, differ from the icosahedral glass model in that they produce Bragg diffraction peaks (as would be observed from ideal periodic crystals). That is, the longitudinal widths of diffraction peaks vanish in the limit of infinitely large samples. The essential difference between the two quasicrystalline descriptions lies in the nature of the growth rules required to generate them.

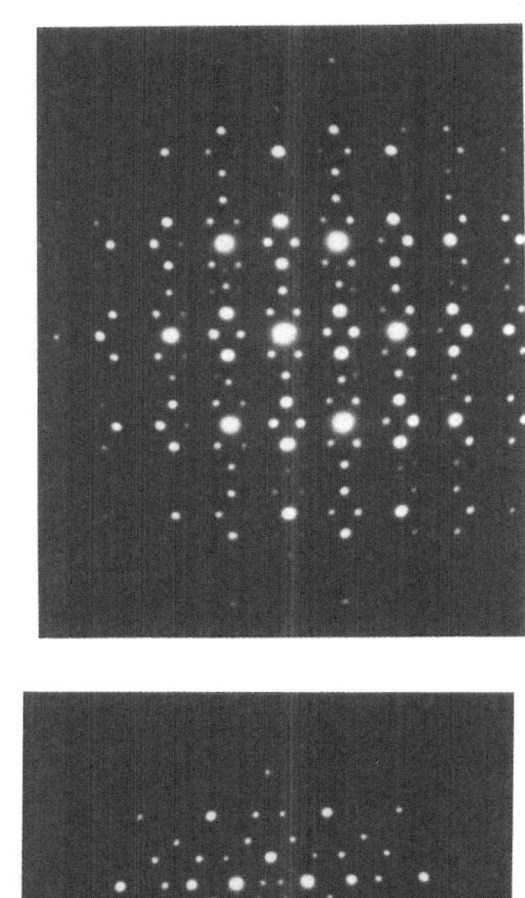

(a)

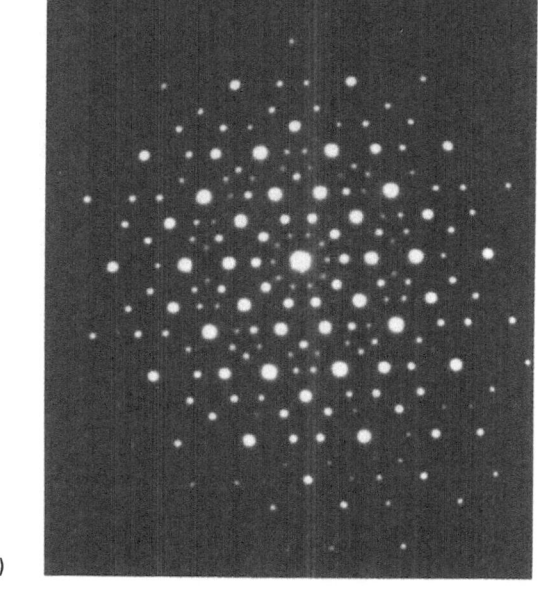

(b)

FIGURE 8.3. Selected area electron diffraction patterns of the two and fivefold planes in the simple icosahedral phase of Al-Mn [4]. Note that the strong peaks along this direction have $\tau^3$ inflation symmetry. The entire diffraction pattern is invariant under inflations or deflations by $\tau^3$.

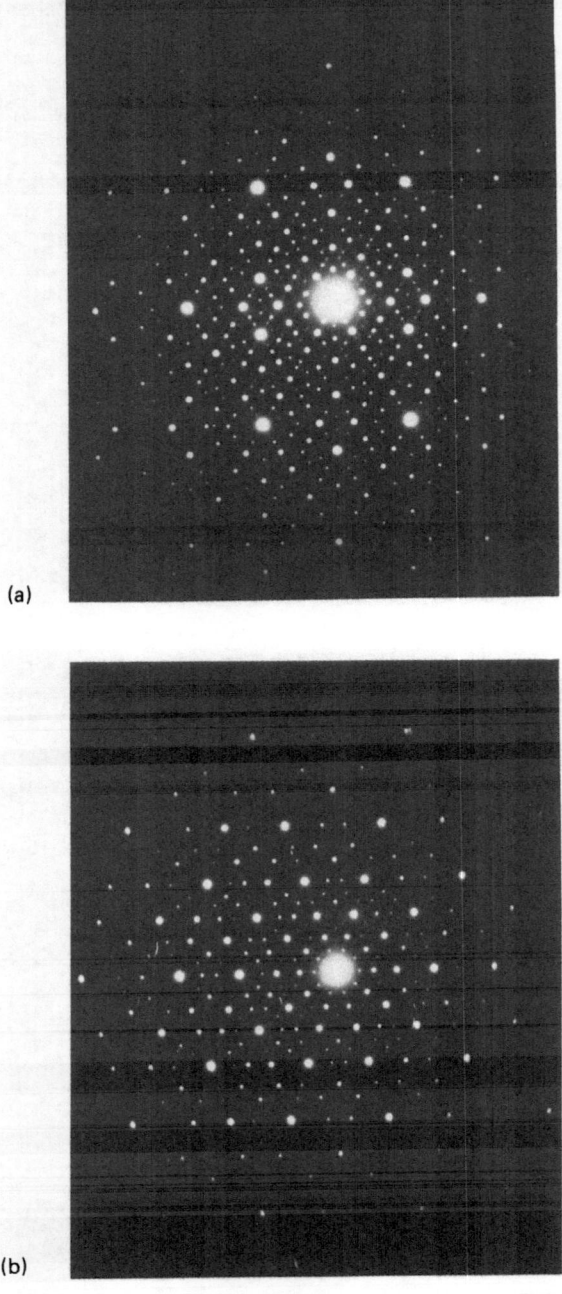

(a)

(b)

FIGURE 8.4. Selected area electron diffraction patterns of the two- and five-fold planes in the face-centered icosahedral phase of Al-Cu-Fe [4]. Note that the strong peaks along this direction have $\tau$ inflation symmetry. The entire diffraction pattern is invariant under inflations or deflations by $\tau$.

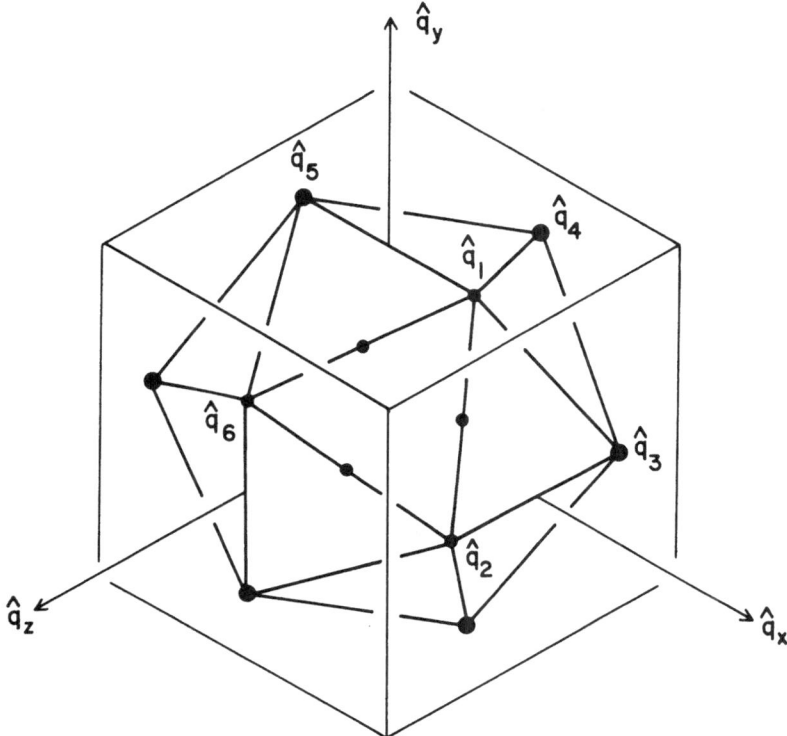

FIGURE 8.5. The vertex basis vectors of an icosahedron.

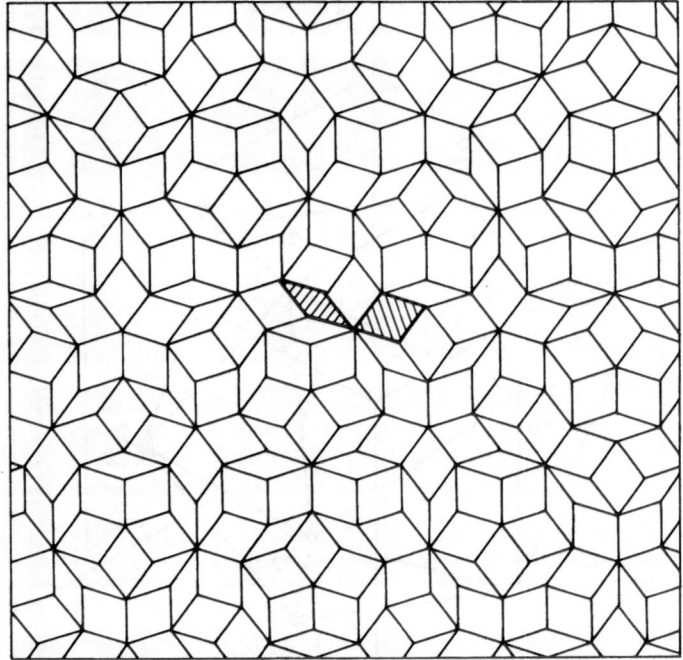

FIGURE 8.6. A two-dimensional Penrose tiling constructed from a "fat" and "skinny" rhombus. The analogous three-dimensional tilings may be constructed from obtuse and acute rhombohedra.

## 8.3.1    Perfect Quasicrystalline Tilings

There are now a number of well-developed mathematical techniques for generating perfect quasilattices with icosahedral, octagonal, decagonal, and dodecagonal rotational symmetries [5]. Perhaps the best known, but least general, icosahedral quasicrystalline structures are the Penrose tilings [6–8]. These tilings are generated by imposing restrictive global matching rules that specify how one constructs a quasiperiodic stacking of two rhombohedral unit cells in three dimensions (or two rhombuses in two dimensions as shown in Figure 8.6 for a decagonal quasicrystal). Octagonal quasilattices can be generated by a quasiperiodic packing of a rhombus and a square unit cell; dodecagonal quasicrystals from a rhombus, square, and regular hexagon. Implicit in this class of models for quasicrystalline structures is the notion that they are the lowest-energy (ground-state) arrangement of the atoms. The diffraction pattern from perfect tilings yields perfectly sharp Bragg peaks.

## 8.3.2   Entropically Stabilized Structures

An obstacle to the physical realization of perfect quasicrystalline structures is the fact that they require nonlocal growth rules to maintain perfection [9,10]. Physically, growth processes are local in the sense that the probability of growth at some particular site depends largely on the local environment of that site. In this view, mistakes in the choice of which tile to place at a particular site are unavoidable, destroying the perfection of the quasicrystalline tiling.

If the mistakes are not particularly ruinous, the quasicrystalline structure may be entropically stablized, at elevated temperatures, with respect to competing crystalline phases [11–14]. At low temperature, these "random tiling" models predict a phase transformation to a stable crystalline phase. While these tilings are not perfectly ordered, in the sense of a Penrose tiling, there is long-range positional order (in three dimensions) [15], and the diffraction pattern consists of Bragg peaks in addition to some diffuse scattering.

## 8.3.3   Quasicrystalline Tilings as Higher-Dimensional Periodic Structures

Of course, it is not immediately obvious why aperiodic lattices such as the Penrose tilings should produce Bragg diffraction. One way of seeing this is provided by a description of these structures in terms of a $d$-dimensional cut through a higher $D$-dimensional periodic lattice. Although this approach has been exhaustively treated in other reviews [16], it is useful to describe it here, since some of the terminology will be required in later sections.

The one-dimensional Fibonacci sequence of atoms in Figure 8.1 may be constructed by following the prescription diagrammed in Figure 8.7. Consider a square lattice of points in two dimensions. Now introduce a second set of axes rotated by some angle, $\theta$, with respect to the first. The axis marked $x_\parallel$ represents the one-dimensional physical subspace of the structure, while $x_\perp$ corresponds to a complementary space. The pointlike atomic basis in the one-dimensional physical space results from the intersection of the $x_\parallel$ axis with the atomic basis in two dimensions. Therefore, the atomic basis in two dimensions must consist of a set of line segments, or more generally, curves [17]. One possible atomic basis for the two-dimensional lattice is represented as a set of line segments, of length $L$, perpendicular to $x_\parallel$. If $\tan(\theta)$ is irrational, the density along $x_\parallel$ is aperiodic. For $\tan(\theta) = \tau^{-1}$, the sequence of long and short lengths between atoms along $x_\parallel$ forms a Fibonacci sequence.

The additional degree of freedom allowed by the complementary space, $x_\perp$, is important for many of the subtle points in the description of quasiperiodic structures. Consider a rigid shift of the $x_\parallel$ axis along the $x_\perp$ direction (dashed line in Figure 8.7). We are still left with a Fibonacci sequence of

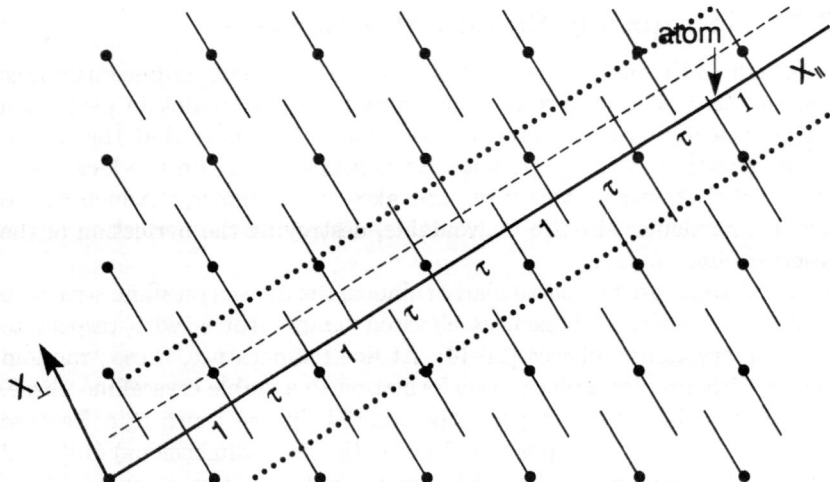

FIGURE 8.7. The one-dimensional Fibonacci sequence of Fig. 8.1 represented as a cut along $x_\parallel$ through a two-dimensional periodic lattice. The line segments parallel to $x_\perp$ describe the atomic basis in two dimensions. A rigid shift along the $x_\perp$ direction (the dashed line) preserves, but rearranges the Fibonacci sequence. The dotted lines parallel to $x_\parallel$ represent the acceptance volume of two-dimensional lattice points that can be projected down to the $x_\parallel$-axis.

long and short segments, but the local sequence of displacements at any particular place in the chain has changed. Shifts along the $x_\perp$ direction correspond to a local rearrangement of the "building blocks." In the context of Landau theory, this rearrangement results from a change in the relative phases of the mass density waves generating the structure, and so the hydrodynamic modes associated with shifts along $x_\perp$ have been termed "phason modes" [18]. The hydrodynamic modes associated with displacements along $x_\parallel$ are, of course, the well-known phonon modes.

For our purposes here, we are more interested in a description of the quasilattice itself than the specific atomic decoration of the quasilattice (although this is an important issue that has received considerable attention over the past few years). Therefore, we do not require any specific knowledge of the atomic basis in the two-dimensional parent structure, but instead concentrate on specifying which lattice points in the higher-dimensional lattice can be *projected* down onto $x_\parallel$ subspace [19]. The prescription for this procedure is also shown in Figure 8.7. Each point in the one-dimensional quasilattice is characterized by a set of two coordinates $(x_\parallel, x_\perp)$. Only those two-dimensional lattice points that fall within some range of $x_\perp$ from the $x_\parallel$ axis (an acceptance region or volume denoted by the dotted lines in Figure 8.7) are allowed lattice points in the one-dimensional structure. If all values of $x_\perp$ were allowed, the physical density in one dimension would be infinite. To produce an ideal one-dimensional

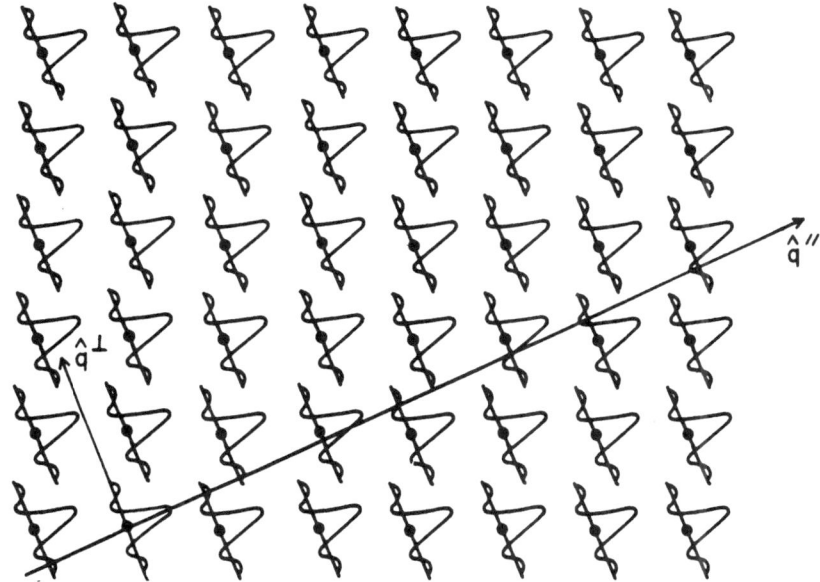

FIGURE 8.8. The Fourier transform of Fig. 8.7. The intensity of the diffraction peaks in real space ($q_{\shortparallel}$-axis) depends on the proximity of the two-dimensional reciprocal lattice point to the $q_{\shortparallel}$-axis.

Fibonacci lattice, the acceptance region also has a slope of $1/\tau$ and is perfectly collinear with $x_{\shortparallel}$.

The diffraction pattern of the one-dimensional Fibonacci sequence (Figure 8.2) results from the intersection of the Fourier transform of the basis of the two-dimensional crystal with the corresponding $q_{\shortparallel}$ reciprocal space axis, as shown in Figure 8.8. The transform consists of a series of $\delta$-functions along $q_{\shortparallel}$ with an intensity given by:

$$I(q_\perp) = \left[\frac{\sin(q_\perp L)}{q_\perp L}\right]^2.$$

Therefore, we see that the intensity of the diffraction peak will depend on its value of $q_\perp$.

Extending this discussion to higher dimensions is straightforward, but difficult to visualize. Imagine starting with a five-dimensional space. If a two-dimensional plane (representing the two-dimensional physical space) is perpendicular to the body diagonal of the 5-cube, the edges of the 5-cube will project to five line segments of equal length, equally spaced around a circle. The resulting pattern is actually a Penrose tiling. This tiling produces a pattern of sharp diffraction spots, and the symmetry of the diffraction pattern reflects the symmetry of the underlying five-dimensional space. In particular, the body diagonal of the 5-cube is a fivefold symmetry axis, just as the body diagonal of a 3-cube is a threefold axis. Consequently, the diffraction pattern is also fivefold symmetric.

The actual icosahedral alloys under consideration here have more than one fivefold axis. Indeed, icosahedral symmetry is defined by the presence of six different fivefold axes. Consequently, the projected structure that defines such a quasicrystal starts in a six-dimensional space, and hence the six miller indices required to describe the diffraction peak positions.

As was true in our one-dimensional example, two coordinates $(\mathbf{R}_\parallel, \mathbf{R}_\perp)$ are required to label atomic positions or lattice points, and each reciprocal lattice point is associated with values for both $\mathbf{G}_\parallel$ and $\mathbf{G}_\perp$. For the icosahedral reciprocal lattice, the $\mathbf{G}_\perp$ of any diffraction peak can be related to its $\mathbf{G}_\parallel$:

$$\mathbf{G}_\parallel = \sum_i n_i e_i^\parallel$$

$$\mathbf{G}_\perp = \sum_i n_i e_i^\perp.$$

The $e_i^\perp$ are simply a rearrangement of the $e_i^\parallel$, so that $e_1^\perp = -e_1^\parallel$, $e_2^\perp = e_2^\parallel$, $e_3^\perp = e_4^\parallel$, $e_4^\perp = e_6^\parallel$, $e_5^\perp = e_3^\parallel$, and $e_t^\perp = e_5^\parallel$.

# 8.4   Icosahedral Glass Structures

The motivation for introducing a third, more severely disordered model for the icosahedral alloys was provided by the first high-resolution X-ray diffraction measurements on rapidly quenched $Al_6Mn$ by Bancel et al. [20]. The powder diffraction scans for both the icosahedral phase and the recrystallized orthorhombic phase are shown in Figure 8.9. One of the important observations that can be made from the top panel is the large apparent breadth of the diffraction peaks for the icosahedral phase. Given that this alloy was produced by rapid solidification, the disorder signalled by broadened diffraction peaks may not seem too surprising. However, two specific points about this disorder stimulated the emergence of the icosahedral glass model. First, taken as a measure of the positional correlation length $\xi_0$, the peak widths correspond to an average grain size of 100–300 Å. This length scale is significantly smaller than the range of bond-orientational order ($\approx 1 - 10 \ \mu$m) measured by electron diffraction for this sample. Second, diffraction peak broadening by conventional mechanisms such as small grain size, defects, or strain has a well-defined systematic dependence on the wave vector $\mathbf{G}_\parallel$ of the diffraction peak. None of the known broadening mechanisms could explain the systematics, or lack thereof, in the pattern of Figure 8.9.

The prototype of the icosahedral glass model was conceived by Schechtman and Blech [21] in their attempt to explain the diffraction pattern from

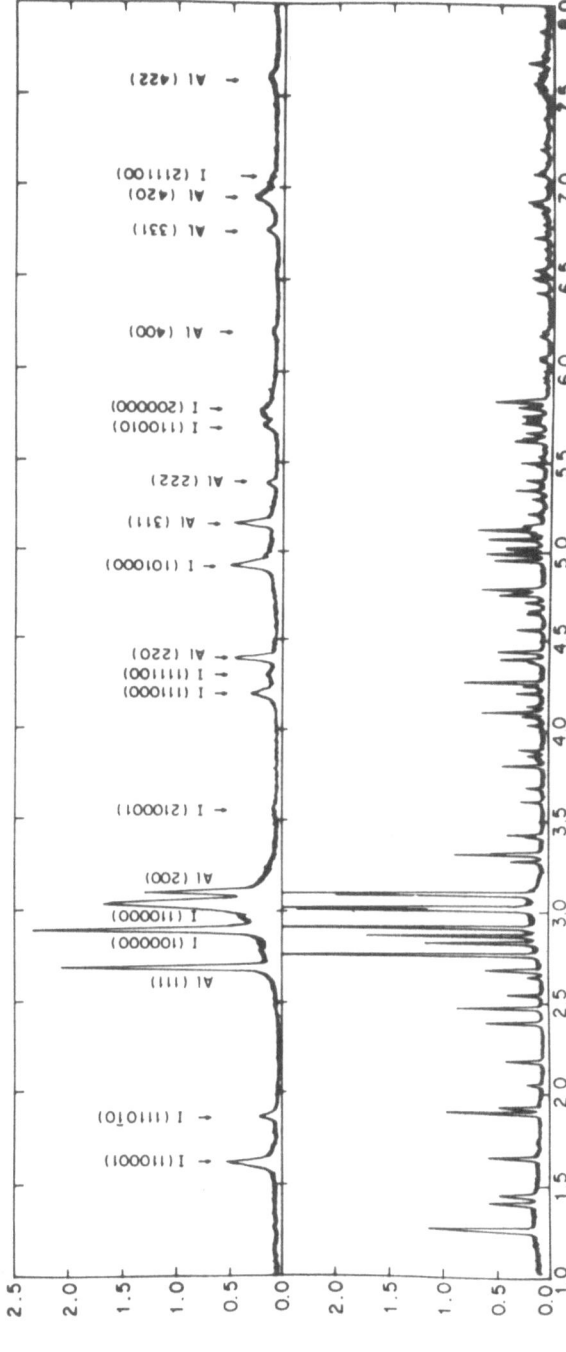

FIGURE 8.9. Top: High-resolution X-ray powder diffraction pattern from rapidly quenched Al₆Mn (after Ref. 20). The icosahedral phase peaks are labeled by a set of six miller indices. FCC Al peaks are also labeled. Bottom: diffraction pattern from the same alloy after heat treatment at 400°C for 3 hr.

the icosahedral phase of Al-Mn. Stephens and Goldman [22,23] further developed this model and specifically considered its relevance to the peak broadening in the diffraction data. Others including Elser [24,25], DiVincenzo [26], and Robertson and Moss [27] have made important contributions to refinements of the model and our understanding of its relation to the quasicrystalline tiling models mentioned above.

As will become evident in the next section, the term icosahedral glass (IG) is a bit misleading since these structures, while very disordered, are far from glassy. However, Peter Stephens and I coined this term to distinguish this class of structural models from the quasicrystalline descriptions. The original icosahedral glass (IG) models may best be described as a disordered packing of icosahedral clusters of atoms assembled in such a way as to maintain the relative orientations of all clusters. This model provides us with an example of the most disordered structures that still yield relatively sharp diffraction maxima with overall icosahedral symmetry. The growth algorithm used to generate the simplest of these structures relies exclusively on the nearest-neighbor environment of a growth site and the propagation of icosahedral bond-orientational order. Unlike the quasicrystalline models described above, the diffraction pattern from an IG is distinguished by an intermixture of sharp and broad diffraction maxima which arise as a consequence of the large inherent structural disorder. More sophisticated growth algorithms allow one to fine-tune the degree of disorder, or eliminate it entirely as one moves from an IG toward the quasicrystalline motifs.

The organization of the remainder of this chapter is as follows. In Section 8.5, I introduce the algorithm employed by Stephens and Goldman [22] to build icosahedral glass structures. This is first illustrated in two dimensions as a decagonal glass, but the basic characteristics of these disordered structures remain the same upon moving into three dimensions. A simple explanation for the presence of sharp diffraction maxima from these models is also presented in terms of the formalism derived by Hendricks and Teller for stacking faults in crystals. In Section 8.6, I review the basic packing algorithms for growing three-dimensional icosahedral glass structures for both the simple icosahedral and the face-centered icosahedral alloys. Section 8.7 compares the scattering patterns from three-dimensional icosahedral glass models with experimental diffraction results.

Before continuing, I would like to point out that an excellent review of the icosahedral glass models has been authored by Stephens [28]. Although overlap in terms of the material covered is unavoidable, I have attempted to focus here on developments since his review.

# 8.5   The Decagonal Glass in Two-Dimensions

## 8.5.1   The Basic Algorithm for Numerical Simulations

The basic algorithm used to generate any icosahedral glass structure must obey two constraints. First, icosahedral bond-orientational order must be maintained across the sample. This can be accomplished by demanding that neighboring icosahedra be joined along a common edge, or at triangular faces (with the faces rotated by 120° relative to each other), or at a vertex. The second constraint is that neighboring clusters must not overlap.

The actual process of aggregating these clusters is illustrated in Figure 8.10 for a decagonal glass in two dimensions. Each pentagon represents a cluster of atoms which has pentagonal symmetry. One pentagon is designated the seed and has five edges available to attach the next pentagonal cluster. Pentagons are attached to a randomly selected edge, provided that each new pentagon does not overlap any pentagon already present. For example, after one pentagon is attached to the seed, eight edges of the two pentagons are now available to accommodate a third cluster. The further addition of new pentagons makes new attachment sites available, but also eliminates some existing sites due to the constraint against overlap. The process of random attachment continues until all available sites within some specified region are occupied.

The finished product of 1033 pentagons, displayed in Figure 8.10, illustrates two general characteristics of the IG structures. First, we see that bond-orientational order is maintained across the sample since each pentagon has either the same orientation as the seed, or its inverse. Second, the structure is not space-filling; the tiling pattern is interrupted by gaps or tears of various degrees. The histograms of $x$ and $y$ coordinates of the centers of pentagons along the edges of the figure emphasize the absence of coherent rows of clusters in this sample.

In order to make contact with diffraction experiments, consider the structure factor $S(\mathbf{Q})$, which is proportional to the scattered intensity from an array of point scatterers placed at the centers of each pentagon,

$$S(Q) = \left| \sum_j \exp(i\mathbf{Q} \cdot \mathbf{R}_j) \right|^2 .$$

The index $j$ runs over all $N$ pentagons in the sample. For a periodic structure, $S(\mathbf{Q})$ consists of a series of Bragg diffraction peaks with intensities proportional to $N^2$. These occur for wave vectors $\mathbf{Q}$, where all unit cells scatter coherently. In a finite sample of characteristic dimension $L$, the Bragg peaks have a half-width-at-half-maximum (HWHM) of $\approx \pi/L$. Figure 8.11 shows a map of one quadrant of $S(\mathbf{Q})$, for the structure shown

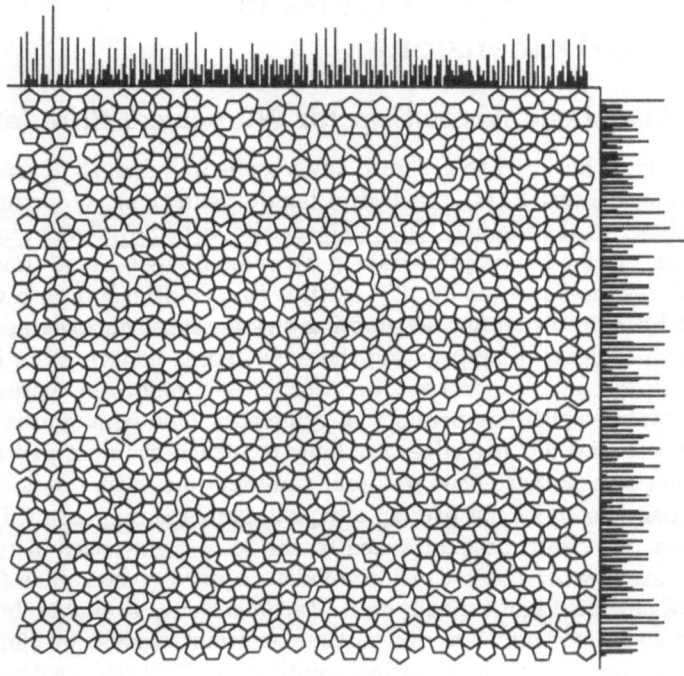

FIGURE 8.10. A sample of 1033 pentagons packed in a square of edge-length $40a$, where $2a$ is the center-to-center separation of connected pentagons. The packing algorithm is described in the text. The histograms along the axes denote the centers of each pentagon.

in Figure 8.10, where each dot represents a value of $\mathbf{Q}$ for which $S(\mathbf{Q})$ exceeds $0.005N^2$. This threshold is somewhat larger than the incoherent background of $\approx N$, but significantly smaller than $N^2$, the maximum possible value of $S(\mathbf{Q})$. The full and dashed lines in the map illustrate the invariance of the diffraction pattern under rotations by $36°$. Therefore, we see that this simple decagonal glass produces relatively localized scattering with overall decagonal symmetry.

How well defined are the diffraction maxima? In order to display the size and shape of these maxima in $S(\mathbf{Q})$, the bottom panels of Figure 8.11 display $S(\mathbf{Q})$ along the $Q_x$ and $Q_y$ directions. The intensities of the strongest peaks are approximately 60% of $N^2$, and the half-widths of these peaks are sample-size-limited ($\approx \pi/L$). Therefore, the strongest peaks in this simulation are as intense and as sharp as those due to coherent diffraction from a perfectly ordered quasicrystal (or crystal) of the same dimensions. In the simulations, the strongest peaks scale like $N^2$ because the "coherence length" ($\xi_0$) for those peaks is larger than the sample size. As the simulation sample size is increased beyond $\xi_0$ for a particular peak, its width remains constant and intensity scales like $N$. The pattern also contains clusters of weaker peaks, such as the group at $10.2/a$ along $Q_x$. If one averages $S(\mathbf{Q})$ over symmetry-equivalent directions, or performs an ensemble average over several simulations, these clusters tend toward a broad, smoother peak which is Lorentzian. However, as explained by Garg and Levine [29], no self-averaging of these clusters of peaks is observed from any single sample since the relative phase of the scattering from any given domain varies randomly throughout the sample. With an X-ray source of high enough spatial and temporal coherence, this speckle could be observed.

## 8.5.2   Understanding the Sharp Diffraction Maxima: The HT Approach

At first it may seem surprising that such a random aggregation of atomic clusters yields a diffraction pattern of relatively sharp peaks. The key to understanding this result lies in the constraints imposed on intercluster distances by the requirement of bond-orientational order. Consider the cluster of pentagons in Figure 8.12, which illustrates how two pentagons may be attached along their common edge. Along the $\hat{x}$ direction (referred to the axes of Figure 8.10), there are only five displacements between the centers of adjacent pentagons, $\pm x_1$, $\pm x_2$, and $0$, where

$$x_1 = 2a \sin\left(\frac{\pi}{10}\right),$$

$$x_2 = 2a \cos\left(\frac{\pi}{10}\right),$$

$$\frac{x_2}{x_1} = \tau = \frac{1 + \sqrt{5}}{2}.$$

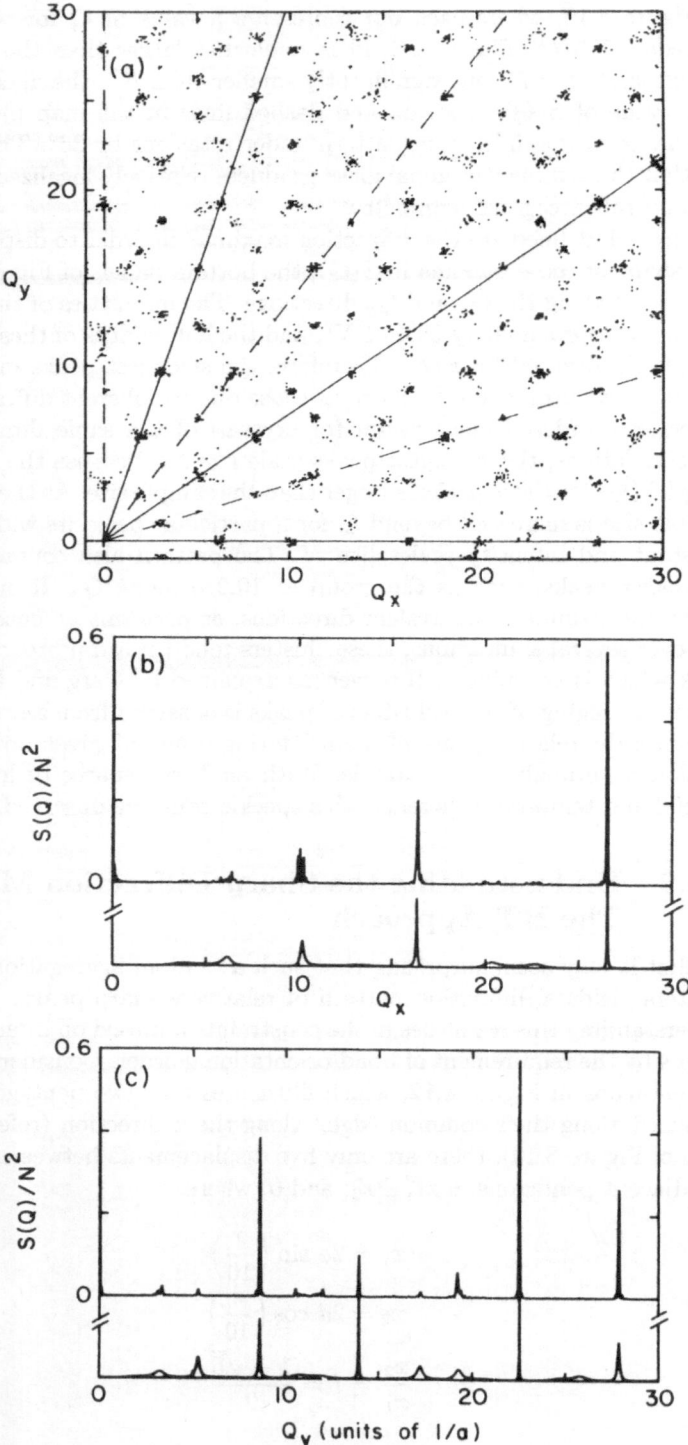

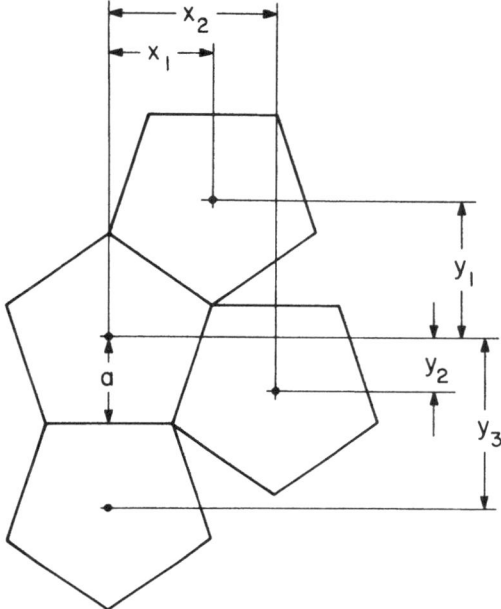

FIGURE 8.12. Fundamental distances separating the centers of pentagonal clusters packed together to maintain bond-orientational order.

The projection of pentagon centers along the $\hat{x}$ direction may now be viewed as a sequence of steps $\pm x_1$, $\pm x_2$, and 0. This small set of fundamental displacements may be contrasted with the displacement vectors that would be allowed for packing hard spheres of radius $a$. In the absence of any requirement for bond-orientational order, the projection of the separation of adjacent sphere centers along any direction takes on all values between 0 and $a$.

Given a random sequence of displacements chosen from this finite set, it becomes clear that the relatively sharp diffraction maxima will occur at wave vectors

$$Q_x \approx \frac{2\pi m}{x_1} \approx \frac{2\pi n}{x_2}$$

with $m$ and $n$ integers. Since $x_2/x_1 = \tau$, the sharpest diffraction maxima are expected with $n/m$ as rational approximants of $\tau$ (i.e., $\frac{3}{2}$, $\frac{5}{3}$, $\frac{8}{5}$, etc.).

FIGURE 8.11. Top: the scattering intensity in reciprocal space for the decagonal glass of Fig. 8.10. Bottom: scans along the $Q_x$ and $Q_y$ directions of the top panel. The upper trace is taken directly from the numerical simulations, while the lower trace is calculated from the Hendricks–Teller formation described in the text.

Along the $\hat{y}$ direction, we note that a displacement of $y_1$ can be followed by $y_2$ or $y_3$, but not $y_1$. Therefore, the sequence of displacements is more restricted, and so we choose our set of fundamental displacement vectors to be $y_1 + y_2$, $y_1 + y_3$, and $y_3 - y_2$. In a similar manner, we can choose a finite set of fundamental displacement vectors along any direction by a suitable combination of those along $\hat{x}$ and $\hat{y}$.

The qualitative explanation above can be placed on a more quantitative basis by using the formalism developed by Hendricks and Teller (HT) [30] to discuss the scattering patterns from partially ordered layer lattices. Essentially, the problem at hand is analogous to the presence of stacking faults in crystals (such as the Bi-based high-$T_C$ superconductors, or graphite intercalation compounds). Given a set of layer spacings, $d_1 \ldots d_N$, which occur with some probability $f_j$, the scattered intensity per layer can be written as:

$$\frac{S(Q)}{N} = \frac{1 - C^2}{1 - 2C \cos\overline{\phi} + C^2}$$

$$C = \sum_j f_j \cos(Qd_j - \overline{\phi})$$

$$\sum_j f_j \sin(Qd_j - \overline{\phi}) = 0.$$

Here we consider the scattering power from each of the $d_N$ layers to be identical. The diffraction patterns along $Q_x$ and $Q_y$ which result from assuming that each of the fundamental displacement vectors along $\hat{x}$ or $\hat{y}$ occurs with equal probability are displayed in the bottom panels of Figure 8.11, and closely mimic the trends observed in the simulations. As $N \to \infty$, the HT description predicts peak intensities proportional to $N$.

It is necessary to point out here that while the HT description provides us with some insight as to why sharp diffraction maxima are observed, it is inherently unphysical since it does not explicitly exclude overlaps between adjacent clusters. In the simulations, two pentagons can be separated by less than $2a$ along the $\hat{x}$ direction only if they are also separated by the appropriate distance along $\hat{y}$. In the absence of additional growth rules, the growth process can be viewed as a branched, self-avoiding random walk in two dimensions, while the HT description must be viewed as a one-dimensional random walk.

# 8.6   The Icosahedral Glass in Three Dimensions

We now turn our attention to three-dimensional packings of objects with icosahedral symmetry. Several simple solids have icosahedral symmetry; e.g., the icosahedron with 20 triangular faces, the pentagonal dodecahedron with 12 pentagonal faces, and the rhombic dodecahedron with 30 golden rhombuses. Although it is easiest to visualize the packing of hard solids, it should be emphasized that the physical basis for this model is the packing of clusters of atoms that have icosahedral symmetry.

Such atomic clusters do, in fact, exist. Interestingly enough, they are found in large unit cell structures such as $\alpha$-Al-Mn-Si [31] and R-Al-Li-Cu [32], which are cubic phases respectively close in composition to the Al-Mn-Si and Al-Li-Cu icosahedral alloys. This observation in itself provides further motivation for consideration of the icosahedral glass model.

Before we proceed to describe specific features of each of these modes of packing, a few general comments are in order. First, as was true for the decagonal glass described above, we require that bond-orientational order be maintained throughout the sample so that a unique set of twofold, three-fold, and fivefold axes may be defined. For the numerical simulations, the algorithm used to pack icosahedra is essentially the same as that used for the decagonal glass. Starting with a seed, a new icosahedral unit is attached to a randomly chosen icosahedron of the existing structure as long as there is no overlap. The attachments continue until a cube of some edge length $L$ can accommodate no further units. The length scale of the simulation is defined by the center-to-center separation ($2a$) of two connected icosahedra. The diffraction patterns from these computer-generated structures are calculated by assuming a point scatterer, of scattering length 1, placed at the center of each icosahedral cluster. Therefore, we are probing the network of connected icosahedra and not the specific atomic decoration of these clusters.

The basic packing algorithms make no provision for relaxation of the atomic clusters so that once a cluster is attached to the existing aggregate, it is frozen in place. A more realistic, dynamic growth algorithm has been developed by Elser [25] and will be discussed briefly in Section 8.7.2. Furthermore, the icosahedral glass model makes no statement about how these icosahedrally symmetric clusters of atoms form from the melt.

Typically, in these simulations, 50 to 60% of the available volume is consumed by the clusters. This could be compared with the similarly defined packing fraction of icosahedral clusters in the crystalline analogs of the icosahedral phase. For the cubic $R$-phase of Al-Li-Cu, for instance, the packing fraction is approximately 82%. The rather large difference in this comparison arises from the presence of gaps in the IG structure as shown for the decagonal glass in Figure 8.10. We suppose that these gaps are filled with atoms in some unspecified manner. For instance, these gaps might contain highly disordered glassy material.

## 8.6.1 IG Structures Related to the SI Alloys

Historically, the icosahedral glass structures were first proposed to explain the relatively sharp peaks found in the diffraction patterns from the simple icosahedral alloys. Therefore, in this section, I discuss those packings which reproduce the diffraction patterns of the SI alloys. Although the vertex-packing models were the first to be considered by Stephens and Goldman [22], the face-packing algorithm produces results which can be interpreted in terms of the arrangement of atomic clusters in crystalline analogs of the simple icosahedral alloys. Therefore, we first examine the diffraction pattern that results from structures assembled by attaching icosahedral clusters along their threefold axes.

A unique set of two-, three-, and fivefold axes can be defined for the entire sample of icosahedra if the triangular faces which connect nearest-neighbor icosahedra are rotated by 120° with respect to each other. As pointed out by Stephens [28], the 10 pairs of vectors that describe the displacements between adjacent icosahedra can be constructed from linear combinations of the vertex vectors $e_i^\parallel$ defined in the introduction. The displacements along any particular direction (such as a fivefold axis) required to calculate the scattering function via the Hendricks–Teller formalism can be calculated from a simple dot product between that direction and each of these vectors.

The top panel of Figure 8.13 shows the most intense local maxima of $S(\mathbf{Q})$ in one quadrant of a twofold plane for a face-packed icosahedral glass. The units of $Q_x$ and $Q_y$ are $1/a$, and the area of each spot in the diffraction simulation is proportional to the peak intensity of the diffraction maxima. The relative positions and relative intensities of diffraction spots are in very good agreement with the experimental twofold pattern of Figure 8.3. In fact, by a peak-to-peak match of the calculated and experimental peak positions for the rapidly quenched Al-Mn alloy studied by Bancel et al. [20], we can determine the size of the clusters in our simulation. The strong diffraction peak along the fivefold axis (the fundamental 100000 in the notation of Bancel et al.) has a wave vector of 2.896 Å$^{-1}$. The identification of this peak with the strong diffraction peak marked "a" in Figure 8.13 at $15.86/a$ demands that $2a = 10.95$ Å.

Nearly icosahedral clusters of 54 atoms are found in the cubic crystalline phase of $\alpha$-Al-Mn-Si [33]. This structure may be loosely described as a BCC packing of icosahedral clusters connected along the cubic (111) directions by octohedra of Al atoms. Actually, the clusters are a bit distorted by the cubic environment, and the corner and body center clusters are slightly different, so that the structure is actually simple cubic with a "two-cluster" basis. The lattice constant of $\alpha$-Al-Mn-Si is 12.68 Å. Therefore, the intercluster distance is 10.98 Å, in remarkable agreement with the cluster separation determined above. The same relation between the icosahedral phase and cubic $R$-phase of Al-Li-Cu is found. Matching the simulated and experimental icosahedral diffraction patterns requires an intercluster separation of 12.01 Å, compared to the separation of 12.03 Å in the body-centered cubic $R$-phase [34,35]. These results suggest that the icosahedral alloys might simply be viewed as resulting from a terrible job of packing

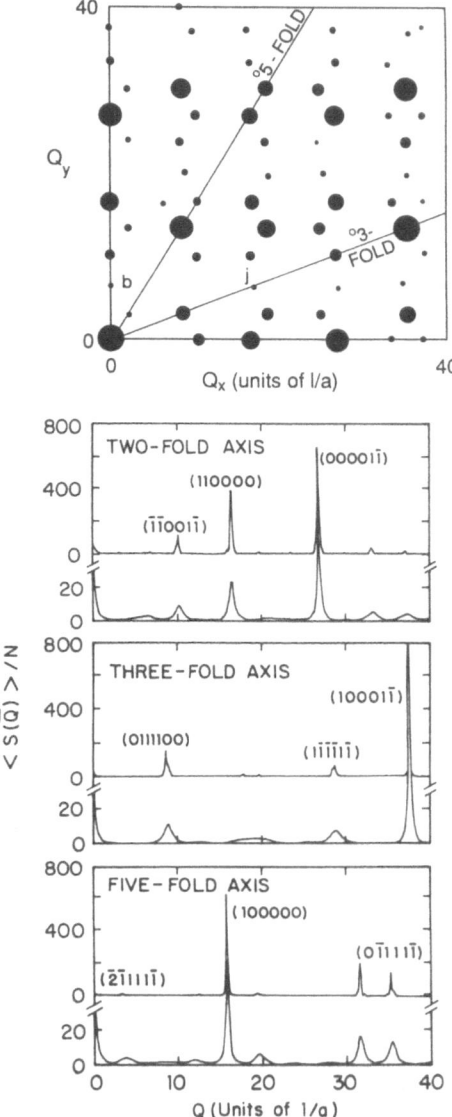

FIGURE 8.13. Top: the face-packed icosahedral glass structure factor in one quadrant of a twofold plane. The size of the spots denotes the relative intensity of the diffraction spots. Bottom: the structure factor calculated along the two-, three-, and fivefold axes for the numerical simulations (upper trace) and Hendricks–Teller model (lower trace).

the atomic clusters in $\alpha$-Al-Mn-Si or R-Al-Li-Cu together.

When we return to the calculated diffraction pattern in Figure 8.13, the top traces of the bottom panel show calculated scans along the high-symmetry two-, three-, and fivefold axes. The bottom traces show $S(\mathbf{Q})$ calculated using the appropriate displacements along these axes in the HT expression discussed in Section 8.5.2. Again, as was true for the decagonal glass, the strongest diffraction maxima have a width which is sample-size-limited (in this particular case, $L = 20a$, or about 900 icosahedra). Larger-scale simulations for box sizes of edge length $100a$ (on the order of 500 Å using the scale determined above) still evidence only sample size broadening in the strongest peaks. Weaker peaks in Figure 8.13, however, clearly display broadening characteristic of icosahedral glass structures. We will return to study the widths of these peaks in more detail in Section 8.7.

An alternative algorithm that produces diffraction patterns in qualitative agreement with the experiment packs icosahedra together vertex to vertex. The possible interconnections between clusters are restricted to the set of displacement vectors $\pm 2a\hat{e}_1^{\parallel}$, $\pm 2a\hat{e}_2^{\parallel}$, $\pm 2a\hat{e}_3^{\parallel}$, $\pm 2a\hat{e}_4^{\parallel}$, $\pm 2a\hat{e}_5^{\parallel}$, and $\pm 2a\hat{e}_6^{\parallel}$. Here $2a$ is the center-to-center separation of two connected icosahedra and the $\hat{e}_i^{\parallel}$ are the vertex vectors defined in the introduction.

The diffraction pattern calculated from this assemblage is shown in Figure 8.14, for one quadrant of a twofold plane. Again, choosing the appropriate scale factor between the simulation and experiment, we find very good agreement between the patterns, both in the positions of the peaks and their relative intensities. The intercluster separations calculated for the vertex-packing model, however, are 19.48 Å for the Al-Mn alloy and 21.35 Å for Al-Li-Cu. No known alloy of Al-Mn or Al-Li-Cu can be found with vertex-packed icosahedral clusters. Recently, however, Levine et al. [36] have discovered a series of crystalline phases in the Ti-Mn-Si alloys which are closely related to the Ti-Mn-Si icosahedral phase [37]. Three distinct lattices with cubic, orthorhombic, and hexagonal symmetry can be described in terms of the packing of icosahedral clusters along the face, vertex, and edge directions, respectively.

## 8.6.2    IG Structures Related to the FCI Alloys

Prior to 1988, all of the known icosahedral alloys produced diffraction patterns similar, except in scale, to those shown in Figure 8.3. In 1988, a group at Tohoku University discovered the first of several face-centered

FIGURE 8.14. Top: the vertex-packed icosahedral glass structure factor in one quadrant of a twofold plane. The size of the spots denotes the relative intensity of the diffraction spots. Bottom: the structure factor calculated along the two-, three-, and fivefold axes for the numerical simulations (upper trace) and Hendricks–Teller model (lower trace.

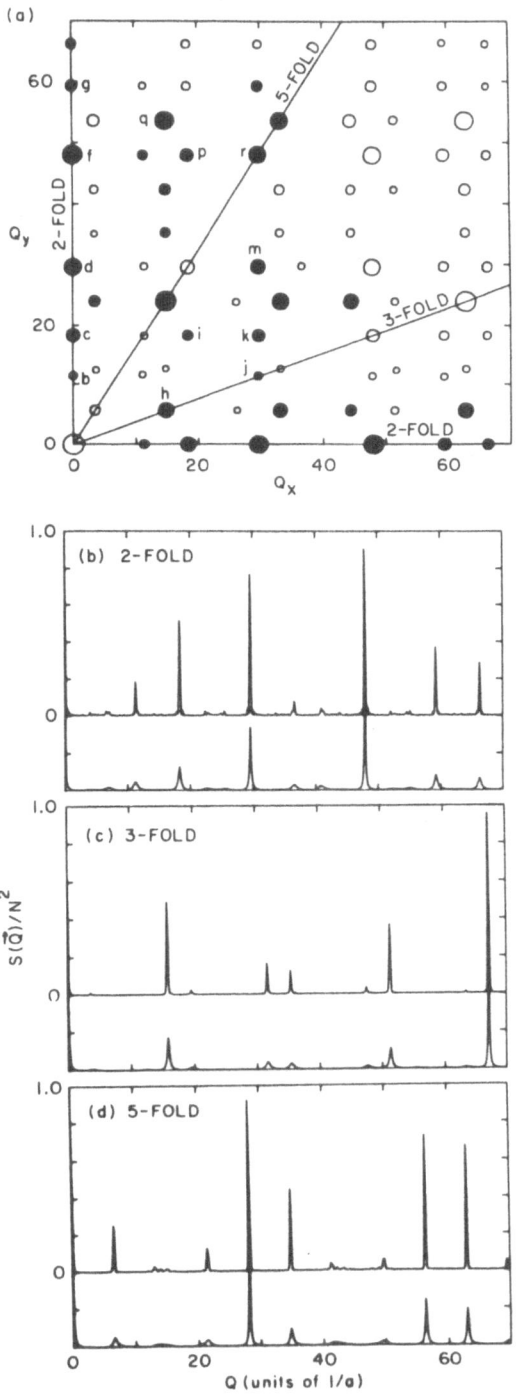

icosahedral phases in a rapidly quenched alloy of Al-Cu-Fe [38]. This discovery has had a profound impact on the field since all of these alloys exhibit resolution-limited peak widths in their diffraction patterns [39,40]. Clearly, then, the icosahedral glass model is not a good starting point for the description of the structure of the FCI phase. Nevertheless, it is instructive to consider icosahedral glass models which can produce FCI diffraction patterns because the basic algorithm—the aggregation of icosahedral clusters—may be applicable to the formation and growth of these alloys.

Shechtman and Blech originally considered icosahedra connected along edges and calculated a diffraction pattern consisting of sharp spots (within their computational resolution) in a pattern with overall icosahedral symmetry. Figure 8.15(a) shows $S(\mathbf{Q})$ calculated in one quadrant of a twofold plane for a sample of edge-packed icosahedra. The top traces of Figures 8.15(b) through (d) display $S(\mathbf{Q})$ from the numerical simulations along the high-symmetry twofold, threefold, and fivefold axes outlined in Figure 8.15(a). As was true for the pentagon packing in two dimensions, as well as the vertex- and face-packing schemes described above, we can define a set of fundamental displacement vectors between adjacent edge-packed icosahedra and calculate the scattering function using the HT formalism. The resultant $S(\mathbf{Q})$ is shown as the bottom trace in Figures 8.15(b) through (d).

Initially, the edge-packing algorithm was rejected since all of the icosahedral alloys studied were SI. Note that the strong peaks along the fivefold axis in Figure 8.15 form a $\tau$ inflation series. Therefore, edge-packed icosahedral glass is properly classified as a face-centered icosahedral structure.

Several recent electron and X-ray diffraction measurements [41,42] have provided strong evidence that the FCI alloys result from chemical ordering on a SI quasilattice [43]. For instance, one could imagine decorating adjacent vertices of a Penrose tiling with different atoms. Similarly, if there are minor differences in the composition between connected icosahedral clusters in a face-packed icosahedral glass, the SI diffraction pattern will develop superlattice spots at the FCI positions. It is perhaps significant that even some SI alloys, like Al-Li-Cu, exhibit some tendency toward chemical ordering, as evidenced by diffuse scattering at the FCI-allowed (but SI-forbidden) reciprocal lattice points [44].

# 8.7   The Success and Failure of the Icosahedral Glass Model: Comparison with Experiments

One of the most beautiful aspects of the icosahedral glass models, from an experimentalist's point of view, is that they provide quantitative predictions, with no adjustable parameters, that can be tested. Although the

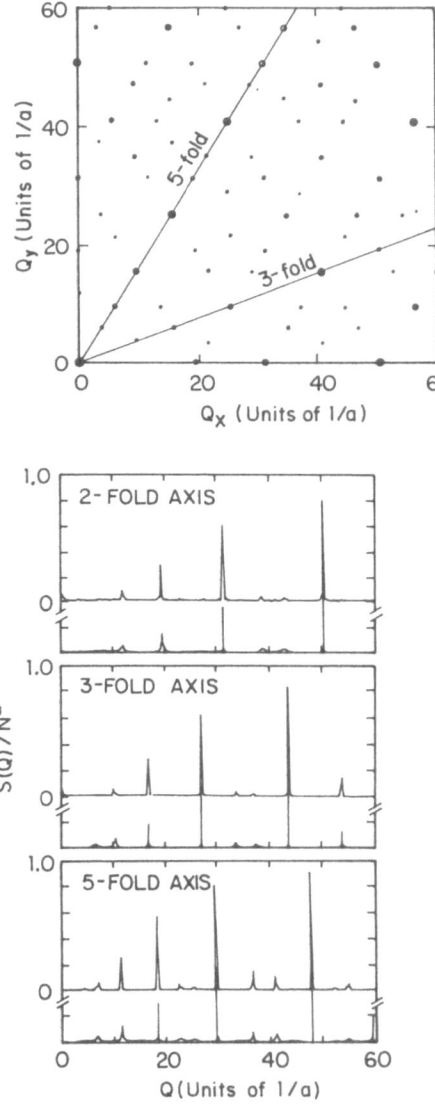

FIGURE 8.15. Top: the edge-packed icosahedral glass structure factor in one quadrant of a twofold plane. The size of the spots denotes the relative intensity of the diffraction spots. Bottom: the structure factor calculated along the two-, three-, and fivefold axes for the numerical simulations (upper trace) and Hendricks–Teller model (lower trace).

icosahedral glass model ultimately falls short in the face of experimental data from the FCI alloys, it has found many successes in explaining the diffraction patterns from the SI alloys. Even here though, modifications of the basic algorithm are necessary in order to obtain good agreement with the experimental data.

## 8.7.1    Peak Broadening in Icosahedral Alloys

Although I have made only qualitative statements above about the peak broadening evident in diffraction simulations from icosahedral glass models, the results of these simulations can be quantitatively compared to experiment. One of the key ingredients in this comparison arose from the realization by Horn et al., in diffraction studies of icosahedral Al-Mn [45], that there were indeed systematic trends to be found in the diffraction peak broadening.

Mechanisms such as finite grain size broadening and strain broadening have distinct signatures in diffraction data. If we denote the average grain size as $L$, the half-widths of peaks in the diffraction pattern are given by $\Delta Q = \sqrt{\left(\frac{\pi}{L}\right)^2 + (\Delta q_{res})^2}$, where $\Delta q_{res}$ is the instrumental resolution. We have already seen that finite grain size contributes to the widths of the sharpest peaks in the icosahedral glass simulations. The additional contribution to the peak width is independent of the wave vector of the diffraction peak. However, peaks of different symmetries can experience broadening to different degrees by this mechanism if the grain shape is anisotropic. Similarly, inhomogeneous strains in a crystal produce peak broadening that increases with the magnitude of the wave vector of the diffraction peak. That is, $\Delta Q$ increases in proportion to $G_{\parallel}$, if we use the notation developed in Section 8.3.

Neither of these trends was observed in diffraction measurements of the simple icosahedral alloys. Instead, it was found that $\Delta Q$ increased almost linearly with $G_{\perp}$. This implies that the dominant strain mechanism in SI alloys originates in the phasonlike, rather than phononlike, degrees of freedom. Furthermore, this "phason-strain" mechanism [46], as it came to be known, appeared to be universal to the icosahedral alloys since the same systematics, and nearly the same magnitude, of broadening were found in slowly cooled icosahedral alloys of Al-Li-Cu [47]. In fact, all SI alloys studied to date exhibit this same peak broadening. By explicitly introducing phason strain [48], the quasicrystalline models for the icosahedral phase can produce diffraction peak broadening which scales linearly with $G_{\perp}$.

It is interesting that none of the FCI alloys produce diffraction peak broadening with these systematics. Figure 8.16, e.g., shows that in the FCI Al-Cu-Ru alloy, the peak widths increase linearly with $G_{\parallel}$ rather than $G_{\perp}$ [39]. Therefore, the dominant disorder mechanism here is normal lattice strain rather than phason strain. The striking difference between the systematics of diffraction peak broadening in the FCI and SI alloys is not presently understood.

Returning to the SI alloys, we see that Figure 8.17 also compares the peak widths obtained from diffraction measurements on Al-Li-Cu [49] with the results of the numerical simulations for the face-packed icosahedral glass. Although the simulations do a reasonable job of predicting the peak widths for small values of $G_\perp$, the broadening in the glass model scales roughly as $G_\perp^{2.5}$, rather than the linear dependence observed experimentally.

Why should the peak widths of an icosahedral glass rely on $G_\perp$ at all? Divincenzo [26] and Elser [24,25] have recast the icosahedral glass model in terms of the cut and projection scheme discussed in Section 8.3.3. Recall that in Figure 8.7, a perfectly ordered one-dimensional quasiperiodic tiling results from a smooth acceptance volume which is parallel to the physical subspace $x_\parallel$. If this acceptance region follows a random path through the 2D lattice, there will be a random sequence of long and short displacement vectors between points in the one-dimensional quasilattice. This is simply the one-dimensional analog of the Hendricks–Teller model discussed before. The detailed dependence of diffraction peak widths on $G_\perp$ is determined by how strongly the root mean square $x_\perp$ of the acceptance strip diverges with the size of the sample. For example, if the acceptance region executes a random walk along $x_\perp$, $\langle x_\perp \rangle \sim x_\parallel^{1/2}$, and the diffraction peak widths grow like $G_\perp^2$ [26]. For a linear $G_\perp$ dependence, $\langle x_\perp \rangle$ must increase linearly with sample size [24]. For the numerical packing simulations, the description of the acceptance volume is further complicated by the existence of gaps in the tiling (Figure 8.10). Tiles separated by a gap correspond to large separations in the relative $x_\perp$ coordinates of the tiles, and so the acceptance volume is punctuated by discontinuities.

## 8.7.2    Modifications of the Growth Algorithm

The range of positional order in an icosahedral glass may be increased by incorporating some preference for particular cluster configurations in the basic face-packing algorithm described in Section 8.6. The basic face-packing algorithm only specifies the set of possible displacement vectors between nearest-neighbor icosahedral clusters of atoms. One could also continue on to specify a set of acceptable and unacceptable displacement vectors between second nearest neighbors, third nearest neighbors, and so on. This has to be done with some care, however, since the spirit of the model demands that the growth rules be local rather than global. The most obvious first-order correction to the packing algorithm is found by studying the arrangements of the same icosahedral clusters in the crystalline analogs of the icosahedral phase alloys.

The icosahedral clusters in the cubic phase of $\alpha$-Al-Mn-Si and R-Al-Li-Cu are connected along their threefold axes (faces) in the same manner

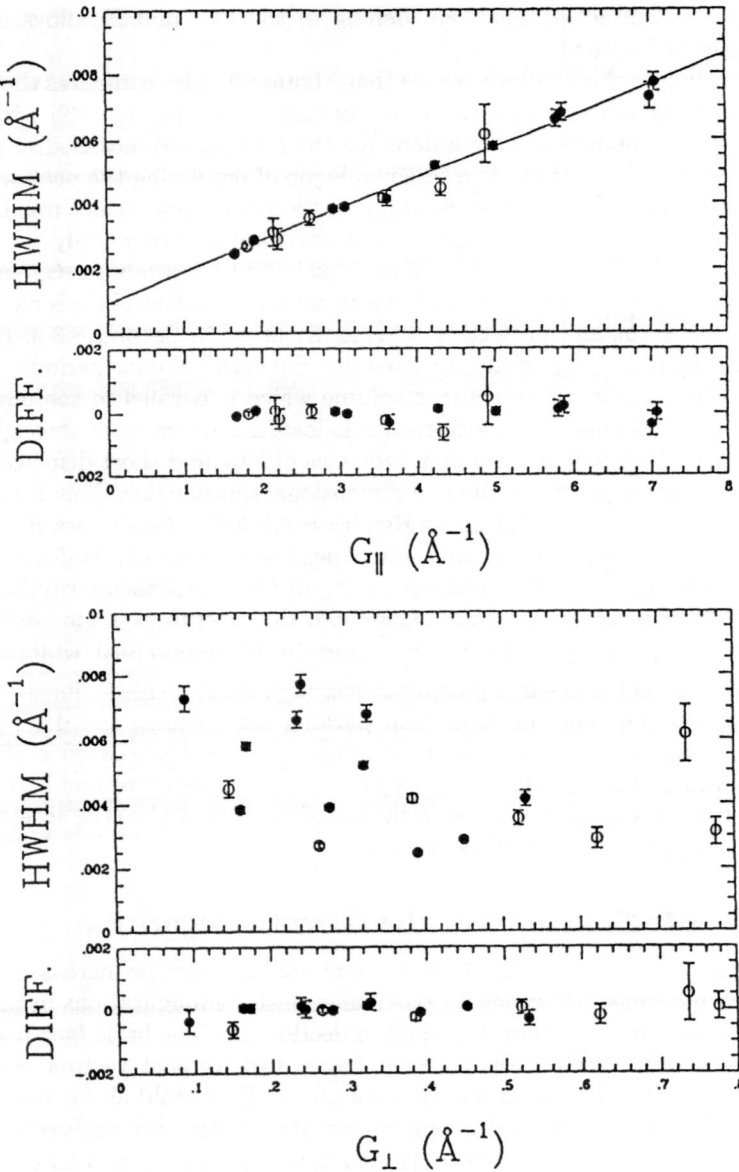

FIGURE 8.16. Results of a high-resolution X-ray study of peak widths and positions in icosahedral Al-Cu-Ru. The filled (open) circles denote diffraction peaks with even (odd) parity. The peak widths scale linearly with $G_{\parallel}$ and have no apparent systematic dependence on $G_{\perp}$.

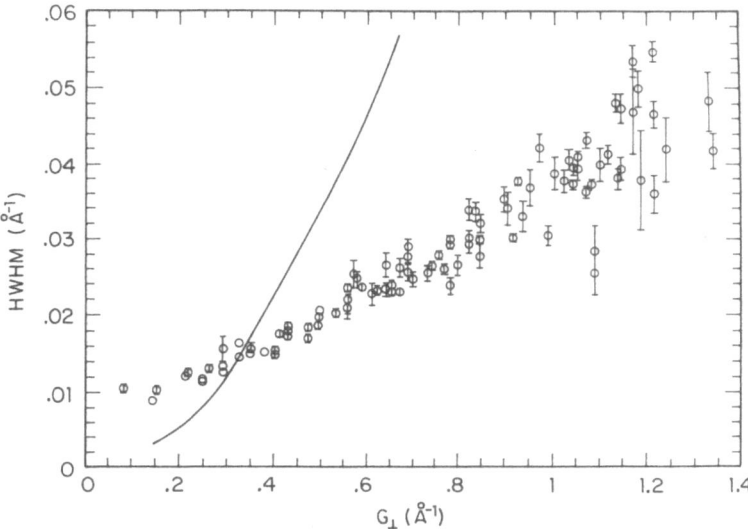

FIGURE 8.17. The $G_\perp$ dependence of peak widths in the icosahedral phase of Al-Li-Cu (after Ref. 49), along with the dependence predicted by the basic face-packing algorithm in Section 7.5.1.

proposed by the IG model. In the cubic phase, however, clusters attach along a subset of eight out of the possible 20 threefold axes of an icosa-hedron, the cubic [111] directions. Each cluster is also coordinated by six second-neighbor clusters along the [100] directions.

By including some preference, in the IG packing algorithm, for two clus-ters to attach to faces of a third which share a common vertex, we favor the next-nearest cluster configurations which are found in the BCC crystalline phase [22]. In the algorithm itself, this is accomplished by increasing the probability for attachment of a new cluster at faces which share a vertex with an occupied face. The result of this "improvement" is a 30% decrease in the magnitude of diffraction peak widths, which now increase like $G_\perp^2$ rather than $G_\perp^{2.5}$.

An alternative route to producing better-ordered icosahedral glasses is to incorporate more realistic growth processes. For instance, the basic packing algorithm allows the attachment of new icosahedral clusters to the existing assembly at any available face. In this sense, the process shares some sim-ilarities with diffusion-limited aggregation models for fractal growth. One can constrain the assembly to grow in concentric shells, so that all available sites within some radius of the initial seed must be filled before attachments at a larger radius are allowed. This results in denser structures with higher connectivity, but no real decrease in diffraction peak widths beyond that found for the first modification described above [28].

Robertson and Moss [27], in extended studies of icosahedral glass (or as they prefer—random cluster) growth algorithms, found that a suitable combination of both of these modifications is successful in obtaining diffraction peak widths with the experimentally observed linear dependence on $G_\perp$. Furthermore, the magnitude of the diffraction peak widths obtained from their simulations are in closer agreement with those found for the SI alloys. The key ingredients of their algorithm are:

(1) The restriction of local attachments to a subset which includes only those which produce second-neighbor cluster configurations found in the cubic $\alpha$-Al-Mn-Si alloy or a perfect quasicrystalline packing (three-dimensional Penrose tiling) of clusters. This constraint appears to discrminate against local configurations which produce large voids in the network.

(2) The structure is grown in concentric shells of a significantly smaller radius (on the order of 0.2 cluster diameters) than that described in Ref. 28.

The networks which result from this growth algorithm have a higher connectivity (each cluster is coordinated by an average of 6 other clusters) and higher density (by about 50%) than the face-packing algorithms described above.

Elser has developed a dynamic growth algorithm for the formation of the icosahedral phase based on the formation and aggregation of icosahedral clusters at the solid/liquid interface [25]. In addition to the essential constraints regarding bond-orientational order and overlap of the "static" algorithms described above, clusters may adjust their positions at the growth interface to maximize the connectivity and density of the structure. This step seems to be crucial to reducing the density of tears, or gaps, in the IG network and results in diffraction peak widths which scale like $G_\perp$ in agreement with experiments.

## 8.7.3   Diffuse Scattering from the Icosahedral Alloys

In addition to the relatively sharp peaks observed in the diffraction patterns from the SI alloys, there is significant diffuse scattering. Odd streaks and shapes, also arranged in patterns with overall icosahedral symmetry, have been noted in electron diffraction studies of rapidly quenched Al-Mn [50], FeTi$_2$ [51], Ti-Mn [37], Ti-Cr-Si [52], and Al-Li-Cu [44,53].

While many of the details of the diffuse scattering vary from system to system, and even among samples of the same alloy, there are features which are common to all of the alloys that have been studied. For instance, in the twofold planes, rings or arcs of diffuse scattering are observed along the fivefolds axes (see Figure 8.18). The centers of these rings are not elements

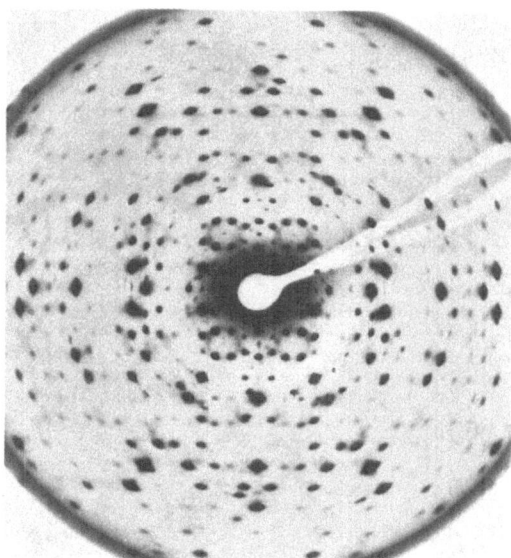

FIGURE 8.18. Monochromatic X-ray precession photo of the twofold plane in Al-Li-Cu (after Ref. 53). The centers of the rings along the fivefold axis correspond to SI-forbidden reciprocal lattice points.

of the SI reciprocal lattice, but instead correspond to positions where strong FCI diffraction peaks are expected, such as $1/\tau \cdot G_{100000}$ and $\tau \cdot G_{100000}$.

The diffuse rings in Al-Li-Cu have been studied in detail by single grain X-ray and neutron scattering techniques [44]. Figure 8.19 shows diffraction scans taken through the center of the diffuse ring along the directions of the three mutually perpendicular twofold axes. All three scans exhibit the same basic features: the maxima at $\zeta = \pm 0.37$ Å$^{-1}$ that define the shell of diffuse scattering and a central diffuse peak at $\frac{1}{\tau} G_{100000}$. Both the position and size of the diffuse ring in the twofold plane agree with the X-ray precession photographs of Figure 8.18. In fact, these data show that the ring in the twofold plane is actually part of a three-dimensional spherical shell of scattering. However, the central peak observed here is not evident in Figure 8.18 and is not always seen in diffraction patterns from other samples of Al-Li-Cu. As pointed out by Henley [43], this diffuse central peak can arise from short-range chemical ordering on a SI lattice—different atomic decorations of adjacent vertices of a Penrose tiling, e.g., or the decorated face-packed icosahedral glass described above.

The widths of both the diffuse shell and the central peak are $\approx 0.1$ Å$^{-1}$ (HWHM). This provides us with a measure of the length scale ($\xi = \frac{1}{HWHM} \approx 10$ Å) of the correlations which produce these features. Within the context of the icosahedral glass model, this suggests that only correla-

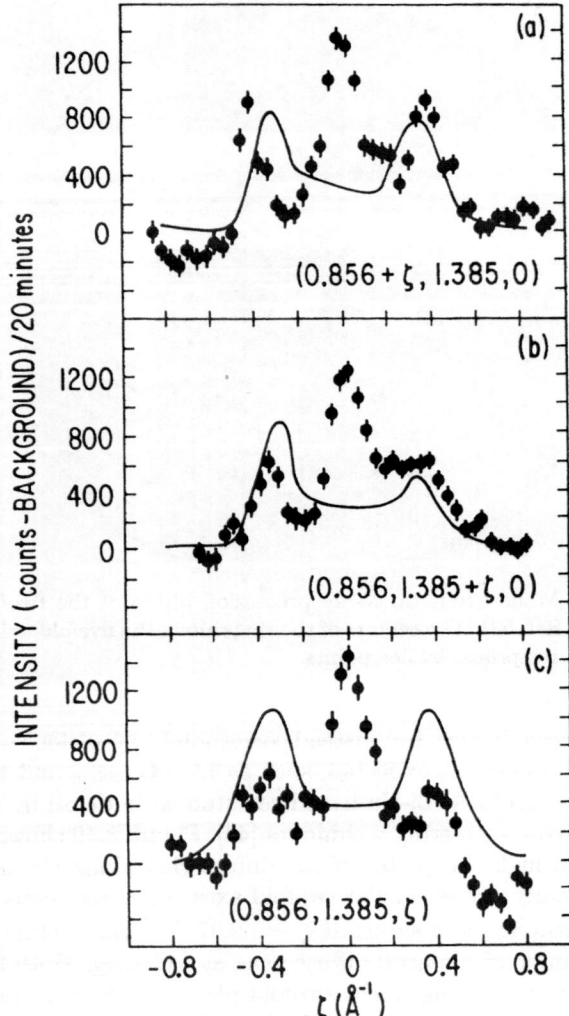

FIGURE 8.19. X-ray diffraction scans (filled circles) taken through the diffuse scattering in the twofold plane parallel to the twofold axes [panels (a) and (b)], and perpendicular to the plane parallel to a twofold axis [panel (c)]. The solid lines are calculated from simulations for a face-packed IG as described in the text (after Ref. 44).

tions between adjoining icosahedral clusters are necessary to explain both the diffuse shell and the peak at $\frac{1}{\tau}G_{100000}$.

The disorder that produces the diffuse scattering can arise from randomness in the manner in which two icosahedral clusters are connected. For instance, in the $R$-phase of Al-Li-Cu, the icosahedral clusters are packed along (111) directions and therefore connect to only eight of the 20 faces (threefold directions) of an icosahedron. One can introduce disorder by allowing icosahedral clusters to join to *any* of the 20 faces with some probability $P_j$. The diffuse scattering can be calculated from an average over the scattering due to the possible two-cluster configurations:

$$S_{\text{diffuse}}(\mathbf{Q}) = f_{\text{cluster}}^2 \langle |1 + P_j e^{i\mathbf{Q}\cdot\mathbf{R}_j}|^2 \rangle.$$

The $\mathbf{R}_j$ are the 20 intercluster vectors along threefold axes, and $f_{\text{cluster}}^2$ is the cluster form factor describing the scattering pattern due to the arrangement of atoms within a single cluster.

In Figure 8.20(b), I show $f_{\text{cluster}}^2$ calculated in one quadrant of a twofold plane using the atomic positions determined by Guryan et al. [35] for the $R$-phase of Al-Li-Cu. Note that this form factor has a strong maximum in the region of the diffuse scattering observed in the icosahedral phase of Al-Li-Cu [Figures 8.20(a) and 8.18]. In Figures 8.20(c) and (d) I show $S_{\text{diffuse}}(\mathbf{Q})/f_{\text{cluster}}^2$ and $S_{\text{diffuse}}(\mathbf{Q})$ for one quadrant of a twofold plane calculated with $P_j = \frac{1}{20}$ for all $j$. Figure 8.20(c) corresponds to the intensity pattern obtained from the configurational average of the possible two-point correlations (centers of the icosahedra) and is independent of the specific decorations of the icosahedral clusters. This pattern will be compared below with the scattering from an undecorated face-packed icosahedral glass. Figure 8.20(d) is the product of Figure 8.20(c) with $f_{\text{cluster}}^2$ shown in Figure 20(b). The important feature here is that the maximum in $f_{\text{cluster}}^2$ at $\frac{1}{\tau}G_{100000}$ now appears as a ring of scattering centered on that wave vector, in qualitative agreement with the present measurements and the precession photographs of Denoyer et al.

The specific model for disorder that was introduced above is really nothing more than the disorder inherent to an icosahedral glass. To illustrate this point, Figure 8.20(e) shows an intensity map of the scattering from an undecorated simulation of face-packed icosahedra. There is a significant "diffuse" background which is very similar to the intensity contours found in Figure 8.20(c). If we decorate this network with the Al-Li-Cu clusters, we obtain the pattern shown in Figure 8.20(f) [the product of Figures 8.20(b) and (e)]. The most intense diffuse scattering found in this pattern (yellow and green regions), beyond that associated with the sharp diffraction maxima, appears as a ring of scattering about $\frac{1}{\tau}G_{100000}$, again quite similar to the experimental pattern of Denoyer et al. We can also compare the scans in Figure 8.19 with simulated scans through the diffuse ring in Figure 8.20(f) (shown as the solid lines). These curves were calculated by averaging

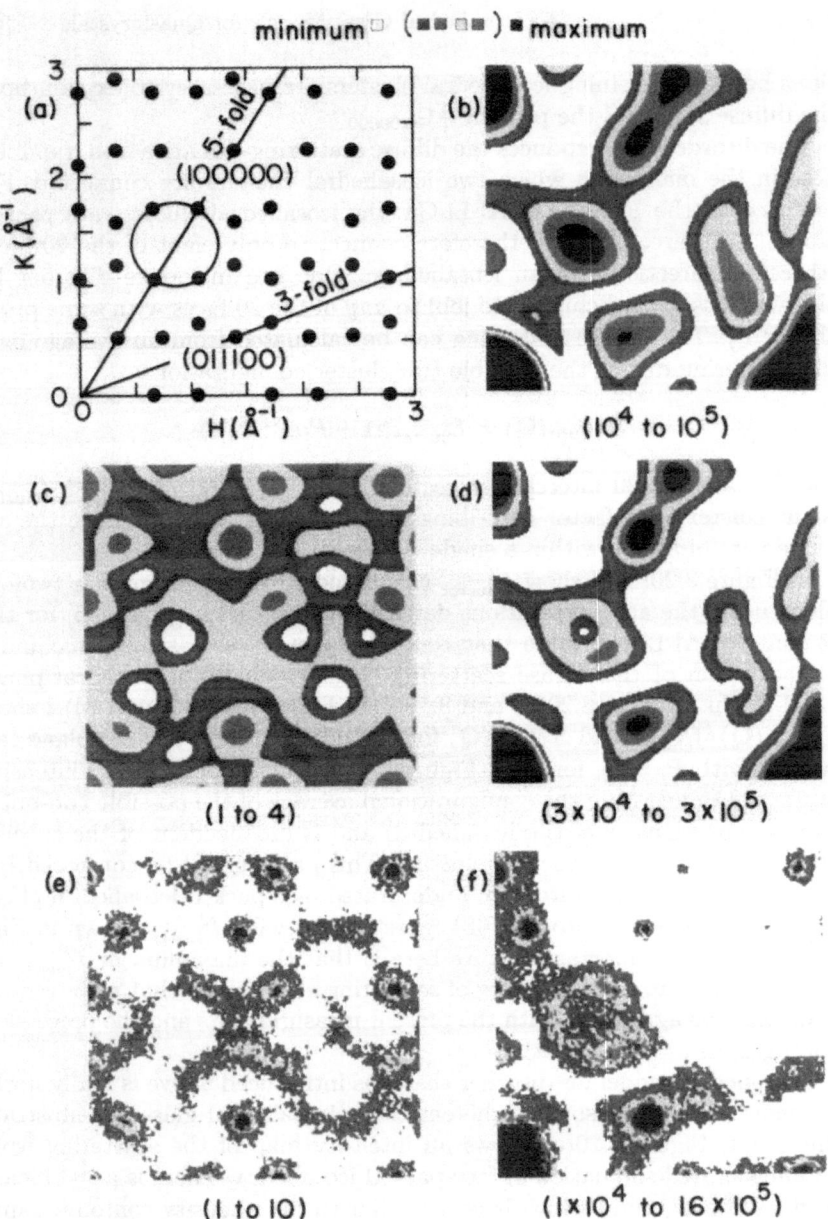

minimum □ (■ ■ ▢ ■) ■ maximum

FIGURE 8.20. (a) A schematic representation of one quadrant of a twofold plane. Filled circles correspond to the most intense reflections from an undecorated quasicrystal. Also shown is the position of the diffuse ring of Fig. 8.19. (b) through (f) plot intensities on a logarithmic scale over the ranges indicated in parenthesis below each panel. (b) The icosahedral cluster form factor, (c) the two-point structure factor, (d) the two-cluster structure factor, (e) the undecorated face-packed icosahedral glass structure factor in the twofold plane, and (f) the decorated IG structure factor, all as described in the text.

over the scattering patterns from several samples of face-packed icosahedra and have been multiplied by a single scale factor for comparison with the experimental data. The X-ray spectrometer vertical resolution acts to broaden, and therefore weaken, the peaks in Figure 8.19(c), as compared to the calculation. The position, size, and width of this diffuse shell, however, agree reasonably well with the experimental pattern of diffuse scattering. Quantitatively, the disorder intrinsic to the basic face-packed icosahedral glass algorithm overestimates the degree of disorder in icosahedral Al-Li-Cu. If we calculate the ratio of the intensities of the diffuse to Bragg peak scattering observed experimentally, it is smaller by about a factor of 5 as compared to the same ratio calculated from our numerical simulations.

How does the central peak arise in this model? In the decorated face-packed icosahedral glass of Section 8.6, the FCI structure was generated by assigning different decorations to neighboring icosahedral clusters. If this chemical ordering is only short-range, the sharp peaks at the FCI positions will yield to broad diffuse peaks.

Gibbons et al. have considered the applicability of these diffuse scattering calculations to the arcs and spheres found in other SI alloys [54]. While the scattering pattern from the bare IG aggregation [Figure 8.20(e)], shows remarkable similarity to the experimentally observed diffuse scattering patterns from TiMn, all attempts at decorating the icosahedral clusters with atoms resulted in a poorer match with experiment.

# 8.8    Concluding Remarks

The two key ingredients of the icosahedral glass models are the existence of clusters of atoms with (nearly) icosahedral symmetry and the constraint of maintaining bond-orientational order in the process of packing these clusters together into a network. The basic algorithms produce heavily disordered networks which, nonetheless, yield relatively sharp peaks in their diffraction patterns. The sharp peaks arise because of the constraint of maintaining bond-orientational order across the sample; only a finite set of intercluster displacements are admitted. The most attractive feature of these structures, as models for the SI alloys, is the simplicity of the growth rules. Even the most simplistic growth algorithms yield good qualitative, and in some cases, reasonable quantitative agreement with experimental diffraction measurements.

*Acknowledgments.* All of my work on the icosahedral alloys has benefited from a close collaboration with P.W. Stephens. I would also like to thank P. Bak, P. Bancel, P. Gibbons, C. Guryan, P. Heiney, K.F. Kelton, S.C. Moss, G. Shirane, and M. Widom for useful and enjoyable interactions over the past few years. Ames Laboratory is operated for the U.S.D.O.E. by Iowa State University under Contract W-7405-ENG-82.

# References

[1] D. Shechtman et al., *Phys. Rev. Lett.* **53**, 1951 (1985).

[2] C. Kittel, *Introduction to Solid State Physics*, Wiley, New York, 1986, p. 9.

[3] For example, see D.P. Shoemaker and C.B. Shoemaker, in *Aperiodicity and Order, Vol. 1: Introduction to Quasicrystals*, edited by M.V. Jarić, Academic, Boston, 1988, pp. 2–40.

[4] These electron diffraction patterns were kindly provided by K.F. Kelton and P. Gibbons at Washington University.

[5] For the interested reader, the description and evolution of quasicrystalline structures may be found in the proceedings of three international conferences on quasicrystals held in Les Houches (1985), Beijing (1987), and Vista Hermosa, Mexico (1989). In addition, several series of reprints and review essays are available: C.L. Henley, *Comm. Condens. Matt. Phys.* **23**, 59 (1987); P.J. Steinhardt and S. Ostlund, eds., *The Physics of Quasicrystals*, Plenum, New York, 1987; M.V. Jarić, *Aperiodicity and Order*, Vols. 1–3, Academic, Boston, 1988–1989.

[6] R. Penrose, *Bull. Inst. Math. Appl.* **10**, 266 (1974); M. Gardner, *Sci. Amer.* **236**, 110 (1977).

[7] A.L. Mackay, *Physica* **114A**, 609 (1982); P. Kramer and R. Neri, *Acta Cryst. A* **40**, 580 (1984).

[8] D. Levine and P. Steinhardt, *Phys. Rev. Lett.* **53**, 2477 (1984).

[9] For example, see Ref. 6.

[10] To some degree, this obstacle can be overcome by including specific defects in the tilings termed "decapods"; G.Y. Onoda et al., *Phys. Rev. Lett.* **60**, 2653 (1989); **62**, 1210 (1989). However, this is still a point which generates strong debate; see M.V. Jarić and M. Ronchetti, *Phys. Rev. Lett.* **62**, 1209 (1989).

[11] M. Widom, K.J. Strandburg, and R.H. Swendsen, *Phys. Rev. Lett.* **58**, 706 (1987).

[12] C.L. Henley, *J. Phys. A* **21**, 1649 (1988); in *Quasicrystals and Incommensurate Structures in Condensed Matter*, edited by M.J. Yacaman et al., World Scientific, Singapore, 1990, p. 152.

[13] M. Widom, D.P. Deng, and C.L. Henley, *Phys. Rev. Lett.* **63**, 310 (1989).

[14] K.J. Strandburg, L.H. Tang, and M.V. Jarić, *Phys. Rev. Lett.* **63**, 314 (1989).

[15] L.H. Tang, *Phys. Rev. Lett.* **64**, 2390 (1990).

[16] For a general discussion, see P. Bak and A.I. Goldman, in *Aperiodicity and Order, Vol. 1: Introduction to Quasicrystals*, edited by M.V. Jarić, Academic, Boston, 1988, pp. 143–170 and references therein.

[17] P. Bak, *Phys. Rev. Lett.* **56**, 861 (1986).

[18] For a review of the hydrodynamic modes in quasicrystals, see T.C. Lubensky in Ref. 16, pp. 199–280.

[19] P.A. Kalugin, A.Y. Kitaev, and L.C. Levitov, *JETP Lett.* **41**, 145 (1985); V. Elser, *Phys. Rev. B* **32**, 4892 (1985); *Acta Cryst.* **A42**, 36 (1986); M. Duneau and A. Katz, *Phys. Rev. Lett.* **54**, 2688 (1985).

[20] P. Bancel et al., *Phys. Rev. Lett.* **54**, 2422 (1985).

[21] D. Shechtman and I. Blech, *Metall. Trans. A* **16**, 1005 (1985).

[22] P.W. Stephens and A.I. Goldman, *Phys. Rev. Lett.* **56**, 1168 (1986); **57**, 2331 (1986); **57**, 2770 (1986).

[23] A.I. Goldman and P.W. Stephens, *Phys. Rev. B* **37**, 2826 (1988).

[24] V. Elser, in the *Proceedings of the XVth International Colloquium on Group Theoretical Methods in Physics*, edited by R. Gilmore and D. Feng, World Scientific, Singapore, 1987, p. 105.

[25] V. Elser, in *Aperiodicity and Order, Vol. 2: Extended Icosahedral Structures*, edited by M.V. Jarić and D. Gratias, Academic, Boston, 1989, pp. 105–136.

[26] D.P. DiVincenzo, *J. Phys.* **47** (Paris), Colloq. C3, 237 (1986).

[27] J.L. Robertson and S.C. Moss, preprint, 1990.

[28] P.W. Stephens, in Ref. 25, pp. 37–104.

[29] A. Garg and D. Levine, *Phys. Rev. Lett.* **60**, 2160 (1988).

[30] S. Hendricks and E. Teller, *J. Chem. Phys.* **10**, 147 (1942).

[31] V. Elser and C. Henley, *Phys. Rev. Lett.* **55**, 2883 (1985); P. Guyot and M. Audier, *Phil. Mag. B* **52**, L15 (1985).

[32] M. Marcus and V. Elser, *Phil. Mag. B* **54**, L101 (1986).

[33] M. Cooper and K. Robinson, *Acta Cryst.* **20**, 614 (1966).

[34] E. Cherkashin et al., *Soviet Phys. Cryst.* **8**, 681 (1964).

[35] C. Guryan et al., *Phys. Rev. B* **37**, 8495 (1988).

[36] L.E. Levine, P.C. Gibbons, J.C. Holzer, and K.F. Kelton, preprint, 1990.

[37] K.F. Kelton, P.C. Gibbons, and P.N. Sabes, *Phys. Rev. B* **38**, 7810 (1988); P.C. Gibbons et al., *Phil. Mag. B* **59**, 593 (1989).

[38] T. Ishimasa et al., *Phil. Mag. Lett.* **58**, 157 (1988).

[39] C. Guryan et al., *Phys. Rev. Lett.* **62**, 2409 (1989).

[40] P. Bancel, *Phys. Rev. Lett.* **63**, 2741 (1989).

[41] J. Devaud-Rzepski et al., *Phil. Mag. B* **60**, 855 (1989).

[42] S. Ebalard and F. Spaepen, *J. Mater. Res.* **4**, 39 (1989).

[43] C.L. Henley, *Phil. Mag. Lett.* **58**, 87 (1988).

[44] A.I. Goldman et al., *Phys. Rev. Lett.* **61**, 1962 (1988).

[45] P.M. Horn et al., *Phys. Rev. Lett.* **57**, 1444 (1986).

[46] For example, see Ref. 18.

[47] P. Heiney et al., *Science* **238**, 660 (1987).

[48] J.E.S. Socolar and D. Wright, *Phys. Rev. Lett.* **59**, 221 (1987).

[49] C. Guryan, Ph.D. dissertation, SUNY at Stony Brook (1990).

[50] N.K. Mukhopadhyay et al., *Phil. Mag. Lett.* **56**, 121 (1987).

[51] C. Dong et al., *Scripta Metal.* **20**, 1155 (1986).

[52] X. Zhang and K.F. Kelton, *Phil. Mag. Lett.* (accepted).

[53] F. Denoyer et al., *J. Phys.* **48**, (Paris), 1357 (1987).

[54] P.C. Gibbons et al., in Ref. 12, p. 516.

# Index